# Theory of Resonances

Reidel Texts in the Mathematical Sciences

A Graduate-Level Book Series

# Theory of Resonances

*Principles and Applications*

by

V. I. Kukulin, V. M. Krasnopol'sky
*Moscow State University, U.S.S.R.*

and

J. Horáček
*Charles University, Prague, Czechoslovakia*

Kluwer Academic Publishers

Dordrecht / Boston / London

**Library of Congress Cataloging-in-Publication Data**

Kukulin, V. I.
    Theory of resonances.

    (Reidel texts in the mathematical sciences)
    Includes index.
    1. Few-body problem.   2. Resonance.
I. Krasnopol'sky, V. M.   II. Horáček, J.
III. Title.   IV. Series.
QC174.17.P7K85 1988      530.1'4      88–12649
ISBN 90–277–2364–8

Published by Academia, Publishing House of the Czechoslovak Academy of Sciences, Prague
in co-edition with
Kluwer Academic Publishers, P.O.Box 17, 3300 AA Dordrecht, The Netherlands.
Kluwer Academic Publishers incorporates the publishing programmes of D. Reidel, Martinus
Nijhoff, Dr. W. Junk and MTP Press

Sold and distributed in the U.S.A. and Canada
by Kluwer Academic Publishers,
101 Philip Drive, Norwell, MA 02061 U.S.A.

Sold and distributed in Albania, Bulgaria, China, Czechoslovakia, Cuba, German Democratic
Republic, Hungary, Mongolia, Northern Korea, Poland, Rumania, U.S.S.R., Vietnam,
and Yugoslavia
by Academia, Publishing House of the Czechoslovak Academy of Sciences, Prague, Czechoslovakia.

Sold and distributed in all remaining countries
by Kluwer Academic Publishers Group,
P.O. Box 322, 3300 AH Dordrecht, The Netherlands.

Scientific Editor
Prof. Ing. Jiří Formánek, DrSc.,
Corresponding Member of the Czechoslovak Academy of Sciences

Reviewer
Prof. Ing. Milan Noga, DrSc.

# Contents

# Preface

It is well known that any development means specialization and differentiation. The soundness of this idea has again been substantiated by the development of quantum physics and by its applications to an ever growing number of topical problems in almost all branches of physics. This process of the differentiation of science separates the specialities, which were associated or indisciplinary as recently as thirty years ago, and makes them so far apart that people working in one of the branches are no longer interested in the problems of, and do not even understand the questions raised by the formerly associate field. We can only counter this unfortunately unavoidable specialization trend by a methodological integration, that is, by developing sufficiently universal and unified methods of studying the phenomena from different domains of physics. An excellent, but unfortunately rare, example of such a methodological unification in present-day theoretical physics is Landau and Lifshitz's ten-volume course of theoretical physics.

The development of the quantum physics of few-body systems may also be regarded as a very successful effort to counter such specialization. This quantum-physics approach, which was initiated in the early sixties by Faddeev in his mathematical works, and by Weinberg, Simonov, and many others in their pioneer studies, gradually expanded to absorb and integrate more and more fields of nuclear physics, atomic and molecular physics, and elementary particle physics in their theoretical and experimental aspects. At present, this trend has reached the status of a new and highly dynamic branch of science which unites, just on the basis of unified approaches and methods, the experts in different fields of physics. (We can get an idea of the development of this branch from the proceedings of the various international conferences on the relevant topics held regularly during the last 25-years).

Another good crystallization centre of various fields of microphysics is the physics of resonance states and processes. Equally, as once vibration theory became a fruitful interdisciplinary science of general laws of vibrational processes in different branches of physics, so the theory of resonance phenomena may well become, when developed appropriately, a science of general laws of formation and decay of long-lived states in molecules, atoms, nuclei, and condensed matter and under hadronic collisions. This tendency can be seen clearly nowadays and has resulted in the first interdisciplinary international conference on the methods and models in physics and theory of resonance processes (see Lecture Notes in Physics, vol. 211: *Resonances-Models and Phenomena*, edited by S. Albeverio, L. S. Ferreira, and L. Streit, Springer, Berlin, 1984).

In this book we shall attempt, as far as we are able, to contribute to the process of integration of methods in the field of resonance physics. Since this field is represented by numerous works from most branches of present-day physics, we have abandoned outright the idea of reviewing even the most fundamental works. It is difficult even to classify this vast mass of works as typifying the methods used, because a great many of the works may be regarded as synthetic, that is they use several different methods.

Fortunately, the basic elements, which are like the building blocks of every edifice, are few so that they can well be described in a moderate volume. We have chosen just this way; that is, we have tried to describe the basic elements of the theory of resonance states and processes along with typical examples of their combination into a completed "construction". Furthermore, we have limited ourselves to the sphere of our professional interests, which is still very large, namely, the resonance states in few-body systems (mainly in atomic and nuclear physics).

The feeling of great surprise which we experienced on studying numerous works from the field of atomic, molecular and nuclear physics has also given a strong impetus to writing this book, because their authors invented again and again, for their particular uses the methods elaborated properly long ago in a neighbouring branch of physics. Avoiding specific examples, we shall still show that many articles published, say, in the Physics Review, Series A are methodically almost an exact replica of earlier works published in Series C, and vice versa, often without any references.

The domain of the physics of few-body systems characterized by methodical unity of approaches seems to us to be just a good basis for the appropriate integration of various branches of resonance theory.

At present we have numerous excellent books on scattering theory (which include the theory of many-particle scattering – see the references in the basic text of the book) where the theory of resonances and resonance processes is described in more or less detail. However, all the books deal practically with only general properties of resonance states, such as the relevance of resonances to the $S$-matrix poles, the time-dependent decay law, etc. At the same time, any particular work has either to find theoretically the *actual* parameters of the resonance states (or the amplitudes of the processes involving the resonances) in terms of one or another dynamic model, or to infer such parameters from experimental data. However, material of this kind, which is of major importance for practical purposes, is absent in the general-purpose books and is disseminated over a great number of original works in *different* branches of physics. Therefore, our aim is to summarize at least a fraction of this vast material, including also our investigations, in a single moderate volume and to explain it in as uniform a language as possible. We leave it to the reader to judge how far we have succeeded in doing that.

The essence of the book is reflected in the table of contents and, in more detail, in the Introduction.

The book is prefaced by a referential chapter aimed at those readers who are eager to start tackling particular problems by sparing them the difficult and tiresome search for the facts and mathematical theorems scattered over the original mathematical and physical literature where various notations and different degrees of generality and rigour are used.

Some of our colleagues and friends who assisted us in writing this book read all or individual chapters and made a number of valuable remarks. We are especially grateful to Vladimir Pomerantsev, a member of our research group at the Moscow State University whose contribution to the elaboration of many results presented in the book can hardly be overestimated. He also rendered great assistance in writing Chapter 4 of the book and looked critically through the material of other chapters. The authors are particularly indebted to Professors J. Formánek and J. Kvasnica of the Charles University, Prague, for their attentive reading of the manuscript and for numerous informative remarks allowing us to improve the text. The authors should also like to thank Professor W. Domcke, Technische Universität München, Professor N. M. Queen, The University of Birmingham and Dr. J. Blank, Charles University Prague for carefully reading parts of the manuscript and pointing out a number of errors.

Since the book is to some extent a first attempt to expound the methods of resonance theory on a uniform basis, it cannot be absolutely free from shortcomings the responsibility for which is borne by the authors alone.

In the techniques and subjects treated there exists a large amount of publications. We would like to apologize to those authors whose work is not directly or sufficiently well treated.

The Soviet authors are also grateful to the administration of Charles University for offering all necessary facilities for their work in Prague which permitted them to finish the book within an acceptable period.

Prague 1986

V. I. Kukulin,<br>
V. M. Krasnopol'sky<br>
*Moscow State University*

J. Horáček<br>
*Charles University*

# Introduction

This book is devoted to the theory of resonance states and processes expressed in terms of nonrelativistic quantum mechanics. There are already many excellent books on quantum mechanics and collision theory where, among other things, a description of general theory of resonance scattering may be found [1–6]. With rare exceptions, however, they usually discuss the parametrization of the scattering amplitude (or of the $S$-matrix) in the near-resonance region, for example the Breit-Wigner parametrization or the appropriate generalization to the case of many levels or many channels. At the same time, the basic problem nowadays is not to get a general parametrization of cross sections but to calculate the parameters of the resonance amplitudes (namely, widths, level shifts, $S$-matrix residues at resonance poles, etc.) proceeding from the fundamental interactions effective in a quantum system. It is the problem of the practical calculation of resonance parameters, however, that is explicitly given little consideration in the general-purpose manuals. They also usually omit the theory of many-body resonance states capable of decaying with emission of two or more particles, although such resonances are extensively used in atomic, molecular, nuclear and particle physics.

On the other hand, present-day pure scientific literature includes many hundreds of works devoted to methods for calculating the resonance parameters and to the general description of many-body resonances. Furthermore, many of the methods which have long been used in nuclear and particle physics are now being reinvented to describe atomic collisions, and vice versa. Therefore the aim of this book is to fill this gap between the textbooks on quantum mechanics and theory of resonance processes on the one hand and the modern-day journal publications on the other, including works from various fields of quantum physics, molecular, atomic, and nuclear physics, several sections of the book dealing with original results of the authors. In choosing the material for the book, we were guided by the desire to include the most general methods and approaches which cannot be found in the standard textbooks, but are widely used in the relevant studies. Moreover, when appropriate, we have included the results of particular numerical calculations for illustrative models bearing in mind that one simple example can well prove to be more convincing than several pages of arguments (as to the exact proofs [35], they impose highly constraining restrictions on the interactions used in most cases).

In a number of cases where lack of space prevented us from giving a detailed derivation we have tried to guide the reader by presenting basic relations and referring him to works where a completed derivation may be found. Sometimes the reader himself will probably be able to complete the derivation, which may be regarded as a good exercise. Finally, bearing in mind the main purposes of this textbook, we end each chapter with a list of references where most of the basic works on a given subject may be found.

**Contents of the book**. Now we shall briefly outline the contents of this book. The first Chapter of the book is auxiliary and deals with subsidiary knowledge from various areas of mathematics, such as the theory for functions of complex variables, methods of analytic continuation, estimation of divergent integrals, concise theory for the Padé approximants, and other information necessary for the subsequent sections of the book to be understood. Chapter 1 also gives necessary information about the Faddeev equations and the theory of three-particle scattering which underlie three-particle (and, generally, many-particle) resonance theory treated in the subsequent chapters of the book. (A comprehensive discussion of the basic mathematical facts, theorems, and equations concerning the modern-day scattering theory in few-body systems may be found in the latest monograph [36]).

Chapter 2 includes general information about resonance states and processes necessary both for understanding of how to detect resonances experimentally in the scattering and reaction cross sections and for learning some basic methods to describe resonance processes (the Kapur-Peierls formalism, etc.). On comparing Chapter 2 with Chapter 4 the reader can readily make sure that the Kapur-Peierls formalism and the Feschbach projection operator approach proper and its various modifications are essentially based on a single fundamental idea, namely, the decay channels of resonance states are artificially made closed, whereupon the pure stationary state is calculated. After that, the decay channels are opened again and the true quasi-stationary state is calculated (usually by the perturbation method) as a state obtainable from the initial (stationary) state when the interaction with the open channels of the continuum is included. The basic difference between the above mentioned approaches is solely that the Kapur-Peirls method is formulated in the configuration space, and the Feshbach approach in the Hilbert space (which is most probably preferable in many problems). Many other approaches (for example, the $R$-matrix method) are based on the same general idea.

Chapter 3 gives a general approach to resonance theory based on the Hilbert-Schmidt method from the theory of integral equations. This approach, developed mainly by Simonov et al., makes it possible, in a uniform way and using unified language, to formulate numerous results obtained by various methods in terms of the resonance state theory. This approach is notable for a good universality, because the given pattern permits a direct generalization in the case of three-particle resonances and allows a simple computational algorithm. The chapter describes also the readily usable methods for calculating the Hilbert-Schmidt eigenvalues and eigenfunctions proposed by the authors.

Chapter 4 describes one of the most extensively used methods to treat resonance states, namely, the projection operator formalism proposed by Feshbach, and its generalization to the many-particle resonance theory carried out by V. Pomerantsev and one of the authors of this book (V.I.K). The last section of the chapter describes the application of the Feshbach projection operator formalism to the calculation of the QBSEC-type metastable states excited in nuclei under elastic and inelastic scattering of nucleons of low and medium energies.

Chapter 5 deals mainly with the behaviour of the resonance poles of the $S$-matrix in various interaction models when the parameters of the interaction Hamiltonian vary. Most attention is paid to the ingenious approach developed by the authors and based on the analytic continuation of the $S$-matrix singularities in the coupling constant. In terms of this approach the resonance and virtual states are inferred from the bound states when the latter are analytically continued in the coupling constant. This approach differs from many other methods by its simplicity and wide universality because it may be applied to calculating both the resonance states proper and the amplitudes of one- and many-particle reactions involving resonances. The chapter presents numerous examples illustrating the application of the approach to various problems.

Chapter 6 discusses a method proposed by the authors and intended for actual determination of the resonance parameters (width, energies, and vertex constants) from phase shift analysis of experimental data. Particular algorithms for processing experimental phase shifts are discussed. The nuclear and atomic physics evidence exemplifies the discussion.

Finally, Chapter 7 discusses the applications of variational methods to calculating the resonance states and treats the dilatation method (i.e. the complex scale transformation method) which has enjoyed wide popularity in recent years, especially in atomic and molecular physics, and is used to study the autoionization state and other long-lived states. We are far from nursing the idea of having exhaustively expounded the above two approaches mentioned (that is the variational and dilatation methods) to which numerous publications are devoted (see the list of references to Chapter 7 which is, however, far from being complete). Our aim is more modest, namely to give some idea of the methodology in the context of this book and to cite basic references to the original works and reviews where the reader can find, if required, the necessary information. On the whole, this agrees fully with our general intention to treat the basic concepts and the calculation methods of the present-day theory of resonances in few-body systems in a comprehensible book of moderate volume. The appendix is also of mathematical character and presents some basic concepts of the theory of rigged Hilbert spaces necessary for the material of Chapter 4 to be understood more clearly. It is known (see, for example, [1]) that the theory for resonance states may also be described without using the formalism of rigged Hilbert spaces. However, the use of the language of rigged spaces makes it possible to present this theory in what is probably the most exact and logical way (see, for example, [34]). Our aim, however, is not to consistently use this ingenious formalism in all the applications, but to describe its usage in

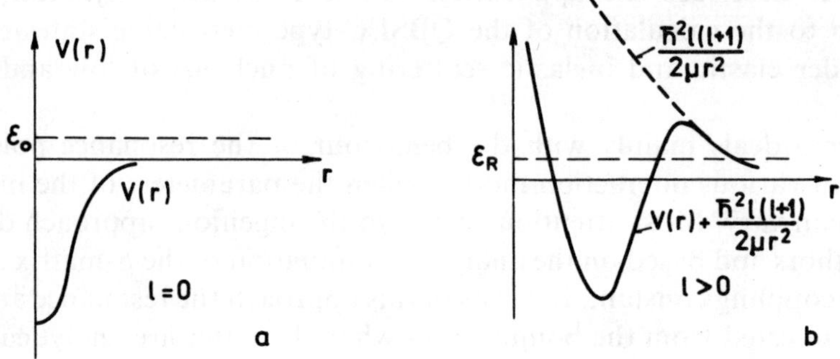


Fig. 0.1

Formation of a confining potential barrier for the superposition of a short-range attractive potential $V(r)$ (Fig. 1a) and a repulsive centrifugal potential $\hbar^2 l(l+1)/2\mu r^2$ for a non-zero particle angular momentum (Fig. 1b), $l > 0$.

constructing a resolvent operator continued to the nonphysical sheet. In the remaining chapters of the book we use a more traditional method, namely, the direct continuation of the matrix elements of operators as ordinary functions of a complex variable.

**Types of long-lived states.** Quantum mechanics deals with numerous types of long-lived (quasi-stationary) states. These states may be classified by the singularities of the scattering matrix, by the mechanism of their production or by the model describing them.

Let us examine briefly the most widespread types of the resonances and the long-lived states of quantum systems. The list of types given below is far from complete, but it shows the diversity of physical phenomena which we are inevitably faced with when we try to get a more or less consistent classification of long-lived states.

(i) *One-particle shape resonances.* This is the simplest and clearest type of long-lived state which arises in the case of potential scattering of a particle by an unexcited target or simply in a potential field in the presence of a confining potential barrier. The energy $E$ of the particle has to be below or of the same order of the height of the barrier (see Fig. 0.1). The particle is captured through tunnelling to the region of strong attraction, thus forming a relatively long-lived (or quasi-stationary) state, whereupon, again through tunnelling, it leaves the inner region of interaction. The potential barrier arises, as a rule, when a strong short-range attraction is applied to a long-range repulsive potential of centrifugal (see Fig. 0.1) or Coulomb type. The lifetime $\tau$ of the quasi-stationary state produced varies inversely with the probability of the decay through tunnelling [1], that is

$$\tau \simeq \hbar/\Gamma ; \qquad \Gamma \sim W_{if} = 2\pi \mid \langle \Phi_i \mid V \mid \Psi_f \rangle \mid^2 \varrho_f \sim P(E) \qquad (0.1)$$

where $P(E)$ is the potential barrier penetration factor, $\varrho_f$ is the state density of decaying particles. Thus $\tau$ is inversely proportional to the potential barrier penetrability $P(E)$. There are numerous examples of shape resonances; for instance, neutron strength function resonance in the optical model of scattering of slow neutrons by nuclei, the well-known resonances in electron scattering by atoms or molecules (e.g. e + $N_2$ scattering at $E_e \simeq 2.3$ eV), and the one-particle resonances arising in mutual scattering of light nuclei of the types $^4$He + $^4$He, $^4$He + $^3$He, $^3$He + n , etc.

(ii) *One-particle virtual states* are similar in their nature to the one-particle shape resonances. Here the confining barrier may be absent, but the interaction potential must show a sharp jump at the boundary of the interaction region. At a low energy of the incoming particle the de Broglie wavelength inside is much smaller than outside the well. As a result, it can readily be shown that inside the well a standing wave is formed which "penetrates" with a small probability through the potential boundary; here the above-barrier reflection factor of the inner wave is close to unity. (A very lucid treatment of one-particle long-lived states can be found in the classical monograph by Blatt and Weisskopf [7]. An example of such a state is the so-called singlet deuteron, i.e. the long-lived state in the n–p system in the $^1S_0$ channel at energy $E_S \simeq -0.066$ MeV.

(iii) *Feshbach resonances.* These states are produced when an incoming particle excites one or several particles of a target (or, more generally, several degrees of freedom of the target) and the particle itself is captured in a long-lived state. The long lifetime arises from the fact that the channels of the direct decay of the intermediate state produced turn out to be closed and the decay of the Feschbach resonance is due to subsequent (total or partial) de-excitation of the target and ejection of the captured particle to the initial channel (elastic resonance scattering) or to the state with a lower positive energy (inelastic resonance scattering). The autoionization states produced in electron scattering by atoms and molecules and the excitation of higher vibrational states in low-energy electron scattering by molecules are very common examples of the Feshbach resonances. In the latter case the production of the vibrational-type Feschbach resonances is stimulated by a shape resonance in the entrance channel. It is the production of this initial relatively long-lived state that permits a light electron to excite vibrations of much heavier nuclei. The Feshbach-type resonances (they are so called because Feshbach was the first to develop a general theory of such states [8]) are an example of the compound states or of the so-called quasi- -bound states embedded in a continuum (QBSEC) in the case where the coupling of inner excitation to the decay channels is relatively weak. This is so in atomic physics where the interaction between electrons is substantially weaker than between nucleons in a nucleus. Thus in the case of electron scattering by atoms, only 1 or 2 atomic electrons are excited, thereby producing an autoionization state which decays mainly through emission of an electron. The autoionization state production is schematically illustrated in Fig. 0.2. The initial stage

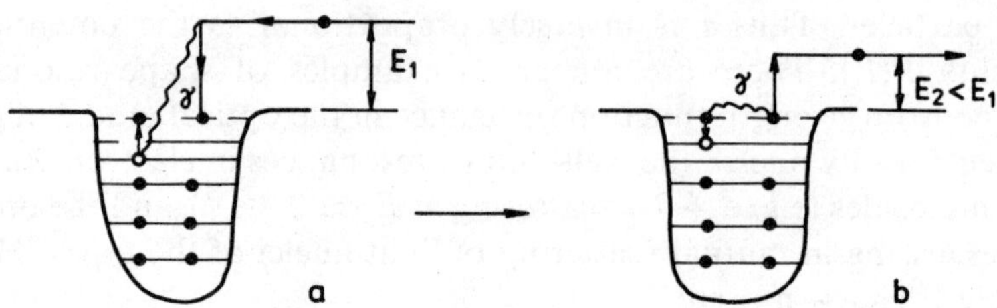


Fig. 0.2

Schematic diagrams illustrating the production mechanism of a typical Feshbach resonance in atomic physics (for example, of an autoionization state). The initial reaction stage is shown in Fig. 2a, and the final stage in Fig. 2b. Wavy lines are electromagnetic interactions.

of the reaction is shown in Fig. 0.2a, and the final stage in Fig. 0.2b.

(iv) *Compound states and QBSECs.* In nuclear physics the interaction between nucleons is strong, so that the incident particle energy dissipation in the target does not stop (except for some special cases) during the first stage shown in Fig. 0.2a. It continues until a very complicated mixture of excited configurations is formed where each particle has but a small fraction of the total energy of the system. As a result, since the number of possible configurations (that is the number of closed channels) of this kind is great, the system can live in such an excited state for a long time. Thus the width $\Gamma$ of such a state turns out to be exceptionally small, but the number of such states in heavy nuclei is enormously high and hence the distances between the levels are very small. Therefore, if the energy spread $\Delta E$ of the incident beam is substantially larger than the distance between the levels of the compound nucleus $D$, that is $\Delta E \gg D$, then, owing to the random character of the phase relations for the excitation amplitudes of various compound states, the amplitude averaged over the energy interval $\Delta E$ is a fairly smooth function of energy and coincides in practice with the one--particle amplitude of scattering by a complex potential which is called nuclear optical potential.

In turn, the Feschbach resonances and the nuclear compound-states a.e particular (though very peculiar) cases of many-channel resonances where the Feshbach resonances arise in atomic physics from "weak" coupling of excited configurations, whereas the nuclear compound states correspond to "strong" coupling[1]).

---

[1]) The terms "weak" and "strong" are used here in accordance with the value of the ratio $\langle V \rangle / D$ which is the ratio of the mean interaction energy to the distance between the levels.

(v) *Many-channel resonances in few-body systems due to strong coupling of channels.* Numerous characteristic types of long-lived states arise in many-channels systems: these include the atomic-physics Feshbach resonances and the compound-nucleus states. Baz [9] predicted another general type of many-channel long-lived state which is due to a near-threshold singularity. Briefly, the Baz idea is that near the threshold of some channel $\alpha$ in a many-channel system the fragments of the system corresponding to this channel prove to be spaced considerable distances apart because of the small binding energy. As a result, because of a small overlap of the wave function in the channel $\alpha$ with the wave functions of other channels, the channel coupling effectively weakens and such a state (if any) becomes a long-lived state. Indeed, in a great many nuclear systems near the respective thresholds a great number of long-lived states can be found (but not always!). Unfortunately, a general and mathematically correct investigation of this interesting problem has not been carried out yet. However, much useful information about many-channel resonances, including near-threshold states, may be found in the fundamental review by Badalyan, Kok, Simonov and Polikarpov [10] and in the physically lucid review by Dalitz [11] devoted to many-channel resonances and elementary particle physics (see also a very voluminous bibliography in Ref. [10]).

The genuine many-channel resonances (the so-called coupled-channel resonances, the CC-resonances) which are due to strong coupling of channels and which disappear when the coupling weakens constitute one of the most interesting types of long-lived states in many-channel systems. Some resonances (if they are identified reliably) which are due to strong coupling of the N–N̄ channel and the annihilation channel may belong to this kind of resonance.

The so-called nuclear quasimolecular resonances, which constitute a new and exceptionally interesting type of nuclear long-lived state discovered twenty years ago [13] and studied intensively nowadays, belong most probably also to many--channel resonances. The nuclear quasi-molecular states arise in the case of scattering of heavy ions of the $^{12}$C + $^{12}$C, $^{12}$C + $^{20}$Ne, $^{28}$Si + $^{28}$Si and even $^{48}$Ca + $^{208}$Pb type [14] at energies (in the centre-of-mass system) of tens of MeV. The basic difficulty in interpreting such long-lived states arises from the high density of nuclear compound states at such excitation energies. For example, in the $^{28}$Si + $^{28}$Si system the density of states $\varrho_f$ at 60–70 MeV excitation energies reached $5 \times 10^6$ MeV$^{-1}$. And since the resonance width $\Gamma$ is proportional to the density of states $\varrho_f$, because

$$\Gamma(E, J) \simeq 2\pi \mid \langle i \mid V \mid f \rangle \mid^2 \varrho_f(E, J) \tag{0.2}$$

where $E$ is the excitation energy and $J$ is the angular momentum of the state, then each long-lived state should simply have been "buried" in this "noise". Meanwhile, the widths of quasi-stationary states observed experimentally are very small compared with those predicated on the basis of (0.2) (and range from 0.1 to 1 MeV). This interesting paradox may be resolved most plausibly by interpreting the observed resonances as nuclear quasi-molecules in which the nuclei are very distant from each other (at the energy near the Coulomb barrier top). Under these conditions, just as in case of the Baz near-threshold states, the channel coupling (that is the coupling of peripheral states with the channels of pure compound states) gets effectively and sharply weaker, so that long-lived states are produced. Unfortunately, the problem cannot be solved completely in this way because the width of the shape-resonances which may be produced by heavy ion scattering is much in excess of the value 0.2 MeV. A correct explanation for the observed phenomena seems to have been given first by Greiner and Scheid [15] and is based on the interpretation of the nuclear quasi-molecular states as simply many-channel resonances in which the entrance channel is coupled through the vibrational and/or rotational excitations of colliding nuclei to a state with a lower kinetic energy of the colliding fragments. As a result, the quasimolecular state turns out to be subbarrier and, hence, relatively long-lived. In spite of the fact that some of the difficulties cannot yet be resolved in terms of this model, the latter makes it possible to answer many questions and has been generally accepted [14].

The rotationally-excited states of van der Waals complexes (diatomic molecule + inert gas atom [24]) and the surface quasi-stationary states of the type of HD-Ag (111) resonances [25] which arise in chemical catalysis (when a wall--scattered molecule of a gas is captured by the surface) are also interesting type . of many-channel resonances.

(vi) *Three-particle near-threshold long-lived states.* These constitute another interesting class of long-lived states where the near-threshold *virtual* states (for which the corresponding $S$-matrix poles are located at negative energies on

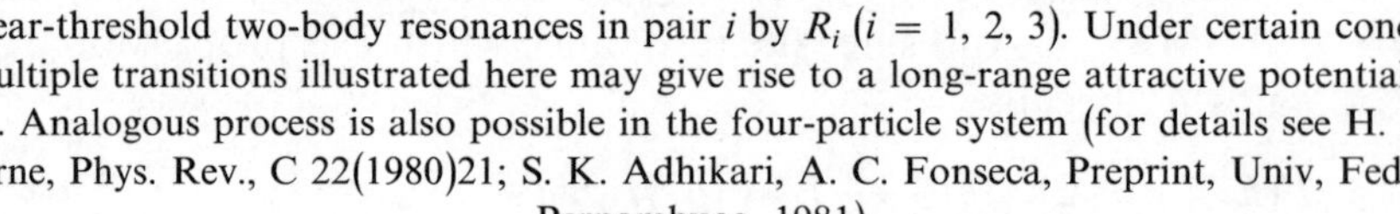

Fig. 0.3

Schematic diagram of the multiple scattering process in a three-body system which plays an important role when near-threshold resonances are present in two-body subsystems. We denote the near-threshold two-body resonances in pair $i$ by $R_i$ ($i = 1, 2, 3$). Under certain conditions the multiple transitions illustrated here may give rise to a long-range attractive potential in the system. Analogous process is also possible in the four-particle system (for details see H. Kröger, R. Perne, Phys. Rev., C 22(1980)21; S. K. Adhikari, A. C. Fonseca, Preprint, Univ, Federal de Pernambuco, 1981).

nonphysical sheets) as well as the resonance states ($S$-matrix poles are located in the complex energy plane, also on nonphysical sheets) are possible. The main condition for the appearance of such long-lived states is the existence of the near-threshold bound, virtual, or resonance states in *two-particle* subsystems. In such cases, *long-range* exchange forces appear in the system even if the *short--range* interactions only enter the interaction Hamiltonian. This long-range interaction is due to multiple transitions of particles between the bound or resonance states of two-particle subsystems (see Fig. 0.3). The so-called Efimov states [16], constitute a characteristic and non-trivial example of such states: the number of these is proportional to the logarithm of the ratio of (large) two--particle scattering length to the radius of the forces. Although pure Efimov states have never been discovered, states naturally related to them are very common among the excited states of light nuclei, for example, states near the *three-particle* threshold [17]:

(1) $^6$Li*; $J^\pi T = 0^+ 1$ state; its excitation energy $E^* = 3.563$ MeV, the location of the $\alpha np$ threshold $E_{\text{th}(\alpha np)} = 3.698$ MeV

(2) $^6$He*; $J^\pi T = 2^+ 1$ state; $E^* = 1.80$ MeV, location of the $\alpha nn$ threshold $E_{\text{th}(\alpha nn)} = 0.83$ MeV;

(3) $^9$Be*; $J^\pi T = \frac{1}{2}^+ \frac{1}{2}$ state; $E^* = 1.68$ MeV, location of the $\alpha\alpha n$ threshold $E_{\text{th}(\alpha\alpha n)} = 1.664$ MeV;

(4) $^{12}$C*; $J^\pi T = 0^+ 0$ state; $E^* = 7.655$ MeV; location of the $3\alpha$ threshold $E_{\text{th}(3\alpha)} = 7.271$ MeV; location of the $\alpha - {}^8$Be threshold $E_{\text{th}(\alpha-{}^8\text{Be})} = 7.367$ MeV.

By their nature, all the above mentioned excited states are three-particle resonances arising from the two-particle near-threshold virtual or resonance states which are present in each pair of particles $\alpha + \alpha (0^+$ state$)$, $\alpha + N(\frac{3}{2}^-$ and $\frac{1}{2}^-$ states$)$ and N–N($^1S_0$ state). It is highly probable that general rules exist (though they have not yet been found) which define the relations between the parameters of two-particle and three-particle resonances in such systems. Although some preliminary studies of the problem have been attempted [18–20], in particular for simple three-particle model systems, generalizations to realistic cases and more comprehensive studies have unfortunately never been completed.

Evidently, the many-particle near-threshold states of *complicated character*, which decay in the three-body channel with a strong (resonance) final-state interaction in all three pairs of ejected particles, are closely related to such resonances. Such states can be found nearly everywhere among the strongly excited states of light nuclei [17]. The methods for calculating such resonances were only developed during recent years. Chapter 4 treats one of the possible approaches developed by one of the authors (V.I.K.). An alternative method using the dispersion N/D technique and applicable rather in the particle physics was developed in Refs. [21, 22] (see the quite comprehensive list of references in review [12]).

In this connection it is appropriate to mention a problem which has been discussed in recent years but which has not yet been completely solved. This concerns the existence of the so-called dibaryon resonances in a two-nucleon system at energies of hundreds of MeV which are associated with high inelasticity (mainly in the NN scattering channels $^1D_2$ and $^3F_3$) and which show themselves near the N–Δ channel threshold in N–N scattering, in the $\pi + d \rightarrow N + N$ reaction, etc. in the corresponding partial waves. Many efforts have been made by various authors to find sufficiently reliable criteria for discriminating between the true dibaryon resonances (as the states arising as a result of rearrangement or excitation of a multiquark system) and the near--threshold anomalies caused by pion production near the threshold and by the formation of the strongly interacting complex of three particles N–π–N. However, the problem cannot be considered as having been solved at present.

Finally, we shall mention a very special but extremely wide-spread type of long-lived quasi-stationary state manifesting itself in many-body systems, condensed media, and crystals, namely,

(vii) *Quasiparticle (collective) excitations or quasiparticles.* They exist in the Bose and Fermi liquids and in other systems. The collective resonances in crystals studied first by Kagan and Afanasyev [23] (see also the book by Pinsker [23]) constitute an interesting new type of long-lived state.

From the above list, which is far from complete, it is clear how immense is the range of problems involving resonance phenomena in quantum physics. Because of the limitations of this book and the interests of the authors, we have treated here only those resonance processes peculiar to few-body systems and relevant essentially to the domain of atomic and nuclear physics. They include the shape resonances and other one-particle near-threshold states, the Feschbach resonances (including autoionization states in atoms), QBSECs, and the theories of three-particle resonances. We shall also discuss some resonance processes in which these states manifest themselves.

Before going to more subtle problems associated with the time-dependent picture of the quasi-stationary state decay, it is expedient to examine briefly the phenomenology of resonance scattering and the manifestation of resonance states in the observed scattering and reaction cross sections (for a more detailed discussion, see Chapter 2). Let us examine first the resonances in pure elastic scattering. We use the standard definition [2] of the partial scattering amplitude $g_l(k)$:

$$g_l(k) = \frac{\exp(2i\delta_l) - 1}{2ik} = \exp(i\delta_l)\sin\delta_l/k \tag{0.3}$$

where the wave vector $k = \sqrt{2\mu E}$; $\mu$ is the reduced mass $(\hbar = 1)$, whence we find the following expression for the partial scattering cross section:

$$\sigma_l(k) = \frac{4\pi(2l + 1)}{k^2}\sin^2\delta_l(k) . \tag{0.4}$$

As we know [2], the phase shift $\delta_l(k)$ near the resonance may be divided (although the division is not unique) into a rapidly changing or resonance phase shift $\delta_R(k)$ and a smooth background shift $\delta_P(k)$, so that $\delta(k) = \delta_R(k) + \delta_P(k)$

(for the sake of brevity, the index $l$ will generally be omitted henceforth). Such a division is convenient because the $S$-matrix unitarity is preserved, since

$$S = S_R \cdot S_P = \exp\{2i(\delta_R + \delta_P)\} \tag{0.5}$$

where $S_R$ and $S_P$ are also unitary.

A properly expressed resonance arises in the case where the resonance phase shift $\delta_R(k)$ changes rapidly (increases) by a value close to $\pi$ within a small energy interval $|E - E_R| \le \Gamma$. The shift $\delta_R(k)$ may be presented as

$$\mathrm{tg}\,\delta_R(k) = \frac{\Gamma/2}{E - E_R} \tag{0.6}$$

which yields after some elementary transformations:

$$\sin^2 \delta_R(E) = \frac{\Gamma^2/4}{(E - E_R)^2 + \Gamma^2/4}. \tag{0.7}$$

In the case where the background scattering is absent, that is $\delta_P = 0$, we find from (0.7) and (0.4):

$$\sigma_l(E) = \frac{4\pi(2l + 1)}{k^2} \frac{\Gamma^2/4}{(E - E_R)^2 + \Gamma^2/4}. \tag{0.8}$$

And this is the famous Breit-Wigner formula describing the shape resonance of the cross section in the discussed case. Here the resonance energy $E_R$ and the width $\Gamma$ are the parameters determining the resonance peak shape in this simplest case (the physical meaning of the width in terms of the dynamic pattern of the resonance formation and decay will be discussed below).

However, the simple parametrization with constant (i.e. energy-independent) $E_R$ and $\Gamma$ is only valid in the case of a sufficiently narrow resonance, which is not very close to the threshold, and in the absence of background scattering. It can readily be shown that the behaviour of the phase shift, such as (0.6), can be a consequence of the presence of an $S$-matrix pole near the real axis $k > 0$ (in its resonance part $S_R(k)$ ):

$$S_R(k) = \frac{k - k_0^*}{k - k_0} \underset{k \to k_0}{\sim} \frac{E - E_R - i\Gamma/2}{E - E_R + i\Gamma/2} \tag{0.9}$$

where $k_0^2/2\mu = E_R - i\Gamma/2$, with such a structure of $S_R(k)$ being necessary for its unitarity to hold (this can be easily verified).

It should be noted that the condition (0.9) is sufficient but not necessary for a resonance (a corresponding counter-example, that is the case of the resonance behaviour of the phase shift $\delta_R(E)$ of type (0.6) *not associated* with a nearby $S$-matrix pole, will be discussed in Chapter 2). However, since the pole origin of the resonance singularities suggests itself most readily and naturally and since

many of the resonance types are inferable from bound states when a perturbation is imposed (for example, the atomic resonances arising in the Stark effect),
while the $S$-matrix poles correspond undoubtedly to the bound states, the near

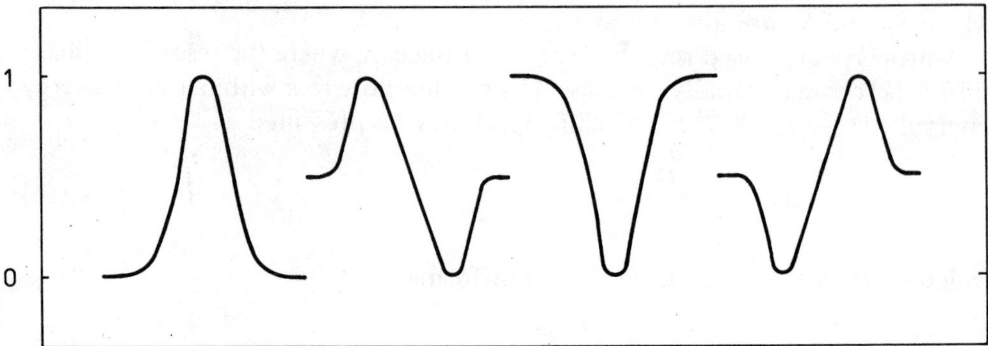

Fig. 0.4

$S$-matrix poles on nonphysical energy sheets are conventionally assumed in
quantum physics to be responsible for the observed resonances. This view-point
has also been accepted throughout this book.

Now we shall discuss the changes in the simple pattern described above which
arise in the realistic situation when $\delta_P \neq 0$, the resonance is not narrow, and,
in addition, inelastic channels for scattering and reaction are open. At $\delta_P \neq 0$,
even in the case of pure elastic scattering, the observed character of the energy
dependence of the cross section may differ strongly from the traditional Breit-
-Wigner form. Fig. 0.4 shows four feasible shapes of the resonance behaviour of
the cross section at different values of the background phase shift $\delta_P(k)$.

As the module of the partial scattering amplitude is

$$|g_i(k)| = |\sin \delta_l|/k$$

it follows from the figure that at the "resonance point" the amplitude module
may take on both the maximum value (i.e. $1/k$) and the minimum (i.e. zero)
value, depending on the background magnitude. In other words, it is the rapid
variations rather then the maximum absolute value of the scattering amplitude
(hence of the cross section) that may be regarded as an indicator of a resonance.
In this case the variations should be due to a rapid increase of the phase shift
within a relatively narrow energy range.

When the reaction channels are open the behaviour of the cross sections can
be even more complicated. Based on the shape of the energy dependence of $\sigma(E)$,
it is difficult to suspect the existence of a resonance (especially in the presence
of a large non-resonance background and of an essential inelasticity). Since the
width $\Gamma$ represents the probability $W$ of the decay (or production) of a given

quasi-stationary state, i.e. $\Gamma = \hbar W$, then, given several possible reaction channels, each channel $\alpha$ must be juxtaposed with its partial width $\Gamma_\alpha$. In this case, if the decay channels are *entirely independent*, then, according to the probability conservation condition, the total resonance width is equal to the sum of the partial widths: $\Gamma = \sum_\alpha \Gamma_\alpha$. In the general case, all $\Gamma_\alpha$ depend differently on energy $E$, that is $\Gamma_\alpha = \Gamma_\alpha(E)$. Under these conditions the total width $\Gamma$ depends on energy in a rather complicated way.

When the background is absent the generalizations of the Breit-Wigner formula in the presence of the inelastic and reaction channels are:

$$\sigma_{l,el} = \frac{\pi}{k^2}(2l + 1)\frac{\Gamma_{el}^2(E)}{(E - E_R)^2 + \Gamma^2(E)/4}, \tag{0.10a}$$

$$\sigma_{l,r} = \frac{\pi}{k^2}(2l + 1)\frac{\Gamma_{el}(E)\Gamma_r(E)}{(E - E_R)^2 + \Gamma^2(E)/4}, \tag{0.10b}$$

$$\sigma_{l,tot} = \frac{4\pi}{k^2}(2l + 1)\{|g_{e,el}|^2 + |g_{l,r}|^2\} =$$

$$= \frac{\pi}{k^2}(2l + 1)\frac{\Gamma_{el}(E)\Gamma(E)}{(E - E_R)^2 + \Gamma^2(E)/4} \tag{0.10c}$$

where

$$\sigma_{l,el} = \frac{\pi}{k^2}(2l + 1)|1 - S_l(k)|^2 = \frac{4\pi}{k^2}(2l + 1)|g_{l,\,el}|^2,$$

$$\sigma_{l,r} = \frac{\pi}{k^2}(2l + 1)(1 - |S_l(k)|^2) = \frac{4\pi}{k^2}(2l + 1)|g_{l,r}(k)|^2,$$

$$\sigma_{l,tot} = \sigma_{l,el} + \sigma_{l,r}$$

and $\Gamma(E)$ is the total width for all the *reaction* channels (that is $\Gamma = \Gamma_{el} + \Gamma_r$). The factorization of the numerator in (0.10) is essentially a consequence of the factorization of the residue of the many-channel $S$-matrix at a resonance pole and means physically an approximate independence of the formation and decay of a given resonance state (that is, the rule for the multiplication of probabilities for the probability of a composite event).

The traditional physical interpretation of a resonance state proceeds from the fact that the spatial part of the total, (i.e. time-dependent) wave function $\Psi(r, t)$ satisfies the Schrödinger equation with a *complex* eigenenergy $z = E_R - i\Gamma/2$ (as was stated by Gamow [26] at the very outset of the development of quantum mechanics):

$$\Psi(r, t) = \Phi(r)\exp\left(-i\frac{z}{\hbar}t\right), \tag{0.11}$$

$$H\Phi(r) = z\Phi(r) \tag{0.12}$$

where $\Phi(r)$ is the complex eigenfunction of a quasi-stationary state (the Gamow wave function). Here, the probability density at a point $r$ is

$$|\Psi(r, t)|^2 = |\Phi(r)|^2 \exp\left(-\frac{\Gamma}{\hbar}t\right) \tag{0.13}$$

that is, it decreases exponentially with time. For the sake of clarity we shall consider a radioactive nuclear decay accompanied by $\alpha$-particle emission. Owing to the condition of conservation of the $\alpha$-particle number in the entire space, the exponential decrease of the number of radioactive nuclei inside a sphere of radius $R$ surrounding the decaying system gives rise to an appropriate $\alpha$-particle flux through the surface of the sphere. In this case, since the number of radioactive nuclei inside the sphere is

$$N(t) \sim \int_R |\Psi(r, t)|^2 \, dr = N_0 \exp\left\{-\frac{\Gamma}{\hbar}t\right\} \tag{0.14}$$

i.e. satisfies the equation

$$\frac{dN}{dt} = -WN$$

(where $W$ is the probability of decay), we easily find from (0.14) that $\Gamma = \hbar W$. If the decay law of some quasi-stationary state is of the form:

$$N(t) = N_0 \exp\left(-t/\tau\right) \tag{0.14a}$$

then the quantity $\tau$, equalling to the time within which the number of radioactive nuclei decrease by $e$ times, is usually called the lifetime of a given quasi-stationary state. In nuclear physics, instead of the lifetime $\tau$, we usually use a quantity proportional to it: $T = \tau \ln 2$, which is called the half-life period of the given radioactive element. It is of interest to find out whence the quasi-stationary quantization conditions arise. For the sake of simplicity we shall consider the case of $S$-waves. The solution $\Phi(r)$ for the radial Schrödinger equation (0.12)

$$\left(\frac{\hbar^2}{2m}\frac{d^2}{dr^2} - U(r) + z\right)\Phi(r) = 0 \tag{0.15}$$

in the asymptotic region $r > R$ behaves, in the general case, as the superposition of two linearly independent solutions $\exp\left(\pm ikr\right)/r$ where $k = \sqrt{(2\mu z/\hbar^2)} = k_1 - ik_2$. However, only the outgoing wave at infinity corresponds to the quasi-stationary state decay (whereas the ingoing wave corresponds to the "forming" state), that is, the boundary condition for the solution (0.15) will be

$$\lim_{r\to\infty}\left(\frac{\Phi'(r)}{\Phi(r)}\right) = i\sqrt{\frac{2\mu z}{\hbar^2}} = i(k_1 - ik_2) \tag{0.16}$$

where $k_1 \geq 0$. In the general case the origin-regular solution for the Schrödinger equation (0.15) is a superposition of the ingoing and outgoing waves. If we limit ourselves only to the outgoing waves, the coefficient at the ingoing wave will vanish. As in the case of bound states, this condition yields a discrete set of eigenenergies which, however, will be complex here: $z = E_R - i\Gamma/2$, i.e. the discrete spectrum at complex energies arises. In turn, the complex eigenvalues determine the energies and widths of the quasi-stationary states.

As follows from (0.16) for $k_2 > 0$, an increasing exponential function, i.e. $\exp(k_2 r)$ enters the asymptotic part of the Gamow wave function, thereby resulting in a divergence of the normalization integrals and in other complications when constructing a resonance state theory. The methods for overcoming these difficulties are now well known and are explained in the subsequent chapters of the book.

Now we go over to the *dynamic* manifestations of the resonance in scattering. Below we shall briefly explain only the basic conclusions, referring the reader to the well-known text-books [1–6] for details. We shall use an approach based on wave packets [1,6] which may give rise to two extreme cases outlined below.

(i) Let us now suppose that the width $\Delta E$ of an incident wave packet is large and much in excess of the resonance width $\Gamma$, that is $\Delta E \gg \Gamma$. As follows from the uncertainty principle, the dimensions $\Delta x$ of the packet are such that

$$\Delta x \sim 1/\Delta p = v_0/\Delta E \qquad (0.17)$$

where $v_0$ is the particle velocity.

The uncertainty in the quasi-stationary state production time is

$$\Delta t_q \sim \frac{\Delta x}{v_0} \sim \frac{1}{\Delta E} \ll \frac{1}{\Gamma} \sim \tau$$

where $\tau$ is the lifetime of the quasi-stationary state. This means that under the condition $\Delta E \gg \Gamma$ the inequality $\Delta t_q \ll \tau$ is valid. In this case the incident wave packet is so well localized in space that the time of the quasi-stationary state production is properly separated from the subsequent decay of the state. In other words, if we want to measure the decay law of a long-lived state, we must use the sufficiently broad (in energy) wave packets satisfying the condition $\Delta E \gg \Gamma$. In the case of properly localized packets we may derive the famous Wigner condition [27] which relates the causality principle to the energy dependence of phase shift. Let us examine a sufficiently narrow incoming wave packet (whose width, however, is much in excess of $\Gamma$):

$$\Psi_{in}(r, t) = \int dk \varphi_l(k)\, e^{-i[kr + W(k)t]} \qquad (0.18)$$

where the spectral function $\varphi_l(k)$ is assumed to be sufficiently monochromatic.

Then the packet (0.18) has a maximum where the phase of the exponential function is stationary, i.e. where

$$\frac{\mathrm{d}\,(kr + W(k)t)}{\mathrm{d}k} = 0 \tag{0.19}$$

i.e. the coordinate of the centre of this maximum is

$$r_c = -\frac{\mathrm{d}W}{\mathrm{d}k}\,t = v_g\,t\,. \tag{0.20}$$

In other words the packet centre has a group velocity $v_g = -\,\mathrm{d}W/\mathrm{d}k$. After scattering, the outgoing wave packet may again be presented in a form analogous to (0.18), but with the phase shifted by $2\delta_l(k)$:

$$\Psi_{\mathrm{out}}(r,\,t) = \int \mathrm{d}k\varphi_l(k)\,\mathrm{e}^{\mathrm{i}\,[kr\,+\,W(k)t\,+\,2\delta_l]}\,. \tag{0.18a}$$

Its maximum is located at the point

$$r_d = -\frac{\mathrm{d}W}{\mathrm{d}k}\,t - 2\frac{\mathrm{d}\delta_l}{\mathrm{d}k}\,. \tag{0.21}$$

i.e. it falls behind the maximum of the corresponding *nonscattered* wave by a distance

$$\Delta r = -2\frac{\mathrm{d}\delta_l}{\mathrm{d}k} = -2v_0\frac{\mathrm{d}\delta_l}{\mathrm{d}E} \tag{0.22}$$

(as has been stated by the scattering theory, the role of a scatterer reduces to production of an additional phase shift $\delta_l(k)$ in the scattered wave). That is, when the phase shift increases with energy, the outgoing scattered wave falls behind the outgoing, but nonscattered, wave by time

$$\Delta t_{sc} = \frac{\Delta r}{v_0} = 2\frac{\mathrm{d}\delta_l}{\mathrm{d}E}\,. \tag{0.23}$$

Using the Breit-Wigner parametrization of the resonance phase-shift, we find

$$\Delta t_{sc} \approx \frac{\Gamma}{(E - E_R)^2 + \Gamma^2/4}$$

i.e., at the point of resonance $E = E_R$ a maximum in the delay time arises:

$$(\Delta t_{sc})_{\max} = 4/\Gamma\,. \tag{0.24}$$

In this case it turns out [1] that the existence of a time delay maximum is by itself sufficient for the Breit-Wigner behaviour of scattering amplitude to occur near the resonances.

A general formula relating the time delay $\Delta t$ to the $S$-matrix behaviour in the one-channel case [1]

$$\Delta t = -i\,\frac{d\ln S}{dE} \tag{0.25}$$

may be generalized directly to the many-channel case [28].

In the case where the phase shift decreases with increasing energy the value of $\Delta r$ given by (0.22) is positive and hence the outgoing scattered wave *leads* in phase with respect to the nonscattered wave. Physically this can be explained by the incident wave reflection from the "surface" of the interaction potential or in the case of a strong attractive potential, by a higher velocity of the waves inside the attractive region compared with free space.

Now we shall use the causality condition to infer the Wigner condition. If the interaction potential is of a finite radius $r_0$, it is evident that

$$\Delta r \leqq 2r_0 \tag{0.26}$$

because the outgoing waves cannot leave the interaction region earlier than an incoming wave arrives. From this, considering (0.22), we find the Wigner condition

$$\frac{d\delta_l(k)}{dk} \geq -r_0 \tag{0.27}$$

which does not curb the phase shift increase rate, but curbs the rate of the phase shift *decrease* by the interaction radius. The condition (0.27) is invalid at low energies where the wave length is significant and the $S$ wave prevails in scattering, while the wave packet cannot be localized in the interaction region. In this case a more exact form of the Wigner condition is

$$\frac{d\delta_0}{dk} \geq -r_0 + \frac{\sin 2(kr_0 + \delta_0)}{2k}. \tag{0.28}$$

At $kr_0 \ll 1$ we get $\sin(kr_0 + \delta) \to k(r_0 - a)$, where $a$ is the scattering length, and (0.28) reduce to the well-known relation

$$\frac{d\delta_0}{dk} = -a,$$

so that the values of $d\delta_0/dk$ for the near-threshold bound states may turn out to be in any way negatively high. It is clear that in the case examined, when $\Delta E \gg \Gamma$, the *observed* cross section *does not exhibit* the Breit-Wigner energy dependence because the interval of the true cross section averaging is $\Delta E \gg \Gamma$, so that in case of a very narrow resonance it does not practically manifest itself in the energy dependence of the cross section.

(ii) Let us now examine the opposite case of $\Delta E \ll \Gamma$, that is when an incident packet has a small energy width with respect to the metastable state width. Here, by varying step by step the incident particle energy we may experimentally trace the energy dependence of the cross section which will obey the formulae (0.10) for different reaction channels. In this case, however, the localization of the incident wave packet is

$$\Delta x \sim \frac{1}{\Delta p} = \frac{v_0}{\Delta E} \gg \frac{v_0}{\Gamma}$$

and, because according to (0.22) the scattered wave lag distance at the resonance point is

$$\Delta r \sim v_0/\Gamma$$

the dimensions of the incident packet in such an experiment is much in excess of the shift of the scattered packet with respect to the unscattered one. This means that under these condition it will be very difficult, or even impossible, to measure the time delay of the scattered packet, that is, the dependence of the metastable state decay.

Let us summarize the above in brief. In the experiments where the energy spread of the incident wave packet $\Delta E \gg \Gamma$, the packet is properly localized spatially, so that we can measure to within a sufficient accuracy the time shift of the scattered wave and hence the law of the metastable state decay. However, in this case it is very difficult, or even impossible, to detect a resonance by measuring the energy dependence of the cross section. And vice versa, in the experiments with narrow wave packets, when $\Delta E \ll \Gamma$, the energy dependence of the cross sections can readily by measured, but the time dependence of the decay is very difficult to find. Thus the two types of experiments complement each other (in the sense of the Bohr complementarity principle).

Finally, we shall very briefly discuss the problem concerning the exact law of quasi-stationary state decay. In a general case the solution may be formulated as follows. If we first form a purely quasi-stationary (i.e. Gamow) state, its decay law will be strictly exponential, as follows from (0.11)–(0.14). However, the problem is usually formulated in the following way: a certain initial state $\Psi_0(r)$ is produced (or specified) which is a superposition of a quasi-stationary state and a particle packet exhibiting a continuous spectrum. In this case, the solution of nonstationary scattering problem $\Psi(r, t)$ is known [5] to be a superposition of a nonresonance wave function $\Psi_n(r, t)$ corresponding to the particles leaving the interaction region immediately after the production of the initial state (i.e. without delay) and a resonance wave function $\Psi_R(r, t)$ taking the form of a travelling wave the leading edge of which moves at velocity $v_R = \sqrt{2E_R/m}$, that is

$$\Psi(r, t) = \Psi_n(r, t) + \Psi_R(r, t)$$

where

$$\Psi_R(r, t) \sim \frac{\Gamma}{2\pi v_k} \exp\left\{ ikr - i\frac{E_R t}{\hbar} - \frac{\Gamma}{2\hbar}\left(t - \frac{r}{v_R}\right)\right\}.$$

Therefore, it is clear that the exponential decay is violated due to the first (nonresonance) term. In particular, it is not difficult to show [1,5] that a departure from the exponential decay law arises either at very short or at very long times $t \gg \tau$. For example, Nicolaides and Beck [29] used a sufficiently general model for a decaying system to show the decay law for long times is of the form

$$|G(t)|^2 \underset{t\to\infty}{\simeq} \exp\left(-\frac{\Gamma}{\hbar}t\right) + \frac{\hbar^2}{\pi^2(4E_R^2 + \Gamma^2)}\frac{1}{t^2}$$

where $G(t)$ is the time evolution amplitude.

In the case of long times the term $t^{-2}$ will dominate and the decay gets nonexponential. A good candidate for the experimental study of departures from the exponential decay law (of an insulated quasi-stationary state) is the decay of the autoionization state of an appropriate atom [29]. If, for example, an atomic autoionization state with energy $E_R \sim 10^{-3}$ eV above the threshold is excited and its lifetimes is $\tau \sim 10^{-11}$s , then, according to the estimates [29], departures of the order of 20 % from the exponential decay law will be observed after a time interval exceeding $18\tau$.

We stress that, *as a whole*, the decay law will be exponential only if the metastable state is populated according to the "Lorentz law", that is, when the probability of a quasi-stationary state with energy E to be produced is given by the Lorentz distribution:

$$W(E) \sim \frac{\Gamma/2}{(E - E_R)^2 + \Gamma^2/4}. \tag{0.29}$$

In the opposite case, the decay law is nonexponential and, according to the Krylov-Fock theorem [30], is fully determined by the energy spectrum of the initial state.

Finally, it should be noted that all facts discussed above are associated with the decay of an isolated resonance or of several resonances totally separated from each other, whereas in the case of a strong overlapping of the levels the decay law gets essentially nonexponential. Discussion of numerous relevant problems may be found in Ref. [31–33].

*References*


1. GOLDBERGER M. L., WATSON K. M., Collision Theory, John Wiley, N. Y. 1964.
2. TAYLOR J. R., Scattering Theory, The Quantum Theory of Nonrelativistic Collisions, John Wiley, N. Y. 1972.
3. LANDAU L. D., LIFSCHITZ E. M., Quantum Mechanics (Nonrelativistic Part) (in Russian), Nauka, Moscow 1974, [English translation: 3rd ed. Pergamon Press, Oxford 1977].
4. BAZ A. I., ZEL'DOVICH Ya. B., PERELOMOV A. M., Scattering, Reactions and Decay in Nonrelativistic Quantum Mechanics, Nauka, Moscow, $2^{nd}$ ed., 1971 (in Russian). [English translation of lst ed.: Israel Program for Scientific Translations, Jerusalem, 1966].
5. FORMÁNEK J., Úvod do kvantové teorie (in Czech), Academia, Prague 1983.
6. NEWTON R. G., Scattering Theory of Waves and Particles ($2^{nd}$ ed.), Springer, New York 1982.
7. BLATT V. F., WEISSKOPF V. F., Theoretical Nuclear Physics, Wiley, N. Y.–London 1952.
8. FESHBACH H., Annals of Physics (N.Y.) *19* (1962) 287.
9. BAZ A. I., Yad. Fiz. (in Russian) *5* (1967) 229. [Sov. J. Nucl. Phys. *5* (1967) 161].
10. BADALYAN A. M., KOK L. P., SIMONOV Yu. A., POLIKARPOV M. I., Resonances in Coupled Channels in Nuclear and Particle Physics, Phys. Reports *82* (1982) No 2, 177.
11. DALITZ R. H., Resonance: Its Description, Criteria and Significance. In: Lecture Notes in Physics *211* (1984) 1.
12. BADALYAN A. M., SIMONOV Yu. A., ECHAYA, (in Russian) – V. *6* (1975) 299 [Sov. J. Part. Nucl. *6* (1975) 119].
13. ALMQVIST E., BROMLEY D. A., KUCHNER J. A., Phys. Rev. Lett. *4* (1960) 515.
14. CINDRO N., POČANIČ D., Resonances in Heavy – Ion Reactions – Structural vs. Diffractional Models. In: Lecture Notes in Physics *211* (1984) 158.
15. GREINER W., SHEID W., J. Phys. (Paris) *C6* (1971) 91:
SHEID W., GREINER W., LEMMER R., Phys. Rev. Lett. *25* (1971) 1 043.
16. EFIMOV V. N., Phys. Lett. *33B* (1970) 563; Yad. Fiz. (in Russian) *12* (1970) 1 080 [Sov. J. Nucl. Phys. *12* (1971) 589]; Nucl. Phys. *A210* (1973) 157.
17. AJZENBERG–SELOVE F., Nucl. Phys. *A320* (1979), *A248* (1975).
18. BRAYSHAW D. D., PEIERLS R. F., Phys. Rev. *177* (1969) 2 539.
19. BADALYAN A. M., SIMONOV Yu. A., Yad. Fiz. (in Russian) *21* (1975) 458 [Sov. J. Nucl. Phys. *21* (1975) 458]; SIMONOV Yu. A., Nucl. Phys. *A266* (1976) 163.
20. FILHO M. P. I., COUTINHO·F. A. B., The Peierls-Brayshaw Resonances, Preprint, Univ. de Sao Paulo 1982.
21. BADALYAN A. M., SIMONOV Yu. A., Yad. Fiz. (in Russian) *21* (1975) 890. [Sov. J. Nucl. Phys. *21* (1975) 458].
22. BADALYAN A. M., SIMONOV Yu., ZhETF (in Russian) *72* (1977) 57 [Sov. Phys. JETP *45* (1977) 29];
SIMONOV Yu. A., ZhETF (in Russian) *69* (1975) 1905 [Sov. Phys. JETP *42* (1975) 966];
See also: In: Proceed. 1977 Europ. Symp. on Few-Part. Problems in Nucl. Phys., Potsdam 1977, p. 81.
23. AFANASYEV A., KAGAN Yu. M., ZhETF (in Russian) *48* (1965) 327 [Sov. Phys. JETP *21* (1965) 215].
PINSKER Z. G., Dynamics of X-ray Scattering in Ideal Cristals, Nauka, Moscow 1974.
24. BASIC Z., SIMONS J., Int. J. Quant. Chem *14* (1980) 467; S. I. Cu, J. Chem. Phys. *72* (1980) 4 772.
25. YU C. F., HOGG C. S., COWIN J. P., WHALEY K. B., LIGHT J. C., SIBENER S. J., Isr. J. Chem. *22* (1982) 305;
MOISEYEV N., The Hermitian Representation of the Complex Coordinate Method: Theory and Application. In: Lecture Notes in Physics *211* (1984) 235.
26. GAMOW G. A., Zs. f. Physik *51* (1928) 204; *52*, 510.
27. WIGNER E. P., Phys. Rev. *98* (1955) 145; Journ. Phys. *23* (1955) 371.

28. SMITH F. T., Phys. Rev. *118* (1960) 349.
29. NICOLAIDES C. A., BECK D. R., Int. J. Quant. Chem. *XIV* (1978) 457.
30. KRYLOV N. S., FOCK V. A., ZhETF (in Russian) *17* (1947) 93.
31. LUBOSHITS V. L., PODHORETSKÝ M. I., Yad. Fiz. (in Russian) *24* (1976) 214 [Sov. J. Nucl. Phys. *24* (1976) 110].
32. LUBOSHITS V. L., Yad. Fiz. (in Russian) *27* (1978) 948 [Sov. J. Nucl. Phys. *27* (1978) 502].
33. LUBOSHITS V. L., Pisma v ZhETF (in Russian) *28* (1978) 32 [JETP Lett. *28* (1978) 30].
34. BÖHM A., Quantum Mechanics in Rigged Hilbert Space – Lecture Notes in Physics *78* (1978) 213.
    GORINI V., PARRAVACINI G., Lecture Notes in Physics *94* (1978) 219.
35. REED M. C., SIMON B., Methods of Modern Mathematical Physics Vol. IV: Analysis of operators, Academic Press, N. Y. 1978.
36. MERKURJEV S. P., FADDEEV L. D., Quantum Scattering Theory in Few-Body Systems (in Russian), Nauka, Moscow 1985.

# Chapter 1

# Mathematical and Quantum-Mechanical Background

In this book we shall use a number of mathematical methods which either are dealt with only in passing in the standard university mathematics course or are not treated at all; for example, the theory of divergent integrals and series, the Padé approximation method, and so on. Accounts of these problems may often be found only in monographs devoted to a narrow range of specialized problems or in original works. We shall also need the basic concepts and equations of the present-day quantum-mechanics theory of many-particle scattering, namely, the Faddeev-Yakubovsky equations which are not usually covered by the standard university courses on quantum mechanics. Therefore we have prefaced the main subject of the book by a concise account of complementary areas of mathematics and quantum scattering theory which provide an indispensable background for understanding the material to be presented.

## 1.1. Some Methods for Regularizing Divergent Integrals

In resonance-state theory we are continually faced with the following form of integrals:

$$I_n(k) = \int_0^\infty r^n\, e^{ikr}\, dr; \qquad n = 0,1 \tag{1.1}$$

where $k$ is a complex value. Similar integrals arise, in particular, when an attempt is made to calculate the normalization of the resonance state function or of the overlap integral of two such functions. If $k$ lies in the upper half-plane $(k = \alpha + i\beta,\ \beta > 0)$, the integrals are properly defined because the integrand $e^{ikr} = e^{-\beta r} \cdot e^{i\alpha r}$ decreases rapidly and oscillates. However, for resonance states where $k = \alpha - i\beta$ (Im $k = -\beta < 0$), the exponential function $e^{ikr} = e^{\beta r} \cdot e^{i\alpha r}$ increases and the integrals $(1.1)^{*)}$ diverge in the upper limit. Therefore expressions of the $(1.1)$ type are inadequate in physical calculations if the

---

*) In the whole text the formulas are denoted as follows. Within a given Chapter (say Chapter 5) they are referred by two numbers, e.g. (3.17). In another Chapter this formula is referred to as (5.3.17).

integrals entering the expressions are to be understood in the conventional sense. However, from the theory of divergent integrals and series [1] we know that many integrals divergent in the conventional sense may be given a definite sense by some regularization, that is, by redefining the integration procedure. To yield reasonable results, a new integration procedure has to satisfy the following requirements:

i) in the region where the integral diverges, the integration must correspond uniquely with a finite expression;

ii) in the region where the integral converges, the integration with regularization has to yield the same result as conventional integration.

The main advantage of a "correct" regularization procedure, which is very important to what follows, is that the regularized integral can be operated on as a conventional convergent integral [1]. This makes it possible to substantially extend the scope of application of the mathematical formalism developed (see below).

Such a procedure may be exemplified by the regularization procedure proposed by Zel'dovich [2], which redefines the (1.1)–type integrals as follows:

$$I_n(k) = \lim_{\varepsilon \to 0} I_n(k, \varepsilon) = \lim_{\varepsilon \to 0} \int_0^\infty e^{-\varepsilon r^2} r^n\, e^{\varkappa r}\, dr\,,$$

$$n = 0, 1; \qquad \varkappa = ik\,. \tag{1.2}$$

The meaning of the regularization is evident. Since the regularizing factor $e^{-\varepsilon r^2}$ decreases at an arbitrary finite $\varepsilon$ more rapidly than the exponential function with an arbitrary exponent $\varkappa$, the integral $I_n(k, \varepsilon)$ converges at $\varepsilon > 0$. In order to examine the behaviour of $I_n(k, \varepsilon)$ at $\varepsilon \to 0$, we replace the variables

$$t = r\sqrt{\varepsilon} - x\,; \qquad x = \frac{\varkappa}{2\sqrt{\varepsilon}}$$

in

$$I_0(k, \varepsilon) = \frac{1}{\sqrt{\varepsilon}}\, e^{x^2} \int_{-x}^\infty e^{-t^2}\, dt = \frac{1}{\sqrt{\varepsilon}}\, e^{x^2} \left[\sqrt{\pi} - \mathrm{Erfc}(x)\right]$$

where

$$\mathrm{Erfc}(x) = \int_x^\infty e^{-t^2}\, dt$$

is the probability integral [3]. By using an asymptotic expression for the function $\mathrm{Erfc}(x)$ (if $\varepsilon \to 0$, then $x \to \infty$) [3], we obtain [4]

$$I_0(k, \varepsilon) = \left(\frac{\pi}{\varepsilon}\right)^{1/2} e^{x^2} - \frac{1}{2x\sqrt{\varepsilon}}\left(1 - \frac{1}{2x^2} + \frac{3}{4x^4} - \ldots\right) =$$

$$= \left(\frac{\pi}{\varepsilon}\right)^{1/2} e^{-\frac{k^2}{4\varepsilon}} + \frac{i}{k} - \frac{2\varepsilon}{k^3} + O(\varepsilon^2) . \tag{1.3}$$

For

$$I_1(k, \varepsilon) = \frac{\partial I_0(k, \varepsilon)}{\partial \varkappa} = \frac{1}{2\varepsilon}\{\varkappa I_0(k, \varepsilon) + 1\}$$

we obtain:

$$I_1(k, \varepsilon) = -\left(\frac{\pi}{4\varepsilon^3}\right)^{1/2} ik\, e^{-k^2/4\varepsilon} - \frac{1}{k^2} + O(\varepsilon) . \tag{1.4}$$

Further, allowing for the fact that at $\mathrm{Re}\, k^2 > 0$, (i.e. at $|\mathrm{Re}\, k| > |\mathrm{Im}\, k|$) the relation

$$\lim_{\varepsilon \to 0} \varepsilon^p \exp\{-k^2/4\varepsilon\} = 0 \tag{1.5}$$

is valid at any real $p$, we obtain the following expressions for the integrals (1.1):

$$I_0(k) = i/k ,$$

$$I_1(k) = -1/k^2 . \tag{1.6}$$

In this way, by using the regularization procedure (1.2), we obtain reasonable finite expressions for the divergent integrals (1.1). The result (1.6) is quite in agreement with the result which we would have obtained if we had calculated the integrals (1.1) in the upper $k$-half-plane and had analytically continued the result to the lower half-plane. This agreement is not accidental, but is one of the necessary conditions of correct regularization.

The regularizing factor $\exp(-\varepsilon r^2)$ is not unique because the same result is obtainable using the regularizing factor [5]

$$\exp(-\varepsilon r^m) \tag{1.7}$$

where

$$1 < m < \frac{\frac{1}{2}\pi}{\mathrm{arctg}\,(-\mathrm{Im}\, k/\,\mathrm{Re}\, k)} . \tag{1.7a}$$

The formula (1.7a) makes it possible to account for the dependence of the selection of regularizing factor on the position of the point $k = \alpha - i\beta$ in the lower $k$-half-plane. Let us rewrite (1.7b) in the form:

$$1 < m < \frac{\tfrac{1}{2}\pi}{\text{arctg}\,(\beta/\alpha)} \tag{1.7b}$$

and recall that the integrand in (1.1) is equal to

$$f(r) = r^n\, e^{ikr} = r^n\, e^{\beta r}\, e^{i\alpha r}\,. \tag{1.8}$$

If the point $k$ lies in the vicinity of the real axis, i.e. $\beta/\alpha \ll 1$ , the range characterizing the increase of the function $f(\mathrm{r})$ is $\varDelta = 1/\beta$ and the oscillation length $\delta = 2\pi/\alpha$ is related to $\varDelta$ as

$$\varDelta \gg \delta\,. \tag{1.9}$$

Since the regularization of the integral is essentially due to the oscillations of the integrand $f(r)$ (in practice, the role of the regularizing factor reduces to just suppressing the function $f(r)$ at great distances), it is evident that the decrease rate of the regularizing factor (that is, the value of the exponent $m$ in (1.7)) is determined by the following qualitative reasoning. A distance at which the integrand is suppressed by the regularizing factor has to be sufficiently great for the function $f(r)$ to show a sufficiently large number of oscillations on this spacing. This imposes an upper limit on the number $m$. For example, at $\beta/\alpha \ll 1$ , and $\varDelta \gg \delta$ we find from formula (1.7b) that $m < M$, where $M = = \varDelta/\delta\pi^2 \gg 1$ , i.e. may be sufficiently high. As the ratio $\beta/\alpha$ increases i.e. as the point $k$ moves away from the real axis, $M$ decreases and at $\beta/\alpha = 1$ we obtain $M = 2$. This means that the Zel'dovich regularization is usable only at $\arg k > -\pi/4$ . At $-\pi/2 < \arg k < -\pi/4$ the regularizing factor must be of the form

$$\exp\left(-\varepsilon r^{1+\gamma}\right) \tag{1.10}$$

where $0 < \gamma < 1$ . If the point $k$ lies in the vicinity of the negative imaginary axis, then $\varDelta \ll \delta$ , that is, the function $f(r)$ varies very strongly on a segment equal to the length of one oscillation and the regularizing factor has to decrease very weakly $(\gamma \to 0)$ in order to cover a sufficient number of the oscillations of $f(r)^1)$. In the extreme $\arg k = -\pi/2, f(r)$ stops oscillating at all and the (1.7) – type regularization due to oscillations becomes impossible. In the case of $\arg k = -\pi/2$ , which corresponds to virtual states, it is necessary either to use an oscillating regularizing factor in which case the regularization is due to the

---

<sup></sup>$^1)$ This leads to the fact that such a procedure is applicable in practice only to sufficiently narrow resonances. In case of broader resonances the integrand decreases very slowly, so that even a regularized integral is difficult to calculate.

oscillations of the factor proper [6], or to make an analytic continuation (see Section 1.3).

Concluding this section, we shall formulate the general requirements on the regularizing factor [1,4–6]. We determine the value $I$ of the definite integral

$$I = \int_0^\infty f(r)\, dr$$

by introducing an auxiliary function $\omega(\varepsilon, r)$ with subsequent movement to a limit of the form

$$I = \lim_{\varepsilon \to 0} \int_0^\infty \omega(\varepsilon, r) f(r)\, dr . \tag{1.11}$$

In the case where the integral exists in the conventional sense, it is required that our definition (1.11) should yield the same value. Here the function $\omega(\varepsilon, r)$ is called the regularizing factor and its derivative

$$\tau(\varepsilon, r) = -\frac{d}{dr}\, \omega(\varepsilon, r)$$

has to satisfy the following conditions:

$$\text{(i)} \quad \int_0^\infty |\tau(\varepsilon, r)|\, dr < M , \qquad \text{where } M \text{ is independent of } \varepsilon ;$$

$$\text{(ii)} \quad \int_0^\infty \tau(\varepsilon, r)\, dr \to 1 \qquad \text{at } \varepsilon \to 0 ;$$

$$\text{(iii)} \quad \int_0^x \tau(\varepsilon, r)\, dr \to 0 \qquad \text{at } \varepsilon \to 0 \ \text{ for any finite } x. \tag{1.12}$$

The Zel'dovich regularizing factor, $\exp\left(-\varepsilon r^2\right)$ and the factor $\exp\left(-\varepsilon r^m\right)$ seem to be particular cases of the function $\omega(\varepsilon, r)$ satisfying the conditions (1.12). It is clear that the regularization (1.11) may be used to calculate not only the (1.1)–type integrals but also more complicated integrals which arise when calculating the matrix elements with resonance state functions.

## 1.2. Padé Approximants and their Applications

Consider the function $f(z)$ which is analytic in the vicinity of the point $z = 0$. The Taylor-series expansion of the function is

$$f(z) = a_0 + a_1 z + a_2 z^2 + \ldots = \sum_{n=0}^{\infty} a_n z^n . \tag{2.1}$$

This series is convergent in a circle $|z| < R$ whose radius $R$ is determined by the nearest singularity of $f(z)$. The simplest approximation for the function $f(z)$ is by a finite number of terms of the series (2.1), i.e. by a polynomial. As the number of terms of this series increases, such an approximation converges only at $|z| < R$ because the polynomial can reproduce zeros of the function, but cannot simulate even the simplest singularity (a first-order pole).

The Padé approximation is a more general and effective method of approximate representation of functions.

The Padé approximant (PA) $f^{[N,M]}(z)$ of the function $f(z)$ is defined as a rational fractional expression of the form [7]:

$$f^{[N,M]}(z) = P_N(z)/Q_M(z) = f(z) + O(z^{N+M+1}) \tag{2.2}$$

where $P_N$ and $Q_M$ are polynomials of degrees $N$ and $M$, respectively. The Padé approximants with $N = M$ are called the diagonal PA. The coefficients of the polynomials $P_N$ and $Q_M$ are determined unambiguously either by the first $N + M + 1$ coefficients of the Taylor series of the function $f(z)$ (2.1) (in this case the PA (2.2) are called PA of type I) or by the values of the function $f(z_i)$ at $N + M + 1$ points $\{z_i\}$ $i = 1, 2, \ldots, N + M + 1$ (such an approximant is called the PA of type II), or finally by the values of the function $f(z_i)$ given with errors $\{\varepsilon_i\} = 1, 2, \ldots, K$ at $K(K > N + M + 1)$ points $\{z_i\}$ (such PA is called the PA of type III).

The general PA form (2.2) leads us immediately to several conclusions:

(i) it is evident that a rational fraction, compared with polynomials, can approximate a much wider class of functions. Since the rational fraction has poles (zeros of $Q_M$ in (2.2)), the function $f(z)$ may also be approximated by such a construction near a finite-order pole. Indeed, if $f(z)$ has a pole of order $m$ at a point $z_0$, the Laurent-series expansion of the function in a minor vicinity of this point gives:

$$f(z) = \sum_{n=m}^{1} \frac{a_{-n}}{(z - z_0)^n} + \sum_{n=0}^{\infty} a_n (z - z_0)^n . \tag{2.3}$$

On truncating the series $(2.3)$ at a certain $n = k$, we obtain

$$f(z) \simeq \frac{a_{-m}}{(z - z_0)^m} + \frac{a_{-m+1}}{(z - z_0)^{m-1}} + \ldots + a_0 + \ldots + a_k(z - z_0)^k =$$

$$= \frac{P_{m+k}(z)}{Q_m(z)}$$

that is, a rational fractional expression of the $(2.2)$ type;

(ii) if the PA $(2.2)$ is Taylor series-expanded, the first $N + M + 1$ terms of the resultant series will coincide with the first $N + M + 1$ terms of the series $(2.1)$ which may be used to construct the PA $(2.2)$. In this case, however, the PA expansion will also contain the terms of all orders higher than $N + M + 1$. In such a way, PA is used to replace the unknown coefficients of the Taylor series by a smooth extrapolation. The unknown coefficients are expressed through the first $N + M + 1$ coefficients of the Taylor series and depend smoothly on $n$. Thus, we see that, in addition to the ability to reproduce the polar singularities of the function $f(z)$, the PA is able to extrapolate, in a smooth manner, the unknown coefficients of the Taylor series using the known coefficients. It is evident in this case that it makes no sense to use PA when approximating functions for which the coefficients of the Taylor series depend irregularly on the number $n$. It should be noted that such functions are very rarely observed in physical applications.

### 1.2.1. Basic Properties of the Padé Approximants

We shall mention here the main properties of the Padé approximants without proofs; for the proofs the reader is referred to the literature [7].

(i) *Uniqueness.* From the definition $(2.2)$ or from the equivalent formula

$$Q_M(z)f(z) - P_N(z) = O(z^{N+M+1}) \tag{2.4}$$

it can be seen that the coefficients of the polynomials $P_N(z)$ and $Q_M(z)$ are determined by the first $N + M + 1$ coefficients of the Taylor series. The coefficients of PA may be found by solving a set of linear equations accurate to within a common multiplier. Consequently, $f^{[N,M]}(z)$ is a unique PA. The polynomials $Q_M$ and $P_N$ are expressed through the coefficients $a_i$ of the Taylor series $(2.1)$ using the following formulas [7]:

$$P_N(z) = \begin{vmatrix} a_{N-M+1} & a_{N-M+2} & \cdots & a_{N+1} \\ \vdots & \vdots & & \\ a_N & a_{N+1} & \cdots & a_{N+M} \\ \sum_{j=M}^{N} a_{j-M}z^j & \sum_{j=M-1}^{N} a_{j-M+1}z^j & \cdots & \sum_{j=0}^{N} a_j z^j \end{vmatrix} \tag{2.5}$$

and

$$
Q_M(z) = \begin{vmatrix} a_{N-M+1} & a_{N-M+2} & \cdots & a_{N+1} \\ \vdots & \vdots & & \vdots \\ a_N & a_{N+1} & \cdots & a_{N+M} \\ z^M & z^{M-1} & \cdots & 1 \end{vmatrix}
$$

(2.6)

For instance, for $N = M = 1$ we obtain:

$$
f^{[1,1]}(z) = \frac{a_0 a_1 + (a_1^2 - a_0 a_2)z}{a_1 - a_2 z} .
$$

(2.7)

For the series

$$
f(z) = \sum_{n=0}^{\infty} z^n
$$

all $a_n = 1$ and the simplest PA

$$
f^{[1,1]}(z) = \frac{1}{1 - z}
$$

(2.8)

immediately gives the exact sum of the series.

(ii) *Invariance of the PA form with respect to rational-fractional transformation of a function.* Let $\hat{A}$ be a fractional rational transformation, i.e.

$$
g(z) = \hat{A}f(z) = \frac{af(z) + b}{cf(z) + d} ;
$$

(2.9)

then,

$$
g^{[N,N]}(z) = \hat{A}f^{[N,N]}(z) = \frac{af^{[N,N]}(z) + b}{cf^{[N,N]}(z) + d} = \frac{\tilde{P}_N(z)}{\tilde{Q}_N(z)} ,
$$

(2.10)

i.e. we get again the diagonal PA $[N,N]$.

(iii) *Invariance of the PA form with respect to conformal transformation of an argument.* Let

$$
z = \hat{B}z' = \frac{az'}{1 + bz'} ; \qquad g(z') = f(\hat{B}z')
$$

(2.11)

Then,

$$f^{[N,N]}(z) = f^{[N,N]}\left(\frac{az'}{1 + bz'}\right) = \frac{P_N\left(\dfrac{az'}{1 + bz'}\right)(1 + bz')^N}{Q_N\left(\dfrac{az'}{1 + bz'}\right)(1 + bz')^N} =$$

$$= \frac{R_N(z')}{S_N(z')} = g^{[N,N]}(z') ,$$

(2.12)

i.e. the diagonal PA form is preserved in the case of the rational-fractional transformation of the argument.

The property (iii) is very important because it is this property that gives rise to the convergence of the PA sequence outside the Taylor-series convergence radius. We shall illustrate it by a simple example.

Let the function $f(z)$ have a singularity at $z = -2$ and all other singularities of the function lie inside a circle of a unit radius with its centre at the point $z = -3$. The Taylor series for $f(z)$ converges in the circle $|z| < 2$ and cannot be used directly to calculate, for example, $f(2)$ and $f(\infty)$. Now we shall replace the variable

$$z = \frac{8y}{1 - 3y} .$$

(2.13)

Conformal mapping (2.13) transforms a part of the $z$ plane outside a unit circle with its centre at the point $z = -3$ into a unit circle with its centre at the coordinate origin in the $y$-plane. Thus, for a function $g(y) = f[8y/(1 - 3y)]$ the Taylor series in the variable $y$ converges in the region $|y| < 1$. The points $z = 2(y = 1/7)$ and $z = \infty(y = 1/3)$ lie inside the convergence circle, so that the values of $f(2) = g(1/7)$ and $f(\infty) = g(1/3)$ may be calculated using the Taylor series for the function $g(y)$. Such series summation method is called the Euler summation method, and the transformation of the type (2.13) – the Euler transformation. We shall now construct PA for $g(y)$

$$g^{[N,N]}(y) = \frac{P_N(y)}{Q_N(y)} .$$

By virtue of (2.12) we get

$$g^{[N,N]}(y) = \frac{P_N(z)}{Q_M(z)} = f^{[N,N]}(z) .$$

(2.14)

Thus, we see that, to calculate $f(2)$ and $f(\infty)$ with the help of the Taylor series, it is necessary to realize the conformal mapping (2.13). To this end we must know the location of the singularities of the function $f(z)$. After that we have to Taylor series-expand $g(y)$ and to sum some segment of the series. If we use PA, then, as can be seen from (2.14), it is sufficient to construct PA $f^{[N,N]}(z)$ using the known Taylor series for the function $f(z)$. According to (2.14), it is equal to $g^{[N,N]}(y)$. Therefore, if the sequence $g^{[N,N]}(y)$ converges for $N \to \infty$ in the convergence circle of the Taylor series for $g(y)$ ($|y| < 1$), then the sequence $f^{[N,N]}(z)$ converges in the entire analyticity region of $f(z)$ ($|z + 3| > 1$). In other words, to calculate $f(2)$ and $f(\infty)$, it is sufficient to compute $f^{[N,N]}(2)$ and $f^{[N,N]}(\infty)$ because $f^{[N,N]}(z)$ will automatically make the conformal mapping (2.13) and, moreover, any transformation of the type (2.11). In this case it is not necessary to know the location of the singularities of $f(z)$ in advance. Thus, the Euler procedure is inherent to the PA form proper.

We have examined this simple example in so much detail to explain a very important property of PA invariance with respect to the conformal (2.11)-type transformation of the argument. This property helps us understand why the convergence region is not a circle, as is the case for the series, and proves mostly to be larger than the Taylor series convergence circle and why the PA are so effective in making the analytic continuation of the function $f(z)$ outside the Taylor-series convergence circle. In fact, if we have proved the convergence of the PA inside the convergence circle of the Taylor series, we shall immediately prove, using the (2.11)-type conformal mapping, the PA convergence in a much wider region. Thus, the function poles are not singular points constraining the PA convergence region

(iv) *Unitarity*. In their application to scattering theory the Padé approximants are often used to approximate the partial-wave $S$-matrix $S(z)$, i.e. a complex--valued function, which is known to satisfy the important unitarity condition [8]:

$$S_l(z)S_l^*(z^*) = 1 .$$

The diagonal PA retain this property of the S-matrix:

$$\{f^*(z^*)\}^{[N,N]} = \{f^{[N,N]}(z^*)\}^* . \tag{2.15}$$

(v) *Convergence of the PA sequence*. The statement that PA $f^{[N,M]}(z)$ converges in the region $z \in D$ to a function $f(z)$ is meant to be

$$f(z) = \lim_{\substack{N \to \infty \\ M \to \infty}} f^{[N,M]}(z) , \qquad \text{for } z \in D . \tag{2.16}$$

A strict proof of the convergence of PA is quite complicated; therefore, rigorous convergence theorems have only been proved in several special cases of the function $f(z)$. Moreover, even a rigorous proof of the fact that PA converges within the Taylor series convergence circle is not available. Baker, Gammel and Wills [46] have only formulated the so called Padé conjecture about the convergence of the Padé-approximant for a function $f(z)$ meromorphic in a circle D including the origin. Here we shall present several cases for which the convergence has been proved rigorously.

*The Stieltjes functions.* Such functions may be defined using the series

$$f(z) = \sum_{j=0}^{\infty} f_j(-z)^j \tag{2.17}$$

which represents a Stieltjes function if and only if there exists a non-increasing function $\varphi(u)$ taking infinitely many values in the interval $0 \le u < \infty$ such that

$$f_j = \int_0^{\infty} u^j \, d\varphi(u) . \tag{2.18}$$

The Stieltjes function may also be defined by the integral form

$$f(z) = \int_0^{\infty} \frac{d\varphi(u)}{1 + uz} . \tag{2.19}$$

The function has a cut along the negative real axis $z(-\infty < z \le 0)$. The PA $f^{[N,M]}(z)$ for the Stieltjes function converges uniformly in the entire cut $z$-plane [7]. In this case the cut is simulated by alternating PA zeros and poles.

The following proposition has been proved for a *meromorphic function*[2]) [9,10]: let $f(z)$ be a meromorphic function in the entire $z$-plane and let $\varepsilon$, $\delta$ and $R$ be three arbitrary positive numbers. Then there exists an integer $N$ such that $|f(z) - f^{[n,n]}(z)| < \varepsilon$ for all $n \ge N$ in the region $|z| < R$, except for the subregion $D_n$ of a measure smaller than $\delta$. Of course, it is impossible to localize the position of the subregion $D_n$ in the region $|z| < R$ .

The proofs of stronger propositions in a general form have failed. However, on the basis of a great number of mathematical examples and numerical calculations, the conjecture has been formulated [7] that a PA converges in the Taylor series convergence region and in an overwhelming majority of cases it converges much more rapidly than the series; according to the property (iii), this means convergence in a much wider region. Examples refuting this conjecture are unknown to the authors.

Concluding this section, we shall examine some examples illustrating the effectiveness of the use of PA instead of the Taylor series.

The Taylor series for the function $f(z) = \sqrt{1 + z}$ is

$$\sqrt{1 + z} = 1 + z/2 - z^2/8 + ... \tag{2.20}$$

---

[2]) The single-valued function $f(z)$ is called meromorphic in the domain $D$ if a finite number of poles are its only singularities in $D$.

Let $S_3(z) = 1 + z/2 - z^2/8$ denote the sum of three first terms of the series. We can construct the simplest $f^{[1,1]}(z)$ Padé approximant using the first three coefficients (see eq. (2.7)):

$$f^{[1,1]}(z) = \frac{4 + 3z}{4 + z} . \tag{2.21}$$

Table 1.1 compares the exact values of $f(z)$ with the sum of the three terms of the Taylor series and with the PA constructed by means of the same three terms at a number of the values of the argument $z$. The convergence region of

Table 1.1

| $z$ | $S_3(z)^*)$ | $f^{[1,\,1]}(z)$ | $f(z) = \sqrt{1 + z}$ |
|---|---|---|---|
| $-0.5$ | 0.718 75 | 0.714 29 | 0.707 11 |
| 0.5 | 1.218 75 | 1.222 22 | 1.224 74 |
| 1.0 | 1.375 00 | 1.4 | 1.414 21 |
| 2.0 | 1.5 | 1.666 67 | 1.732 05 |
| 3.0 | 1.375 | 1.857 14 | 2.0 |
| 4.0 | 1.0 | 2.0 | 2.236 07 |

$^*)$ $S_3(z)$ is the sum of the three first terms of the series (2.20); $f^{[1,\,1]}(z)$ is Padé-approximant (2.21) to $f(z)$.

Table 1.2

| $N$ | $f^{[N,\,N]}(1)$ |
|---|---|
| 1 | 1.4 |
| 2 | 1.413 79 |
| ... | ... |
| 5 | 1.414 213 55˙ |
| Exact value | $\sqrt{2} = 1.414\ 213\ 561\ ...$ |

the series (2.20) is $|z| < 1$ ; therefore, at $z > 1$ $S_3(z)$ has little in common with the corresponding values of $\sqrt{1 + z}$. The PA $f^{[1,1]}(z)$ not only gives a better result for $|z| < 1$, but also approximates properly the function $\sqrt{1 + z}$ beyond the convergence radius. The examined function has a cut in the $z$-plane along the negative real semi-axis from $z = -1$ to $z = -\infty$. The PA (2.21) has a zero at the point $z = -4/3$ and a pole at the point $z = -4$, that is, the zero and the pole lie on the cut. This is not an accidental coincidence because, as the PA order increases, zeros and poles alternate and group on the cut, thereby simulating the latter.

Using the example of the same function we shall examine the convergence of the PA $f^{[N,N]}$ as the order of $N$ increases. We shall consider the PA $f^{[N,N]}(z)$ at the point $z = 1$, $f(z = 1) = \sqrt{2}$.

As can be seen from Table 1.2 the convergence of PA to an exact result is very rapid; at $N = 5$ as many as *eight* significant digits are reproduced correctly.

## 1.2.2. Padé Approximants of Type II (PA II).

In many practical cases we are faced with the problem concerning the approximation of an analytic function specified at a finite number of points. Let $z_1$, $z_2$, ... $z_p$ be $p$ complex numbers and let the analytic function $f(z)$ take on the values $f(z_i) = f_i$ at these points. To approximate $f(z)$ at other points $z$, we may construct a type II PA

$$f^{[N,M]}(z) = \frac{P_N(z)}{Q_M(z)}$$

where $N + M + 1 = p$ and the coefficients of polynomials $P_N$ and $Q_M$ are determined unambiguously by two sets of reference points $\{z_i\}$ and $\{f(z_i)\}$ through the solution of simultaneous linear equations

$$f^{[N,M]}(z_i) = f(z_i), \qquad i = 1, 2, ..., p$$

or, in other words

$$Q_M(z_i)f_i - P_N(z_i) = 0 \ . \tag{2.22}$$

The PA II have the same properties as the PA I [11]. However, it is clear that the PA II rate of convergence will be affected by the location of the reference points $\{z_i\}$ in the analyticity region of the function $f(z)$.

It should be noted that at a great number of reference points (and, hence, at a high order of PA) the set of linear equations (2.22) becomes ill-conditioned. In this case, to construct the PA $f^{[N,M]}$, we may use the technique of continued fractions [12]. This technique makes it possible to construct the PA coefficients on the basis of the values of $\{f_k\}$ and $\{z_k\}$ using the recurrrent formulae (without solving the set (2.22)). Such a procedure is more stable; at very high values of $p$, however, it also becomes sensitive to even minor errors in the initial data and to the rounding-off errors. Furthermore, for reasons associated with a particular characteristic of the problem to be solved, the maximum order of PA must often be *limited*. Here a situation arises where $p > N + M + 1$. In such a case we may proceed in two ways [13] (for simplicity we shall consider $\{f_k\}$ and $\{z_k\}$ to be real):

(i) It is possible to minimize the mean-square deviation of the function $Q_M(z)f(z)$ from $P_N(z)$ at points $\{z_k\}$, that is, to minimize (see (2.4))

$$I = \frac{1}{p} \sum_{k=1}^{p} |Q_M(z_k) f_k - P_N(z_k)|^2 \qquad (2.23)$$

with respect to the variations of the coefficients $\{a_i\}$ and $\{b_j\}$ of the polynomials

$$P_N(z) = \sum_{j=0}^{N} a_j z^j \quad \text{and} \quad Q_M(z) = 1 + \sum_{j=1}^{M} b_j z_j, \text{ respectively. Such a minimization}$$

leads, in the standard way, to a set of linear equations having dimensions of $N + M + 1$: to find $\{a_j\}$ and $\{b_j\}$

$$\begin{cases} \sum_{i=0}^{N} a_i f_{im} - \sum_{j=1}^{M} b_j f_{jm}^{(2)} = f_{0m}^{(2)}; & m = 1, 2, ..., M \\ \sum_{i=0}^{N} a_i z_{il} - \sum_{j=1}^{M} b_j f_{jl} = f_{0l}; & l = 0, 1, ..., N \end{cases} \qquad (2.24)$$

where

$$f_{ij} = \sum_{k=1}^{p} z_k^{i+j} f_k; \qquad f_{ij}^{(2)} = \sum_{k=1}^{p} z_k^{i+j} f_k^2; \qquad z_{ij} = \sum_{k=1}^{p} z_k^{i+j}. \quad (2.25)$$

The coefficients $\{a_i\}$ and $\{b_i\}$ are determined unambiguously from eqs. (2.24);

(ii) it is possible to determine the PA coefficients $\{a_i\}$ and $\{b_i\}$ by equating the first $N + M + 1$ moments of the expression $f(z)Q_M(z) - P_N(z)$ to zero, that is

$$\sum_{k=1}^{p} z_k^l [Q_M(z_k) f(z_k) - P_N(z_k)] = 0; \qquad l = 0, 1, ..., N + M.$$

This requirement leads to a set of $N + M + 1$ equations to find $\{a_i\}$ and $\{b_i\}$

$$\sum_{i=0}^{N} a_i z_{il} - \sum_{j=1}^{M} b_j f_{jl} = f_{0l}; \qquad l = 0, 1, ..., N + M. \qquad (2.26)$$

For the notation of $z_{ij}$, $f_{ij}$ and $f_{0j}$, see (2.25). A detailed comparison between the various methods for constructing PA II is made in Refs. [7, 12, 13].

In the general case the problem of how to approximate the analytical functions in mathematics has many different aspects and is treated in numerous publications (see, for example, [12, 14]). We shall only mention here those aspects of the problem which are important to understanding the material of the subsequent chapters.

### 1.2.3. Padé Approximants of Type III (PA III).

The problem of approximating the analytic function $f(z)$ given at a finite number of points $\{z_k\}$ gets often complicated because the function value is specified at the points within some errors $\{\varepsilon_k\}$. Such a situation arises also when the function to be approximated is a numerical solution of an integral or differential equation (in this case the level of the errors $\{\varepsilon_k\}$ depends on the numerical methods used) and when the function results from physical measurements (in this case the value of $\{\varepsilon_k\}$ is determined by experimental accuracy).

In such cases, the parameters of the approximating functions are found using the $\chi^2$ minimality criterion (for simplicity we shall consider $\{f_i\}$, $\{z_i\}$, and $\{\varepsilon_i\}$ to be real):

$$\chi^2 = \sum_{k=1}^{p} \frac{1}{\varepsilon_k^2} \left| f_k - \frac{P_N(z_k)}{Q_M(z_k)} \right|^2 . \tag{2.27}$$

However, direct minimization of (2.27) with respect to the coefficients of the polynomials $P_N$ and $Q_M$ leads to non-linear expressions. Several methods for avoiding this difficulty have been proposed [15–17]. Here we shall describe briefly one such method [16] which makes it possible to reduce the problem of the $\chi^2$-minimization to successive iterations each of which is a linear problem. The algorithm of the method is very simple. We reduce the expression in (2.27) under the modulus sign to a common denominator and write the ansatz determining $\chi^2_{S+1}$ at the $s + 1$-th step

$$\chi^2_{S+1} = \sum_{k=1}^{p} |Q_M^{(S+1)}(z_k) f_k - P_N^{(S+1)}(z_k)| / [\varepsilon_k^{(S+1)}]^2 . \tag{2.28}$$

where $Q_M^{(0)} \equiv 1$; $\varepsilon_k^{(S+1)} = \varepsilon_k Q_M^{(S)}(z_k)$ and the polynomial $Q_M^{(S)}$ is determined by substituting the coefficients found at the previous step.

The minimization of $\chi^2_{S+1}$ with respect to the coefficients $\{a_j\}$ of the polynomial $P_N^{(S+1)}(z)$ and $\{b_j\}$ of the polynomial $Q_M^{(S+1)}(z)$ leads to a set of linear equations (2.24) where, however, the coefficients are slightly redefined:

$$f_{ij} = \sum_{k=1}^{p} \frac{z_k^{i+j} f_k}{[\varepsilon_k^{(S+1)}]^2}; \quad f_{ij}^{(2)} = \sum_{k=1}^{p} \frac{z_k^{i+j} f_k^2}{[\varepsilon_k^{(S+1)}]^2}; \quad z_{ij} = \sum_{k=1}^{p} \frac{z_k^{i+j}}{[\varepsilon_k^{(S+1)}]^2} .$$

$$\tag{2.29}$$

The effectiveness of the method and the iteration convergence rate for a number of trial functions is illustrated in Ref. [17]. In Chapter 6 this method will be used to derive the parameters of resonances from experimental data (see also ref. [16]).

### 1.2.4. Application of Padé Approximants

At present PA are used in many areas of physics and mathematics. Many articles and monographs describe their applications (see, e.g., [7, 9–11, 13, 15–18, 46]). Here we shall discuss three applications: (i) summation of poorly convergent and divergent series, which is very important when using perturbation theory; (ii) solution of the integral equations of scattering theory; (iii) determination of the zeros, poles and branch points of a function. Another important application of PA is to make analytic continuation of a function outside the convergence circle of the Taylor series. This application was discussed briefly above and will be considered in more detail in Section 1.3.

(i) *Summation of poorly convergent and divergent series.* We are often faced with a situation where a physical problem cannot be solved in a closed analytic form, but the solution may nevertheless be written in the form (2.1) in some parameter $z$ using perturbation theory. However, at a physical value of the expansion parameter (henceforth we shall consider $z_0 = 1$ to be such a physical value), the obtained series either converges poorly or even diverges in many cases. At the same time, in the majority of cases the solution of the problem is an analytic function of $z$ at $z = z_0$ and the divergence of the series is due to a singularity lying within the unit circle, but far from the physical domain of the values of $z$.

As was shown in Section 1.2.1 (see Tables 1.1 and 1.2) the Padé approximants converge more rapidly than the power series in the convergence region of the series and they also converge where the series diverges. Now we shall show that the definition of a generalized sum of a series corresponds to the summation on the basis of the Padé approximants. Let us consider a sequence $S_k$ of partial sums of the series (2.1):

$$S_k = \sum_{i=0}^{k} a_i z^i . \tag{2.30}$$

Let us assume that the partial sum $S_{k+1}$ may be determined in terms of $S_k$ using some functional relation $g$:

$$S_{k+1} = g(S_k) .$$

It may obviously be assumed that the function $g$ exists and is unique.

Then, the sum of series (2.1) is defined to be the solution of the functional equation

$$S = g(S) . \tag{2.31}$$

If the series (2.1) converges, then (2.31) defines the ordinary sum of the series, if (2.1) diverges, then (2.31) defines the generalized sum [1].

We shall present several examples of such summation of series. We shall consider the well-known series, the geometric progression

$$S = 1 + z + z^2 + z^3 + \ldots$$

which may be rewritten as

$$S = 1 + z(1 + z + z^2 + z^3 + \ldots) = 1 + zS .$$

The resultant equation is an equation of the type (2.31) for the geometric progression, whence we get immediately the classical result:

$$S = \frac{1}{1 - z} .$$

As an example of series which do not have sums in the conventioal sense, but have them in the sense of (2.31), we consider the series

$$S = 1 - 1 + 1 - 1 + 1 - \ldots =$$
$$= 1 - (1 - 1 + 1 - 1 + 1 - \ldots) = 1 - S ,$$

i.e., $S = 1/2$
and

$$S = 1 - 2 + 3 - 4 + 5 - 6 + \ldots =$$
$$= 1 - (2 - 3 + 4 - 5 + \ldots) =$$
$$= 1 - (1 - 1 + 1 - 1 + 1 - \ldots) -$$
$$- (1 - 2 + 3 - 4 + 5 - \ldots) = 1 - 1/2 - S ,$$

i.e. $S = 1/4$.

It can be shown [19] that the Padé method is effective for solving the equation (2.31). In other words, for a convergent series the Padé approximants make the convergence more rapid, and for a divergent series they define a generalized sum in the sense of (2.31).

All this evidently applies to numerical series, because any numerical series

$$S = \sum_{i=0}^{\infty} a_i$$

may be considered to be the value of the function $S(z) = \sum_{i=0}^{\infty} a_i z^i$ at the point $z = 1$ .

Consider another interesting example of summation of a series having a zero radius of convergence by the Padé approximation technique. When seeking the energy levels of an anharmonic oscillator [20] in terms of perturbation theory, the series

$$F(z) = \sum_{n=0}^{\infty} n! \, z^n \tag{2.32}$$

arises whose convergence radius is exactly zero. The function $F(z)$ has the following integral representation:

$$F(z) = \int_0^\infty \frac{e^{-x}}{1 - zx} \, dx \; . \tag{2.33}$$

The function $F(z)$ given by (2.33) is called the Euler function. From (2.33) it is clear why the convergence radius of (2.32) is zero. The point $z = 0$ is branch point of the function $F(z)$ and in the $z$-plane there is a cut from the point $z = 0$ to $z = \infty$. From (2.33) it is also seen that the function $F(z)$ can be reduced to a Stieltjes function. Therefore a sequence of diagonal Padé approximants constructed on the basis of the coefficients of the series (2.32) converges to (2.33) in the entire z plane, except for the cut $[0, \infty]$. The cut itself is simulated by the alternating zeros and poles of the Padé approximants. Thus, we see that the Padé method allows one to correctly sum even the series (2.32) having, one would think, purely formal meaning because it does not represent the function $F(z)$ anywhere except for the coordinate origin.

If the methods described above cannot be used, then to sum or accelerate the convergence of the series

$$S = \sum_{i=0}^\infty a_i \; ,$$

we may use the PA II. The method of operations may be as follows. The partial sum

$$S_k = \sum_{i=0}^k a_i$$

may be considered to be a function of the variable $z = 1/k$, i.e. $S_k = F(1/k)$, the exact sum being $S = F(0)$. From the first $N + M + 1$ partial sums $\{S_k\}(k = 1, 2, ..., N + M + 1)$ we may construct the approximant $F^{[N,M]}(z)$ for the function $F(z)$ and determine

$$S \simeq F^{[N,M]}(0) \; .$$

Consider, for example, the series

$$S = \sum_{n=1}^\infty 1/n^2 \; ,$$

$$S_1 = 1 \; ; \qquad S_2 = 1.25 \; ; \qquad S_3 = 1.361 \; ...$$

The approximants constructed on the basis of $S_1$ or $S_1$ and $S_2$ or $S_1$, $S_2$, and $S_3$ yield

$$F^{[0,0]}(0) = 1 \; ; \qquad F^{[0,1]}(0) = 1.5 \; ; \qquad F^{[1,1]}(0) = 1.65 \; ...$$

respectively. The exact result is $F(0) = \pi^2/6 = 1.645\ldots$ A substantial accelerati-
on of the convergence is evident. It should be noted that the PA $F^{[0,1]}$ *does not contain any information* other than the numbers $S_1$, and $S_2$, and $F^{[1,1]}[0]$ other than $S_1$, $S_2$, and $S_3$. Certainly, the use of the Padé method implies some regular behaviour of $S_k$ as a function of $k$.

(ii) *Determination of the zeros, poles, and branch points of an analytic function.* The PA I and PA II may be used to find zeros of an analytic function, its poles, and even its branch points.

First we examine the determination of the zeros of the function $F(z)$:

$$F(z) = 0 \tag{2.34}$$

or of its poles. We replace the equation (2.34) by the equation

$$F^{[N,M]}(z) = P_N(z)/Q_M(z) = 0 \, ,$$

i.e.

$$P_N(z) = 0 \, , \tag{2.35}$$

thereby reducing the problem to searching for the zeros of the polynomial $P_N(z)$. In many cases the zeros of the polynomial $Q_M(z)$ prove to be approximate values of the poles of $F(z)$. The polynomials $P_N$ and $Q_M$ may be constructed from the coefficients of the Taylor series (PA I). For example, $\tan(z)$ has the nearest pole at $z = \pi/2 = 1.5707$ and the nearest zero at $z = \pi = 3.14\ldots$ The approximant of order [0, 1] yields a pole at $z = \sqrt{3} = 1.71$, the [1, 1] approximant has already a pole at $z = 1.581\ldots$ (it differs from $\pi/2$ by 1 %) and a zero at 3.86. The PA [1, 2] has a zero already at $z = 3.24$, and so on. That is, the application of the Padé method makes it possible to find the zeros and poles of an analytic function by means of its truncated Taylor series.

The application of the PA II makes it possible to construct the following simple algorithm for finding the complex zeros of an analytic function [11]. Let the values of the function $F(z)$ at three points $z_1$, $z_2$, and $z_3$ be equal to $F_1$, $F_2$, and $F_3$, respectively. We construct a PA II $F^{[1,1]}(z)$, $F^{[1,1]}(z_i)$ being equal to $F_i(i = 1, 2, 3)$. Let $z_4$ be a zero of $F^{[1,1]}(z)$. Further, using the three points $z_2$, $z_3$, and $z_4$ we construct the next PA in the next step, etc. The position of the zero after the $n$-th iteration is determined by the formula

$$z_{n+3} = \frac{z_{n+2}\dfrac{(z_{n+1} - z_n)}{F(z_{n+2})} + z_{n+1}\dfrac{(z_n - z_{n+2})}{F(z_{n+1})} + z_n\dfrac{(z_{n+2} - z_{n+1})}{F(z_n)}}{\dfrac{(z_{n+1} - z_n)}{F(z_{n+2})} + \dfrac{(z_n - z_{n+2})}{F(z_{n+1})} + \dfrac{(z_{n+2} - z_{n+1})}{F(z_n)}} \, . \tag{2.36}$$

It is possible to show [11] that the convergence of such a procedure is exponential, that is, if in the first step we determine a zero to within accuracy $\varepsilon$, then in the second step the accuracy will be $\varepsilon^3$, in the third step $\varepsilon^9$, etc., i.e. for $\varepsilon = 10\,\%$ the accuracy is $0.1\,\%$ in the second and $10^{-7}\,\%$ in the third step. It should be noted that the described procedure is very simple and linear and is equally simple for finding both real and complex zeros of $F(z)$.[3]

The PA II can be used to construct a universal algorithm for finding a root branch point for a function given *numerically* (or by a numerical algorithm). Let the function $f(z)$ have a root branch point and let the values of the function $y_i = f(z_i)$ and of its derivative $f'_i = df(z)/dz\,|_{z\,=\,z_i}$ be known at a finite number of points $\{z_i\}$, $i = 1, 2, ..., p$ .

We want to find the location of the branch point $z_0$ and the values of the function at this point $y_0 = f(z_0)$. Since $f(z)$ has a root branch point at the point $z_0$, we get

$$y = f(z) \underset{z \to z_0}{\approx} y_0 + a_1 \sqrt{z - z_0} \tag{2.37}$$

at $z \to z_0$ .

The inverse function $z = z(y)$ at $y \to y_0$ is of the form:

$$z \underset{y \to y_0}{\simeq} z_0 + a_2(y - y_0)^2 \tag{2.38}$$

and is an analytic function of $y$ near $y = y_0$ . From (2.37) it is seen that

$$\frac{df}{dz} \underset{z \to z_0}{\simeq} \frac{a_3}{y - y_0}$$

at $z \to z_0$ and, thus, the function $x(y) = 1/(df/dz)$ is of the form:

$$x(y) = 1/(df/dz) \underset{y \to y_0}{\simeq} a_4(y - y_0) \tag{2.39}$$

at $y \to y_0$ , i.e.

$$x(y_0) = 0 \tag{2.40}$$

at the branch point $y = y_0$.

Let us consider now the function $f(x)$. From (2.37) and (2.39) it follows that in the vicinity of $y = y_0$

$$f(x) \equiv f(z(x)) = y_0 + \tilde{a}x , \tag{2.41}$$

---

[3] This procedure may be used to find the poles of the analytic function $F(z)$ too. Such a problem reduces to finding the zeros of the function $1/F(z)$.

i.e. the function $f(x)$ is analytic in the vicinity of $x = 0$ . Considering (2.40), we find from (2.41) that

$$y_0 = f(x = 0) . \tag{2.42}$$

Now, we shall use the above properties to construct the algorithm. We transform the reference set of the derivatives $\{f_i'\}$ into the set $\{x_i\} = \{1/f_i'\}$ ; then, using the sets $\{x_i\}$ and $\{y_i\}$ we construct the Padé approximant $f^{[N,M]}(x) = P_N(x)/Q_M(x)$ $(N + M + 1 = p)$ and find the coefficients of the polynomials $P_N$ and $Q_M$. Since $f(x)$ is analytic in the vicinity of $x = 0$ , the approximate equality

$$y_0 \simeq f^{[N,M]}(0) = P_N(0)/Q_M(0) \tag{2.43}$$

Table 1.3

| Order of | I | II | |
|---|---|---|---|
| the PA | $z_0$ | $z_0$ | $y_0$ |

$$y = 1 + (z - z_0)^{1/2} + 0.1\,(z - z_0);\ z_0 = 0.5$$

| Order of the PA | $z_0$ (I) | $z_0$ (II) | $y_0$ (II) |
|---|---|---|---|
| 1,1 | 0.441 183 21 | 0.440 922 97 | 0.999 315 19 |
| 2,2 | 0.499 952 12 | 0.499 952 07 | 0.999 999 99 |
| 3,3 | 0.499 999 83 | 0.499 999 80 | $1. - 5 \times 10^{-11}$ |
| 4,4 | 0.499 999 99 | 0.500 000 09 | $1. + 5 \times 10^{-11}$ |
| 5,5 | 0.500 000 01 | 0.500 000 06 | $1. - 7 \times 10^{-11}$ |
| Exact value | 0.5 | 0.5 | 1 |

$$y = 1 + 5 \exp \left\{ - (z - z_0)^{1/2} \right\};\ z_0 = 0.5$$

| Order of the PA | $z_0$ (I) | $z_0$ (II) | $y_0$ (II) |
|---|---|---|---|
| 1,1 | 0.443 351 9 | 0.450 763 1 | 5.901 95 |
| 2,2 | 0.500 430 8 | 0.500 426 3 | 5.987 64 |
| 3,3 | 0.499 987 3 | 0.499 987 5 | 5.997 89 |
| 4,4 | 0.500 011 7 | 0.500 011 9 | 5.999 56 |
| 5,5 | 0.499 993 6 | 0.499 992 4 | 5.999 86 |
| 6,6 | 0.500 000 5 | 0.500 002 4 | 5.999 86 |
| 7,7 | 0.500 000 3 | 0.500 000 2 | 5.999 91 |
| Exact value | 0.5 | 0.5 | 6.0 |

turns out at $N, M \to \infty$ to be the exact equality (2.42). Thus, (2.43) determines approximately the value of $y_0$. Further, using the set of reference points $\{y_i\}$ and $\{z_i\}$, we construct the PA $z^{[N,M]}(y) = R_N(y)/S_M(y)$, $N + M + 1 = p$ , where $p$ coefficients of polynomials $R_N$ and $S_M$ are unambigously determined by the sets $\{z_i\}$ and $\{y_i\}$. Using the value $y_0$ obtained from (2.43), we can eventually determine the position of the branch point $z_0$:

$$z_0 \simeq z^{[N,M]}(y_0) = R_N(y_0)/S_M(y_0) \, . \tag{2.44}$$

Table 1.3 presents the calculations carried out for two trial functions $y = 1 + (z - z_0)^{1/2} + 0.1(z - z_0)$ and $y = 1 + 5 \exp\{-(z - z_0)^{1/2}\}$. The first column of Table 1.3 shows the values of $z_0$ obtained from formula (2.44) in the case where the value $y_0$ is given exactly for all orders of the Padé approximants. The second and third columns give the values of $y_0$ and $z_0$ obtained in a self-consistent way using formulae (2.43) and (2.44). It can be seen that the values $z_0$ and $y_0$ are obtainable to a high accuracy already at not very large $N$ and $M$ values.

(iii) *Application of the PA-technique to solving integral equations.* We shall consider a Fredholm type II equation:

$$f(x) = \varphi(x) + \int K(x, y) f(y) \, dy \tag{2.45}$$

(where $x$ and $y$ are one or many-dimensional variables) which describes two- or few-particle scattering, for example the Lippmann-Schwinger or Faddeev-Yakubovsky equations. It is convenient to solve equation (2.45) by iterations. The zero-order approximation is simply equal to the inhomogeneous term

$$f_0(x) = \varphi(x) \, .$$

The first-order solution is

$$f_1(x) = f_0(x) + \int K(x, y) f_0(y) \, dy$$

and so on. Formally, the exact solution may be written in the form of an infinite series

$$f = \sum_{i=0}^{\infty} K^i \varphi \, . \tag{2.46}$$

The series (2.46) is called the Neumann series and converges if and only if the absolute values of all eigenvalues $\eta_n$ of the operator $K$

$$K\psi_n = \eta_n \psi_n \tag{2.47}$$

are smaller than unity, i.e.

$$|\eta_n| < 1 \qquad \text{for all } n \, . \tag{2.48}$$

However, in the majority of cases of physical interest the condition (2.48) is not satisfied and the series (2.46) diverges. Under such conditions we may proceed in the following way. Instead of equation (2.45) we consider the equation

$$f(\lambda, x) = \varphi(x) + \lambda \int K(x, y) f(\lambda, y) \, dy \tag{2.49}$$

where $\lambda$ is a complex parameter. From the theory of integral equations [21] it is known that the resolvent of equation (2.49), as well as the solution $f(\lambda, x)$, is a meromorphic function of $\lambda$. For the sake of brevity we shall henceforth omit all the variables, except for $\lambda$. In the vicinity of $\lambda = 0$, $f(\lambda)$ is an analytic function and may take the form of the Taylor series:

$$f(\lambda) = \sum_{i=0}^{\infty} \lambda^i K^i \varphi,$$ (2.50)

which coincides with the Neumann series for an equation with the kernel $\lambda K$. It is clear that by selecting

$$|\lambda| < |\eta_{max}^{-1}|$$ (2.51)

we get a series (2.50) converging at all energies. However, it is $\lambda = 1$ that corresponds to the solution for the equation (2.45)

$$f = f(\lambda = 1).$$ (2.52)

At $\lambda = 1$, however, the series (2.50) coincides with (2.46) and diverges. This means that some singularities $\lambda_v$ of $f(\lambda)$ which are related to $\eta_v$ as

$$\lambda_v = \eta_v^{-1}$$ (2.53)

lie inside a unit circle. This situation is known to us from the previous section where the summation of divergent series was treated. Therefore we may proceed in a familiar way, namely, using the coefficients of the Taylor series (2.50) we construct the Padé approximant

$$f^{[N,M]}(\lambda) = P_N(\lambda) / Q_M(\lambda)$$ (2.54)

and determine the solution of equation (2.45):

$$f \approx f^{[N,M]}(\lambda = 1) = P_N(1)/Q_M(1).$$ (2.55)

The singularities $\lambda_n$ of $f(\lambda)$ correspond to zeros of $Q_M(\lambda)$. In a special case where the kernel $K$ of equation (2.45) is a degenerate kernel of rank $N$, i.e.

$$K = \sum_{i=1}^{N} |\varphi_i\rangle \langle\varphi_i|$$

the diagonal approximant $f^{[N,N]}$ coincides with the exact solution of the equation. This means that we may apply the above-mentioned Padé-summation of the Neumann series to integral equations with compact kernels.

The described method for solving integral equations is used extensively in the scattering theory of two, three, and four particles. For example, it is this method that was used to carry out the earliest and, most complete calculations of three-and-four-particle systems with realistic NN interactions [22].

To complete this section we should point out, as we have already noted when discussing the convergence theorems, that the theory of Padé approximants is more poorly developed than the power series theory. Therefore, the validity of the Padé method in each particular case is frequently not evident beforehand and can only be verified *after* examining the PA convergence, as the PA order increases, on the basis of the appraisal of the locations and motions of the PA zeros and poles as a result of the variations in the PA order.

Hence, it is clear that if $f(z)$ has $M_0$ poles the PA $f^{[N,M]}(z)$ at $M > M_0$ has a greater number of poles than $f(z)$. How can the true and false poles be distinguished from each other? Numerous calculations have shown that the poles of the PA $f^{[N,M]}(z)$ with large residues correspond to the true poles of $f(z)$, whereas the superfluous poles have a small residue, which means that the zeros of the numerator and denominator of $f^{[N,M]}(z)$ almost coincide at this point (for more details see Section 1.3.3).

Furthermore, as the coefficients of the Taylor series (when using the PA I), as well as the values of the function $f(z_i)$ (when using the PA II), are always known to within a limited accuracy, the errors give rise to false singularities of $f^{[N,M]}(z)$. Numerical investigations have shown that, as the PA order increases, the true singularities corresponding to the singularities of $f(z)$ move only slightly whereas the false singularities move rapidly towards the boundary of the analyticity domain of $f(z)$.

## 1.3. Some Methods of Analytic Continuation

In many practical cases we are often faced with the following situation. Let a quantity $y$, which is an analytic function of another quantity $z$ (i.e. $y = f(z)$) at $z \in D$, be specified in a subdomain $D_1$ of the domain $D$ (i.e. at $z \in D_1$ we may measure or calculate $f(z)$). Now let $D = D_1 + D_2$. We have to determine this function for $z \in D_2$ if $f(z)$ is known at $z \in D_1$. Such a procedure to determine $f(z)$ in the entire domain $D$ on the basis of its values in the subdomain $D_1$ is called analytic continuation.

Since nearly all of the existing theories for resonance states are based on one or another type of analytic continuation (of $S$-matrix, scattering amplitude, Jost function and others, or of the dynamic, e.g. Lippmann-Schwinger or Faddeev, equations) to the non-physical energy sheet, we shall devote a special section to a brief description of the most extensively used analytic continuation methods.

### 1.3.1. Analytic Continuation by Means of Power Series

Let the function $f(z)$ be an analytic function in the domain $D$. Further, let this function be specified by a Taylor series in the vicinity of the point $z_0$ (see Fig. 1.1):

$$f_1(z) = \sum_{n=0}^{\infty} a_n(z - z_0)^n .$$

$$(3.1)$$

Since the series (3.1) converges in a circle $D_1$ whose radius does not exceed the distance from the point $z_0$ to the nearest point at the boundary of the domain $D$ (from the nearest singularity of the function $f(z)$), i.e. at $|z - z_0| < R_1$, the

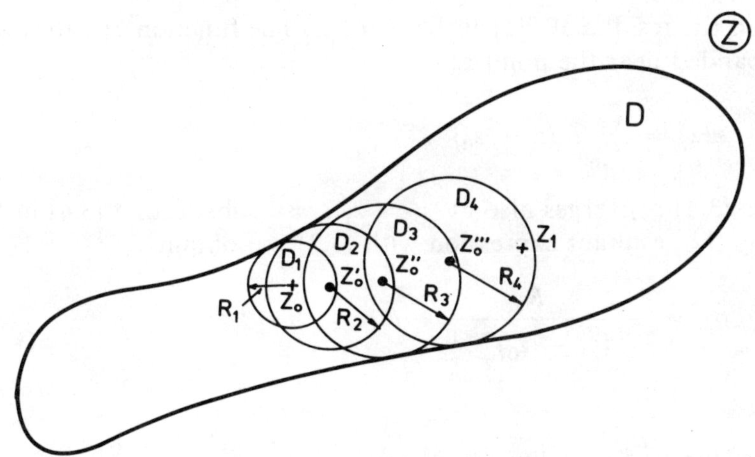



Fig. 1.1
Analytic continuation by means of power series.

function $f(z)$ will only equal $f_1(z)$ at $z \in D_1$. If we want to find, for example, $f(z_1)$ we cannot use the expansion (3.1). To find $f(z_1)$, we have to select [23] a point $z_0'$ in the domain $D_1$ and Taylor series-reexpand the function $f_1(z)$ at the point $z_0'$

$$f_2(z) = \sum_{n=0}^{\infty} b_n(z - z_0')^n .$$

It is quite obvious that $f_2(z) = f_1(z)$ at $z$ corresponding to the intersection of the domains $D_1$ and $D_2$ and that the function $f_2$, which is the analytic continuation of $f_1(z)$ outside $D_1$, determines the analytic function $f(z)$ in $D_2$. In the example shown in Fig. 1.1 the function $f(z_1)$ has to be found by repeating three times the above described operation of the Taylor series reexpansion of the function. In such a way we obtain the analytic continuation of the function to the domain $D_4(z_1 \in D_4)$. This is just the procedure of analytic continuation by means of a series.

The fact that the power series convergence circle is constrained by the nearest singularity strongly limits in practice the direct use of power series for analytic continuation. A number of methods have been proposed for avoiding this difficulty [24], of which the Hadamard method seems to be the most useful. Let

$$f(z) = \sum_{n} a_n(z - z_0)^n \tag{3.2}$$

be a Taylor series for a function $f(z)$ whose nearest singularity (a simple pole) is located at a point $z_1$. Then, $f(z)$ may be written as

$$f(z) = \frac{R_1}{z - z_1} + \varphi(z) , \tag{3.3}$$

where $R_1$ is the residue of $f(z)$ at the point $z_1$. The function $\varphi(z)$ may be Taylor series-expanded near the point $z_0$:

$$\varphi(z) = \sum_{n=0}^{\infty} \tilde{a}_n (z - z_0)^n . \tag{3.4}$$

The series (3.4) converges also at $z = z_1$. By substituting (3.4) in (3.3) and comparing the resultant expression with (3.2), we obtain

$$a_m = - \frac{R_1}{(z_1 - z_0)^{m+1}} + \tilde{a}_m . \tag{3.5}$$

From (3.5) we get:

$$\lim_{m \to \infty} \frac{a_m}{a_{m+1}} = \lim_{m \to \infty} (z_1 - z_0) \frac{\tilde{a}_m (z_1 - z_0)^{m+1} - R_1}{\tilde{a}_{m+1}(z_1 - z_0)^{m+2} - R_1} = z_1 - z_0 \tag{3.6}$$

where we have used the following relation resulting from the convergence of the series (3.4):

$$\lim_{m \to \infty} \tilde{a}_m (z_1 - z_0)^m = 0 . \tag{3.7}$$

The expression (3.6) determines the location of the pole of $f(z)$

$$z_1 = z_0 + \lim_{m \to \infty} \frac{a_m}{a_{m+1}} . \tag{3.8}$$

The residue at the pole can be determined by rewriting (3.5) as

$$R_1 = -a_m (z_1 - z_0)^{m+1} + \tilde{a}_m (z_1 - z_0)^{m+1} . \tag{3.9}$$

Further, using (3.7), we obtain the following expression from (3.9):

$$R_1 = - \lim_{m \to \infty} a_m (z_1 - z_0)^{m+1} . \tag{3.10}$$

Thus, the relations (3.8) and (3.10) determine the location of the nearest pole of $f(z)$ and the residue at this pole through the Taylor series coefficients for $f(z)$ at the point $z_0$. After finding $z_1$ and $R_1$, we may use formula (3.5) to obtain $\tilde{a}_n$, the Taylor series coefficients for $\varphi(z)$. This method may be generalized to the case of several singularities [24].

The Hadamard method is the technique for dividing the function $f(z)$ specified by the Taylor series coefficients into polar parts and a smooth function $\varphi(z)$. The method demonstrates a very important fact. The Taylor series coefficients carry much more information about the function than we might expect. They carry information about the behaviour of the function not only inside but also outside the convergence circle and even about its singularities. However, the information carried by the series coefficients is latent and the Hadamard method is one of the ways of extracting it. Another way, the method of conformal transformation, has been described in Section 1.2.1. It was shown there how a change of variables may be used to extend the Taylor series convergence circle in practice up to the entire analyticity region of the function $f(z)$. However, to make such a conformal transformation it is necessary to know *the location* of the function singularities *beforehand*. The Padé method is to some extent devoid of this shortcoming.

### 1.3.2 Analytic continuation by means of Padé approximants.

In section 1.2.1 we have shown that the Padé approximants realize the conformal transformation of the variable $z$ (the Euler transformation),

$$z = \frac{ay}{1 + by}$$

*automatically,* i.e. without first determining the location of singularities of $f(z)$. This means that the use of the Padé approximation technique makes it possible to find the necessary parameters $a$ and $b$ systematically and most effectively, proceeding from the coefficients $a_n$ of the Taylor series $(2.1)$ for the function $f(z)$ (PA I) or from the function values at a finite number of points ( PA II or PA III). In this way, the use of the PA allows us to continue the function in practice to the entire analyticity domain. Owing to the presence of the PA poles, the PA proper not only provide for an effective analytic continuation but also make it possible to investigate the character and the location of the singularities of a function to be continued. As we noted in Section 1.2, the PA poles reproduce the poles of the function while the alternating zeros and poles simulate the cuts. In many cases the PA II or PA III turn out to be more suitable for analytic continuation than the PA I. In particular, the PA II ensure a smooth interpolation of the function $f(z)$ between the points $z_i$ which is often more effective than the Lagrange interpolation formulae. Furthermore, owing to the fact that PA reconstruct the $f(z)$ singularities, they are used in an effective way to continue $f(z)$ outside the region of its definition (that is, outside the localization region of the points $z_i$). Numerous applications of the PA I–III [7, 11, 13, 15–18, 22] have shown that the Padé approximation is the optimal method for analytic continuation.

### 1.3.3. Stability of Analytic Continuation

The most serious problem of analytic continuation concerns the stability of the result with respect to minor variations of initial data. Errors are always present in the initial data because the Taylor series coefficients or the values of the

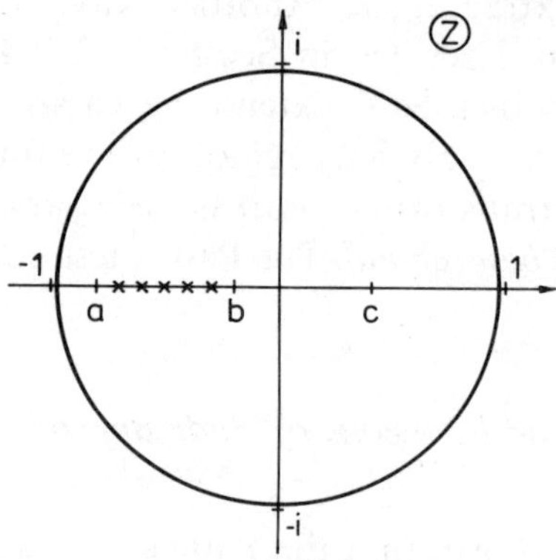

Fig. 1.2

function at definite points are always calculated within a certain finite accuracy. Let us examine an example [25]. Let $f(z)$ be an analytic function in a unit circle $D$ and let $f(z)$ be defined at a number of points located on the interval $[a, b]$ (see Fig. 1.2) within an error not exceeding $\varepsilon$. Let also the analytic continuation be from the data in the interval $[a, b]$ to a point $c$.

We shall take $f(z_i)$ to be the values of the function in the interval $[a, b]$ and $f_1(z)$ to be an analytic function in a unit circle and assume that

$$|f_1(z_i) - f(z_i)| \le \varepsilon ; \qquad z_i \in [a, b] .$$

As $f_1(z)$ describes properly (within the error $\varepsilon$) the function $f(z)$ in the interval $[a, b]$, it may be taken to be a solution. What will be then the error at the point $c$? It is evident that at any $A > 0$ the function

$$f_2(z) = f_1(z) + \varepsilon \exp \{A(z - b)\}$$

will also describe $f(z)$ properly in the interval $[a, b]$. The difference $|f_2(c) - f_1(c)|$ at the point $c$ may, however, take any high values (depending on $A$). Consequently, to obtain a stable analytic continuation it is necessary to select the functions exibiting the necessary analytic properties. The Padé approximant exibiting the necessary analytic properties is a convenient tool which makes it possible to carry out a stable analytic continuation for a large class of functions $f(z)$. This question has been investigated by several authors (see e.g. [26, 27]).

We shall discuss the problem here using a simple example. We shall again deal with the simplest series $\sum_{n=0}^{\infty} z^n$. Now, however, let the series coefficients contain random noise

$$\tilde{f}(z) = \sum_{n=0}^{\infty} \left(1 + \varepsilon r_n\right) z^n \tag{3.11}$$

where $\varepsilon$ is a small number, for example, of the order of $10^{-3}$; $r_n$ is a random complex number, such that $|r_n| < 1$.

From the mathematical point of view the entire curve $|z| = 1$ is a singularity of the function $f(z)$ which, therefore, cannot be continued analytically outside this line. On the other hand, it is intuitively clear that the main part of $f(z)$ must be an analytic function $1/(1 - z)$ which has a pole at $z = 1$, whereas its non-analytic part must be small, (of the order of $\varepsilon$). We can write down the PA $\tilde{f}^{[N,N]}(z)$ for the function $\tilde{f}(z)$ and study their zeros and poles at different $N$. The calculations have yielded the following result [27]:

(i) in the vicinity of $z = 1$ there exists a pole which is stable with respect to the variations of $N$;

(ii) there exists a zero associated with this pole. The zero is located at a distance of the order of $1/\varepsilon$ from the pole, and its location changes as $N$ varies. The zero conforms to the fact that at $z \to \infty$ (i.e. at $1/(1 - z) \to 0$) the diagonal PA $\tilde{f}^{[N,N]}(z) \to$ const.

(iii) All other zeros and poles are located as doublets (one pole + one zero) in the vicinity of the circle $|z| = 1$. The distance between them and, hence, the residue at the pole are of the order of $\varepsilon$. The locations of the doublets are extremely unstable with respect to the variations of the Padé approximant order $N$.

Thus, if we calculate the value of $\tilde{f}(z)$ using the PA $\tilde{f}^{[N,N]}(z)$ we shall obtain a result very close to $1/(1 - z)$ to within an accuracy $\varepsilon$ even at $z$ beyond the unit circle, excluding those values of $z$ which lie in the vicinities of the doublets. However, since the location of the doublets also changes as $N$ varies, we may also obtain a correct result in these regions if we change $N$. The correctness of the result is verified by its minor changes as $N$ varies further.

The conclusions drawn from the above simple example proved to be true for a large class of functions [26, 27]. The PA are stable with respect to random errors and they suppress, rather than enhance, the noise. On the other hand, the character of the errors in the initial data is of great importance. For, as shown by the examples considered in the literature (unfortunately, a strict rigorous theory has never been developed), even minor systematic errors may *radically change* the conclusions arrived at above.

Our own experience shows that the PA are quite stable with respect to the random errors which usually arise in computer calculations and that they make it possible to carry out stable analytic continuation to quite large distances on the basis of the data calculated within a moderate accuracy.

However, some analogue of the Wiener theorem from the time series theory [28] occurs in this case, that is, for a fixed accuracy of the initial data, there exists an optimal order of the PA which gives the best-accuracy extrapolation at a given point. When the PA order exceeds the optimal level, the accuracy of the prediction deteriorates. At the same time, as the accuracy of the initial data improves, the optimal order increases.

### 1.3.4. Analytic Continuation of Contour Integrals

Most of the complex-valued functions with which we were faced in our analysis and in the applications may be represented in the form of singular contour integrals (Cauchy-type integrals). In the theory of resonances a special role is played by a spectral decomposition of the Hamiltonian resolvent. The Hamiltonian $H$ of a physical system with resonance states must have a continuous spectrum. Therefore, the matrix element of its resolvent $G(z) = (z - H)^{-1}$ may be presented as (see Section 4.3).

$$\langle \varphi \mid G(z) \mid \psi \rangle = \sum_i \frac{\langle \varphi \mid F_i \rangle \langle F_i \mid \psi \rangle}{z - \lambda_i} +$$

$$+ \int_\Lambda \frac{\langle \varphi \mid R(\lambda) \rangle \langle L(\lambda) \mid \psi \rangle}{z - \lambda} \, \mathrm{d}\lambda \qquad (3.12)$$

where the integral is over the continuous spectrum of $H$. The shift of the integration contour offers ample possibilities of making an analytic continuation of the functions represented by the contour integrals. Following [43], we shall examine a theorem of the Cauchy type integral continuation

$$F(z) = \frac{1}{2\pi i} \int_C \frac{f(\xi)}{\xi - z} \, \mathrm{d}\xi . \qquad (3.13)$$

If the contour $C$ divides a plane into parts, then in each of them we have an analytic function. In any case the contour forms a boundary of the analyticity domain.

The problem of analytic continuation of the Cauchy-type integral through the integration contour is solved by the following *theorem*:

Let $f(\xi)$ be analytic in the domain $D$ containing a part $C_0$ of the contour $C$ and let $C_0$ divide the domain $D$ into two subdomains $D^+$ and $D^-$ lying respectively on the left and on the right of $C_0$ (Fig. 1.3). Let also $F^+(z)$ and $F^-(z)$ be functions represented by the integral (3.13) in the domains $D^+$ and $D^-$, respectively. Then the function $F^-(z) + f(z)$ gives the analytic continuation of the function $F^+(z)$ into the domain $D^+$ through the arc $C_0$.

We present here a proof based on the replacement of the integration contour because it may be used as an example for analogous examinations of other integrals. We denote by $C^*$ a contour differing from $C$ in that $C_0$ is replaced by

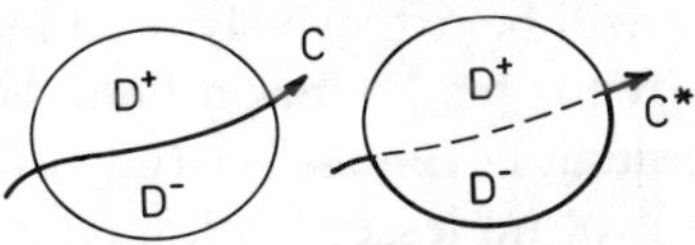


Fig. 1.3

a part of the $D$ boundary (see Fig. 1.3), and by $L^-$ the entire boundary of $D^-$. It is clear that at any $z \notin C_0$, $z \in D$ we have

$$\frac{1}{2\pi i} \int_{C^*} \frac{f(\xi)}{\xi - z}\, d\xi = \frac{1}{2\pi i} \int_{C} \frac{f(\xi)}{\xi - z}\, d\xi + \frac{1}{2\pi i} \int_{L^-} \frac{f(\xi)}{\xi - z}\, d\xi . \qquad (3.14)$$

The integral on the left-hand side of (3.14) defines the function $F(z)$ analytic in the domain $D$ because the points of the contour $C^*$ are absent there. At $z \in D^+$ the first integral on the right-hand side of (3.14) is equal to $F^+(z)$ by definition, while the second integral is zero according to the Cauchy theorem. Therefore,

$$F(z) = F^+(z), \qquad z \in D^+ . \qquad (3.14a)$$

At $z \in D^-$ the first integral is equal by definition to $F^-(z)$, while the second integral equals $f(z)$ according to the Cauchy integral formula

$$F(z) = F^-(z) + f(z), \qquad z \in D^- . \qquad (3.14b)$$

Thus, the function $F(z)$ analytically continues $F^+(z)$ from the subdomain $D^+$ to the entire domain $D$, whence we obtain that the discontinuity across the contour $C_0$ equals

$$F^+(z) - F^-(z) = f(z), \qquad z \in C_0 \qquad (3.15)$$

Applying this theorem to the matrix element (3.12), we obtain

$$\langle \varphi \,|\, G_+^{II}(z) \,|\, \psi \rangle = \langle \varphi \,|\, G_-(z) \,|\, \psi \rangle -$$

$$- 2\pi i \,\langle \varphi \,|\, R(z) \rangle \langle L(z) \,|\, \psi \rangle \qquad (3.16)$$

where $\langle \varphi \,|\, G_+^{II}(z) \,|\, \psi \rangle$ is the continuation of the matrix element specified at the upper rim of the cut $\varLambda$ to a second, nonphysical, sheet; $G_-(z)$ is a resolvent operator taken on the lower rim of the cut $\varLambda$, i.e. on the physical sheet.

## 1.4. Scattering Theory in a Three-Body System and the Faddeev Integral Equations

The material of this section will be extensively used in the subsequent chapters, so the reader is advised to focus his attention here. We preface this section by a very brief historical introduction. Extensive study of the quantum-mechanical three-body problem which is of no lesser (and maybe even of a greater) importance than the well-known classical three-body problem was initiated by L. D. Faddeev's fundamental works published in the early sixties [29]. The works were not only the first to formulate the integral equations describing all possible processes of scattering and rearrangement in a three-body system, but also (and this is of major significance) proved rigorously the uniqueness of the solution of the equations derived and, examined in detail the most important of their properties. An enormous number of studies were carried out later by many people all over the world. They were aimed at generalizing, specifying, and modifying the Faddeev equations, at developing practical effective methods to solve them numerically, and at applying them to numerous special systems ranging from atomic and molecular physics to elementary particle physics. Within the last twenty years the studies have given rise to a new and extended field of quantum physics, so-called physics of few-body systems. Representative conferences on the subjects pertaining to this field are held annually.

The aim of this section is to treat in a simple, lucid, and compact way, the basic facts and equations of three-particle scattering theory which are necessary for the material of the subsequent chapters to be properly understood. A more comprehensive discussion of the above mentioned problems may be found in the well-known handbooks [30–32, 39, 44] and in the numerous reviews cited therein. To begin with, we refer the readers particularly to a very lucid book [30] by E. Schmid and H. Ziegelmann or to the lectures [44].


### 1.4.1. The Features of the Three-Body Problem as Compared with the Two-Body Problem

The Lippmann-Schwinger (LS) equation for the two-particle Green functions $g(z)$ $(\text{Im } z \neq 0)$

$$g(z) = g_0(z) + g_0(z)\, V g(z) \tag{4.1}$$

describing the scattering in the two-body system (after going over to a centre-of--mass system and separating the centre-of-mass variables) is known [8, 45] to be a Fredholm-type integral equation and, therefore, may be solved directly by the known methods (they are numerous [21, 30]).

The situation is different with the systems of three and more particles. For the sake of clarity we shall use some graphic illustrations. Iterations of equation (4.1) give rise to the series

$$g = g_0 + g_0 V g_0 + g_0 V g_0 V g_0 + \cdots \qquad (4.2)$$

which may be depicted with the help of graphs (see Fig. 1.4).

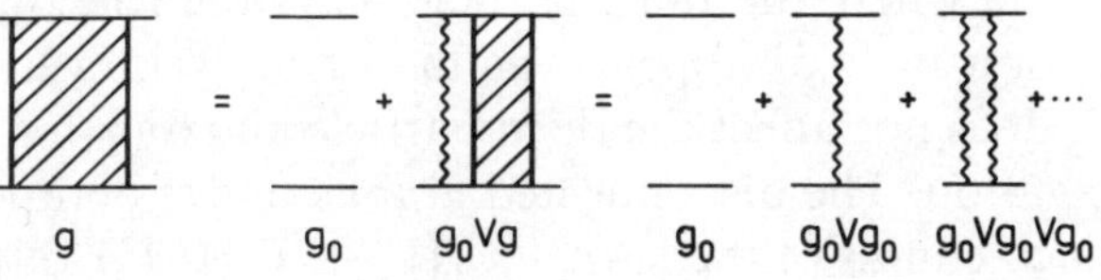

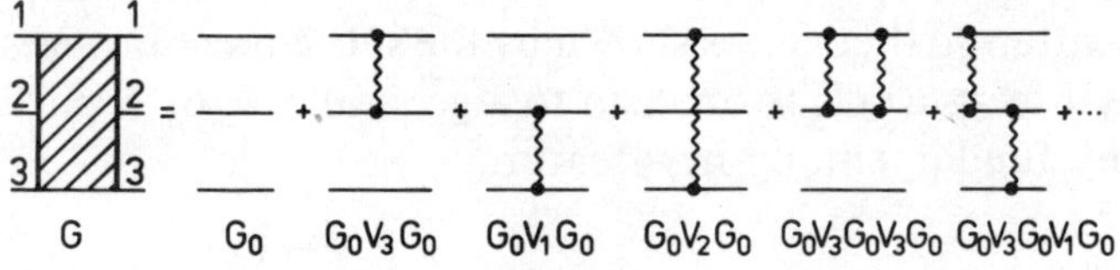



Fig. 1.4

From the graphs it follows that, except the inhomogeneous term, all the graphs are connected and correspond to an iterated kernel series all terms of which are nonsingular.

By analogy with eq. (4.1) we write down now the three-particle Lippmann-
-Schwinger equation

$$G = G_0 + G_0 V G , \qquad (4.3)$$

where

$$V = V_{12} + V_{13} + V_{23} = \sum_{i=1}^{3} V_i .$$

The corresponding three-particle iteration series will be fully similar to eq. (4.2), i.e.

$$G = G_0 + G_0 V G_0 + G_0 V G_0 V G_0 + \cdots \qquad (4.4)$$



Fig. 1.5

Again we depict graphically an iteration series corresponding to (4.4) (see Fig. 1.5). Here, side by side with the fully connected graphs (the sixth term on the right-hand side of eq. in Fig. 1.5), there are also disconnected terms (for example, the fifth term on the right-hand side of eq. in Fig. 1.5). The explicit form of the fifth term in Fig. 1.5 is

$$T_5 = \langle k_3, q_3 \mid T_5 \mid k_3', q_3' \rangle = \langle k_3, q_3 \mid G_0 V_3 G_0 V_3 G_0 \mid k_3', q_3' \rangle =$$

$$= A(k_3, k_3' \, ; \, q_3, q_3') \, \delta(q_3 - q_3') \, . \tag{4.5}$$

Here, the Jacobi momentum variables $k_3$ and $q_3$ correspond to the coupling scheme $(1 + 2) + 3$ where the third particle plays the role of a spectator. The Dirac delta function in (4.5) appears, as may readily be verified, because the interaction in $T_5$ does not affect the third particle and $\delta(q_3 - q_3')$ corresponds to its free propagation. The disconnected graphs under iterations arise directly from the disconnectedness of the kernel $G_0(V_1 + V_2 + V_3)$ of the three-particle equation (4.3). This circumstance leads to a non-Fredholm-type integral equation because its kernel contains unavoidable delta functions. Therefore it is clear that the "dangerous" delta functions will not appear only if the kernel (or its first iterations) is connected; more precisely, if it corresponds to the connected graphs. Strictly speaking, the absence of the delta functions in the kernel is not yet sufficient for a respective equation to be of the Fredholm type. However, this is highly probable and is accepted in all the present-day works on scattering theory. The physical cause of the fact that the three-particle Lippmann-Schwinger equation does not have a "good" kernel and, hence, a unique solution is the existence in the three-body problem, in contrast to the two-body problem, of several continua (two-particle ones and a three-particle one) overlapping each other. That is, at the same energies in the three-particle system there are, generally speaking, several interconnected two- and three-particle scattering channels and, hence, the kernels of mathematically correct equations have to include the singularities (poles and cuts) corresponding to each channel, which is certainly not the case in the kernel $G_0(V_1 + V_2 + V_3)$ of the three-particle Lippmann-Schwinger equation. Of course, this circumstance necessitates a division of the complete wave function into the components each of which corresponds to a respective division of the complete system into individual subsystems. Thus, we see that three-particle scattering theory must in principle differ from two-particle scattering theory. As shown by the subsequent progress, the generalization of the three-particle formalism to a general case of $N$ particles does not give rise to any fundamentally new features.

### 1.4.2. The Faddeev Equations and their Structure

To obtain a three-particle integral equation with a "good" kernel it is necessary to exclude the disconnected parts in some way, that is, it is necessary to rearrange the Lippmann-Schwinger equation (4.3) so that we are able to obtain only the connected graphs by iterations. It was Faddeev who found the method for such a rearrangement based on the division of a complete three-particle Green function G into terms corresponding to the first (or to the last) interaction

$$G = G_0 + G^{(1)} + G^{(2)} + G^{(3)} \tag{4.6}$$

where

$$G^{(i)} = G_0 V_i G . \tag{4.7}$$

The division (4.6) (together with the subsequent derivation (4.7)–(4.10)) is now called the Faddeev reduction. Let us substitute the division (4.6) in (4.7)

$$G^{(i)} = G_0 V_i G_0 + G_0 V_i (G^{(i)} + G^{(j)} + G^k); \qquad i, j, k = 1, 2, 3 . \tag{4.8}$$

The disconnected graphs in the series (4.4) have arisen from the terms of the form $G_0 V_i G_0 V_i G_0$ . If we transfer the $i$-th term from the right-hand side of eq. (4.8) to the left, i.e.

$$(1 - G_0 V_i)G^{(i)} = G_0 V_i G_0 + G_0 V_i (G^{(j)} + G^{(k)}) , \tag{4.9}$$

and then invert the operator $(1 - G_0 V_i)$ , we obtain a set of equations where the iteraction in the $j$-th pair is always followed by the $i$-th pair interaction $j \neq i$. Indeed,

$$G_i = G_0 + G_0 V_i G_i \;\Rightarrow\; (1 - G_0 V_i)^{-1} G_0 = G_i$$

and we find then from (4.9):

$$G^{(i)} = G_i V_i G_0 + G_i V_i (G^{(j)} + G^{(k)}) =$$
$$= G_i - G_0 + G_i V_i (G^{(j)} + G^{(k)}), \qquad j, k \neq i \tag{4.10}$$

where

$$G_i = (E - H_0 - V_i)^{-1} .$$

Eqs. (4.10) are the Faddeev equations for the Green function components. If we express the set (4.10) in the matrix form

$$\begin{pmatrix} G^{(1)} \\ G^{(2)} \\ G^{(3)} \end{pmatrix} = \begin{pmatrix} G_1 - G_0 \\ G_2 - G_0 \\ G_3 - G_0 \end{pmatrix} + \begin{pmatrix} 0 & G_1 V_1 & G_1 V_1 \\ G_2 V_2 & 0 & G_2 V_2 \\ G_3 V_3 & G_3 V_3 & 0 \end{pmatrix} \begin{pmatrix} G^{(1)} \\ G^{(2)} \\ G^{(3)} \end{pmatrix}$$

we can see clearly that the diagonal kernels are zeros, thereby giving, through iterations, only the terms corresponding to the connected graphs. The kernel $G_i V_i$ can be rewritten through a two-body $t$-matrix in the $i$-th channel

$$G_i V_i = G_0 t_i$$

and, by writing (4.10) as

$$G^{(i)} = G_0 t_i G_0 + G_0 t_i (G^{(j)} + G^{(k)}) , \tag{4.11}$$

we get, by iterations, the series, depicted grafically in Fig. 1.6, where, expect the inhomogeneous term, all the rest (i.e. iterated) terms correspond to the connected diagrams.

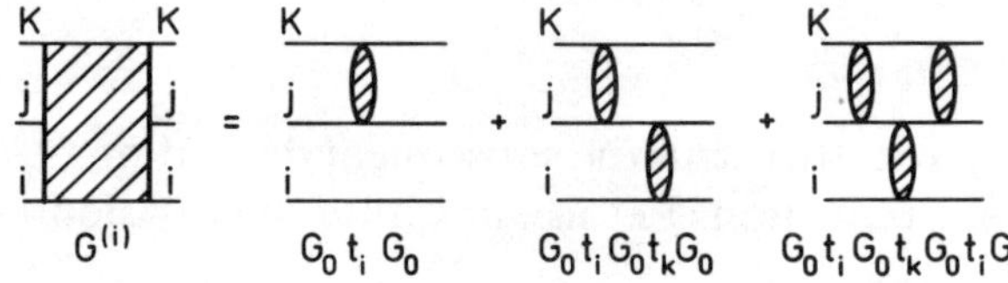

Fig. 1.6

On the basis of the Faddeev equations for the resolvent $G(z)$ we may easily obtain the appropriate equations for the wave function and for the complete scattering matrix. For example, the equations for the wave function are obtainable from the well-known relation [30]

$$|\psi_{\alpha m}^{(\pm)}\rangle = \lim_{\varepsilon \to 0} \pm i\varepsilon G(E \pm i\varepsilon) |\Phi_{\alpha m}\rangle \tag{4.12}$$

where $|\Phi_{\alpha m}\rangle$ is the initial-state function, that is, the product of the bound-state function in channel $\alpha$ with number $m$ by the plane wave for the third particle. Substituting the Faddeev expansion (4.6) for the resolvent $G(E \pm i\varepsilon)$ and using the well-known [30] definitions of scattering wave functions

$$\lim_{\varepsilon \to 0} \pm i\varepsilon G_i(E \pm i\varepsilon) |\Phi_{\alpha m}\rangle = \delta_{i\alpha} |\Phi_{\alpha m}\rangle$$

$$\alpha = 1, 2, 3; \qquad i = 0, 1, 2, 3$$

($i$ is the number of incident particle; $\alpha$ is the label of a bound-state pair) and

$$|\psi_{\alpha m}^{(\pm)}\rangle_i = \lim_{\varepsilon \to 0} \pm i\varepsilon G^{(i)}(E \pm i\varepsilon) |\Phi_{\alpha m}\rangle \, ,$$

we obtain the Faddeev's splitting for the three-body scattering wave function

$$|\psi_{\alpha m}^{(\pm)}\rangle = \sum_{i=1}^{3} |\psi_{\alpha m}^{(\pm)}\rangle_i \, . \tag{4.6a}$$

After that, using the above definitions and the Faddeev equations for the resolvent operator, we can eventually obtain the set of Faddeev equations for the wave-function channel components $|\psi_{\alpha m}^{(\pm)}\rangle_i$ in the case, for example, of the $i$-th particle scattering by the bound state of the rest two:

$$|\psi_{\alpha m}^{(\pm)}\rangle_i = \delta_{i\alpha} |\Phi_{\alpha m}\rangle + G_0(E \pm i\varepsilon)t_i(E \pm i\varepsilon) \left( |\psi_{\alpha m}^{(\pm)}\rangle_j + |\psi_{\alpha m}^{(\pm)}\rangle_k \right) , \tag{4.13}$$

$$i, j, k = 1, 2, 3 \, .$$

Now we shall examine a Faddeev reduction for the system of a heavy core plus two light particles (not necessarily identical). Although the above equations are correct in a general case, the system permits a more convenient Faddeev reduction [31]. Let us denote the core by number 3, and the valent particles by numbers 1 and 2. We divide the total Hamiltonian of the system

$$H = K_1 + K_2 + V_{31} + V_{32} + V_{12} \qquad (4.14)$$

into two parts

$$H = H_0 + V_{12}, \ H_0 = K_1 + K_2 + V_{31} + V_{32} \qquad (4.15)$$

where $K_1$ is the operator of the $i$-th particle kinetic energy.

Now, we introduce the channel resolvents $G_3$ and $G_{12}$ corresponding to the interaction operators $V_3 = V_{13} + V_{23}$ and $V_{12}$, respectively:

$$G_3(E) = (E - K_1 - K_2 - V_3)^{-1}, \ G_{12}(E) = (E - K_1 - K_2 - V_{12})^{-1}.$$

Recalling that $G_3(E)$ is a Green function of two non-interacting particles in the field of a heavy core, we may write the following expression for $G_3(E)$ through the one-particle Green functions $g_1(E)$ and $g_2(E)$:

$$G_3(E) = - \frac{1}{2\pi \, \mathrm{i}} \int\limits_{-\infty}^{\infty} g_1(\varepsilon)g_2(E - \varepsilon) \, \mathrm{d}\varepsilon = g_1 * g_2 \qquad (4.16)$$

where the integral represents the so-called convolution of $g_1$ and $g_2$ and is taken at the upper rim of the cut in the $E$-plane, i.e. at $E + i0$ . The methods for calculating the convolution (4.16) are presented in [33, 34]. Here we shall only give the results using coordinate representation. Then, using the partial wave expansion and neglecting the spin variables, we get:

$$G_{l_1 l_2}(r_1 \, r_2; \, r_1' \, r_2'; \, E) = \int\int \mathrm{d}\Omega_1 \, \mathrm{d}\Omega_2 \, \mathrm{d}\Omega_1' \, \mathrm{d}\Omega_2'$$
$$Y_{l_1 l_2}^{LM*}(\hat{r}_1, \hat{r}_2) \, Y_{l_1 l_2}^{LM}(\hat{r}_1', \hat{r}_2') \, G(r_1, r_2; r_1', r_2'; E) \qquad (4.17)$$

where the spherical tensor

$$Y_{l_1 l_2}^{LM}(\hat{r}_1, \hat{r}_2) = \sum_{m_1 (m_2)} \langle l_1 m_1 l_2 m_2 \mid LM \rangle \, Y_{l_1 m_1}(\hat{r}_1) Y_{l_2 m_2}(\hat{r}_2)$$

is expressed through the spherical harmonics of the vectors $r_1$ and $r_2$ and through the Clebsch-Gordan coefficients. After that, we expand one-particle resolvents $g_1$ and $g_2$ in a spectral series (in the units $\hbar = 2\,m = 1$)

$$g_l(r, r'; z) = \sum_n \frac{R_{nl}(r)R_{nl}^*(r')}{z - E_{nl}} + \int\limits_0^{\infty} \frac{k^2 \mathrm{d}k R_l(k, r) \, R_l^*(k, r')}{z - k^2} \qquad (4.18)$$

where $R_{nl}(r)$ and $R_l(k, r)$ are the wave functions of the discrete spectrum and scattering, respectively. Substituting the expansion (4.18) in (4.16), we find the following representation for $G_{l_1 l_2}(r_1, r_2; r_1', r_2'; E)$

$$G_{l_1 l_2}(r_1, r_2; r_1', r_2'; E) = \delta_{\lambda_1 l_1} \delta_{\lambda_2 l_2} \sum_{\substack{n_1 \lambda_1 \\ n_2 \lambda_2}} \frac{R_{n_1 \lambda_1}(r_1) R_{n_2 \lambda_2}(r_2) R^*_{n_1 \lambda_1}(r_1') R^*_{n_2 \lambda_2}(r_2')}{E - E_{n_1 \lambda_1} - E_{n_2 \lambda_2}} +$$

$$+ \delta_{\lambda_1 l_1} \sum_{n_1 \lambda_1} \int_0^\infty k_2^2 \, dk_2 \frac{R_{n_1 \lambda_1}(r_1) R_{l_2}(k_2, r_2) R^*_{n_1 \lambda_1}(r_1') R^*_{l_2}(k_2, r_2')}{E - E_{n_1 \lambda_1} - k_2^2} +$$

$$+ \delta_{\lambda_2 l_2} \sum_{n_2 \lambda_2} \int_0^\infty k_1^2 \, dk_1 \times \frac{R_{l_1}(k_1, r_1) R_{n_2 \lambda_2}(r_2) R^*_{l_1}(k_1, r_1') R^*_{n_2 \lambda_2}(r_2')}{E - E_{n_2 \lambda_2} - k_1^2} +$$

$$+ \int_0^\infty k_1^2 \, dk_1 \int_0^\infty k_2^2 \, dk_2 \frac{R_{l_1}(k_1, r_1) R_{l_2}(k_2, r_2) R^*_{l_1}(k_1, r_1') R^*_{l_2}(k_2, r_2')}{E - k_1^2 - k_2^2} \equiv$$

$$\equiv G_{bb} + G_{bc} + G_{cb} + G_{cc} \tag{4.19}$$

where $G_{bb}$ correspond to the contribution of the one-particle bound states only; $G_{bc}$ gives the contribution of the states in which the first particle is in the discrete spectrum, and the second in continuum; $G_{cb}$ is the analogous contribution with transposed numbers of particles; $G_{cc}$ correspond to two particles in continuum.

The analytic structure of $G_{l_1 l_2}(E)$ follows directly from the expansion (4.19) and is described in [33, 34] in detail. This analytic structure is very important for understanding and calculating the resonance states in terms of the discussed model because the three-particle resonance poles in the near-threshold region are determined by the branch points on the non-physical sheets, which are due to resonances in each subsystem. Moreover, the Green function $G_{l_1 l_2}(E)$ is a part of the kernel of the three-particle equation (see below) and the calculation method depends on the kernel singularities.

Now we shall consider the three-particle equations describing our model. Carrying out the Faddeev reduction on the analogy of the general case, but using the decomposition (4.15) of the total Hamiltonian, we obtain:

$$G(E) = G_0(E) + G^{(12)}(E) + G^{(3)}(E) \tag{4.20}$$

where

$$G^{(12)}(E) \equiv G_0 V_{12} G; \qquad G^{(3)} = G_0 V_3 G = G_0 (V_{31} + V_{32}) G \, .$$

Now we obtain a set of *two* Faddeev equations for the components $G^{(12)}$ and $G^{(3)}$:

$$G^{(12)} = G_{12} - G_0 + G_{12}V_{12}G^{(3)} , \tag{4.21a}$$

$$G^{(3)} = G_3 - G_0 + G_3V_3G^{(12)} . \tag{4.21b}$$

In the same way, we obtain a set of two equations for the Faddeev components of the total wave function. For example, the equations for the wave function of deuteron scattering in the field of a nuclear core will be [31]:

$$\psi^{(12)}_{\varepsilon_{12}\,p_3} = \varphi^{(12)}_{\varepsilon_{12}\,p_3} + G_{12}V_3\psi^{(3)}_{\varepsilon_{12}\,p_3} , \tag{4.22a}$$

$$\psi^{(3)}_{\varepsilon_{12}\,p_3} = G_3V_{12}\psi^{(12)}_{\varepsilon_{12}\,p_3} . \tag{4.22b}$$

Here in the latter equation we have taken into account that in channel (3) there is no incident wave because the incident wave is only present in channel (12). Multiplying (4.22a) by $V_2$, and (4.22b) by $V_3$, and substituting the obtained homogeneous equation in the inhomogeneous one, we find finally a single integral equation for the transition operator, $T^{(12)}$, corresponding to the scattering of the bound state ("deuteron") by the nucleus:

$$T^{(12)}_{\varepsilon_{12}p_3} = \chi_{\varepsilon_{12}\,p_3} + T_{12}\,\Delta G_3 T^{(12)}_{\varepsilon_{12}\,p_3} \tag{4.23}$$

where

$$T^{(12)}_{\varepsilon_{12}p_3} = V_{12}\psi^{(12)}_{\varepsilon_{12}p_3} ; \qquad \chi_{\varepsilon_{12}p_3} = V_{12}\varphi^{(12)}_{\varepsilon_{12}p_3} ,$$

$$\Delta G_3 = G_3 - G_0 = G_0V_3G_0 .$$

Turning now from the operator equation (4.23) to an equation in the momentum representation, we obtain a Fredholm-type integral equation having a unique solution [31, 39]. In the presence of one-particle resonances and near-threshold bound states in subsystems (13) and (23), nearby singularities which just give rise to the resonance behaviour of the solution for (4.23) appear in the kernel of (4.23) $T_{12}\,\Delta G_3$ (see the discussion and application of the above equations in Chapter 5).


## 1.5. The Hilbert-Schmidt Method in the Theory of Integral Equations


As mentioned in Section 1.3, the integral equations of scattering theory (the Lippmann-Schwinger equation in case of two particles and the Faddeev equations in three-particle case) are Fredholm equations of the second type and may be written as (2.45):

$$f = \varphi + Kf . \tag{5.1}$$

If we introduce the resolvent of such an equation

$$R = (1 - K)^{-1} ,\tag{5.2}$$

the solution $f$ may formally be written as $f = R\varphi$ . As mentioned above, the solution may also be formally presented in the form of a Neumann iteration series:

$$f = \sum_{i=0}^{\infty} K^i \varphi \tag{5.3}$$

and the convergence of this series will be governed by the eigenvalues of the operator $K$:

$$K\chi_n = \eta_n \chi_n .\tag{5.4}$$

The divergence of the series is due to the maximum modulo eigenvalues

$$|\eta_m| > 1$$

(for simplicity we assume that there exists only one such value for $n = m$). If the kernel $K$ of the integral equation is symmetric or Hermitian, that is, if it satisfies the condition

$$K(x, y) = K^*(y, x)\tag{5.5}$$

and equation (5.1) is treated in the Hilbert space $L^2(a, b)$, then it is possible to prove the powerful theorems of the series expansion of the kernel $K$ in terms of its eigenfunctions[4]). The expansions will greatly simplify the solution and the study *of the corresponding equations and* therefore are extensively used in the applications, particularly in scattering theory.

First we shall explain briefly the respective theorems. The reader will find the proofs and details in the textbooks on integral equations and functional analysis (see, for example, [21, 35]). After that, we shall indicate the difficulties arising when applying the Hilbert-Schmidt formalism to quantum scattering theory.

*Theorem 1.* If the kernel $K(x, y)$ satisfies the condition (5.5) and is not zero almost everywhere[5]), then there exists an orthonormal (finite or infinite) sequence of eigenfunctions $\chi_n(x)$ of the operator $K$, corresponding to the eigenvalues $\eta_n$ such that for *arbitrary* function $g(x) \in L^2(a, b)$ there exists the expansion

$$g(x) = h(x) + \sum_n (g, \chi_n) \chi_n(x)$$

convergent in mean. Here $h(x)$ is a function (dependent on $g(x)$) exhibiting the property that $Kh = 0$ , that is, belonging to the null space of the operator $K$.

---

[4]) These theorems were proved first by Hilbert and Schmidt in the beginning of this century.

[5]) In this case the operator $K$ has at least one non-zero eigenvalue and all its non-zero eigenvalues are of a finite multiplicity.

From theorem 1 is follows immediately that

$$Kg(x) = \sum_n \eta_n(g, \chi_n) \chi_n(x) \, . \tag{5.6}$$

The series (5.6) converges in mean for all functions of the type

$$f(x) = K\, g(x) = \int_a^b K(x, y)\, g(y)\, dy \, .$$

Further, we may prove that under the condition

$$\int_a^b |K(x, y)|^2 \, dx\, dy < C$$

(where $C$ is a constant) the series (5.6) converges absolutely and uniformly. We may prove another theorem on expansion.

*Theorem 2.* Any function $K(x, y)$ with a summable square which satisfies eq. (5.5) may be series expanded in terms of the eigenfunctions of the operator $K$ generated by the kernel $K(x, y)$

$$K(x, y) = \sum_n \eta_n \chi_n(x) \chi_n(y) \tag{5.7}$$

($\eta_n$ are the eigenvalues of the operator $K$), the series (5.7) being convergent in mean.

From this theorem it follows that

$$\|K\| = \sum_n \eta_n^2 \tag{5.8}$$

or

$$\|K(x, y) - S_N(x, y)\| = \sum_{n>N} \eta_n^2 \tag{5.9}$$

where $S_N(x, y)$ denotes the $N$-th partial sum of the series (5.7). However, the uniform convergence of the series (5.7) does not always take place, even for the continuous functions $K(x, y)$.

Therefore, we shall present another important theorem on the uniform convergence of the expansion (5.7).

*Theorem 3.* (Mercer's theorem) If the operator $K$ generated by the continuous symmetric kernel $K(x, y)$ is positively defined, that is $(Kf, f) \geq 0$ for arbitrary $f$, or, which is equivalent, all eigenvalues $\eta_n$ are positive, then the series (5.7) converges uniformly.

It should be added that for any continuous symmetric kernel $K(x, y)$ the iterated kernel i.e.

$$K_2(x, y) = \int_a^b K(x, z) \, K(z, y) \, \mathrm{d}z \, ,$$

is positive and symmetric. Indeed,

$$(K_2 f, f) = (K^2 f, f) = (Kf, Kf) \geq 0 \, .$$

Besides, the eigenfunctions $\chi_n(x)$ of the operator $K$ are, as is evident, eigenfunctions of the operator $K^2$ and the eigenvalues are equal to the squares of the eigenvalues of $K$:

$$K^2 \chi_n = \eta_n^2 \chi_n \, .$$

Now we shall briefly touch upon the applicability of these theorems to the quantum scattering theory. Some difficulty arises from the fact that the operator $K$ corresponding to the kernel of the Lippmann-Schwinger equation

$$K(E) = G_0(E) \, V \tag{5.10}$$

is, generally speaking, non-symmetric and non-Hermitian. However, at negative energies the Green function $G_0(E) = (E - H_0)^{-1}$ is a negatively defined operator ($H_0$ is the kinetic-energy operator) and, hence, the operator $K(E)$ may be transformed into a symmetric operator

$$K_s(E) = -[-G_0(E)]^{1/2} V [-G_0(E)]^{1/2} \tag{5.11}$$

which, as evident, is directly related to the initial operator $K$:

$$K(E) = -[-G_0(E)]^{1/2} K_s(E) [-G_0(E)]^{-1/2} \tag{5.12}$$

(the similarity transformation).

On the other hand, it is clear that the eigenvalues of the symmetrized operator $K_s$ coincide with the eigenvalues of the initial operator $K$ and the eigenfunctions of the two operators are related to each other in a simple way. Therefore, taking into account this fact, the whole Hilbert-Schmidt theory at $E < 0$ is applicable to the Lippmann-Schwinger equation. As regards the physical region of the positive energies, several possible approaches may be used. One of them was developed by Sasakawa [36] who suggested replacing the complex-valued operator

$$G_0^{(+)}(E + i\varepsilon) = \int \frac{\psi_0^*(\mathscr{E}) \, \psi_0(\mathscr{E})}{\mathscr{E} - E - i\varepsilon} \, \mathrm{d}\mathscr{E} =$$

$$= P \int \frac{\psi_0^*(\mathscr{E}) \, \psi_0(\mathscr{E})}{\mathscr{E} - E} \, \mathrm{d}\mathscr{E} + i \, \pi \delta(E - \mathscr{E})$$

by the Green function in the sense of principal value

$$G_0^P(E) = P \int \frac{\psi_0^*(\mathscr{E})\,\psi_0(\mathscr{E})}{\mathscr{E} - E}\, d\mathscr{E}$$

which is Hermitian. This operator will be used further in the Lippmann–Schwinger equation instead of the initial Green function $G_0^{(+)}$.

Using now a simple relation which exists between the operators $G_0^{(+)}V$ and $G_0^P V$, we may readily relate the solutions for the initial and symmetrized problems to each other.

The numerical calculations carried out by Sasakawa for a number of particular potentials have demonstrated a more rapid convergence of the series (5.7) for the operator $G_0^P V$ than for the initial operator $G_0^{(+)}V$.

Another approach is that all the operators and the corresponding expansions in eigenfunctions of type (5.7) in the region $E + i\varepsilon$ $(E > 0)$ are considered to be the analytic continuation of the corresponding operators from the region $E < 0$ where all quantities appearing in the scattering theory (Green function, $S$-matrix, eigenvalues and eigenfunctions of the operator $G_0 V$, etc.) are real and well defined. Such a definition not only gives a mathematical method which may be used for the corresponding proofs, but is also a constructive definition underlying the method for the *actual calculations* of eigenfunctions and eigenvalues and also of quantities associated with them, such as $S$-matrix, scattering phase shifts, etc. (see Chapter 3). Because of its convenience and universality this method will be used throughout this book.

The Hilbert–Schmidt expansion is extensively used in scattering theory for various purposes, in particular, to calculate resonance states. One of the very useful applications is associated with the acceleration of convergence or with rearrangement of the Born series (from divergent to convergent at low energies; for detailed description of this approach see Ref. [38]). However, in the subsequent sections we shall be more interested in the analytic properties of the eigenvalues and eigenfunctions of the Hilbert–Schmidt problem as applied to the Lippmann–Schwinger and Faddeev integral equations, because it is these analytic properties that are of paramount importance to the theory of resonance states.

## 1.5.1. *The Hilbert–Schmidt Expansion for the Lippmann–Schwinger Integral Equation*

The Hilbert–Schmidt method, as applied to the Lippmann–Schwinger equation, is described in sufficient detail in Refs. [32, 37, 38]. Therefore, we confine ourselves here to a brief review of the basic ideas and formulae necessary in what follows.

We shall consider the Lippmann–Schwinger equation for the $l$-th partial-wave component of the scattering matrix $t_l(z)$:

$$t_l(z) = V_l + G_0(z) V_l t_l(z) \tag{5.13}$$

where $G_0(z) = (z - H_0)^{-1}$, $V_l$ is the potential in the $l$-th partial wave. For the kernel of this equation $G_0(z)V_l$ we may write the following eigenvalue problem:

$$G_0(z) V_l \psi_{nl}(z) = \eta_{nl}(z) \psi_{nl}(z) \tag{5.14}$$

where $\psi_{nl}(z)$ is an eigenfunction and $\eta_{nl}(z)$ is an eigenvalue. The eigenfunction $\psi_{nl}$ is normalized with weight $G_0^{-1}$:

$$\left(\psi_{nl}, G_0^{-1} \psi_{n'l}\right) = -\delta_{nn'} \tag{5.15}$$

and, together with the eigenfunction $g_{nl}(z)$ of the kernel $V_l G_0(z)$

$$V_l G_0(z) g_{nl}(z) = \eta_{nl}(z) g_{nl}(z) , \tag{5.16}$$

constitutes a biorthogonal set of functions:

$$\left(g_{nl}, \psi_{n'l}\right) = -\delta_{nn'} \tag{5.17}$$

where

$$\psi_{nl}(z) = G_0(z) g_{nl}(z) . \tag{5.18}$$

The main properties of the eigenvalues $\eta_{nl}(z)$ [32, 38] are as follows:

(1)  For a "normal" operator $V_l$, $\eta_{nl}(z)$ will always form a discrete set (this is due to the compactness of the operator $G_0(z) V$ [22] at $z = E + i\varepsilon$ ; non-compact operators also have a continuous series of eigenvalues).

(2)  At any given value of $z$, only a finite number of $\eta_{nl}(z)$ lie outside a unit circle $(|\eta_{nl}(z)| > 1)$. This property is important in investigating the convergence of Born series, because it is $\eta_{nl}(z)$ lying outside the unit circle that lead to the divergence of the series at a given energy $z = E > 0$ .

(3)  $\eta_{nl}(z)$      is real for      $z < 0$ .

(4)  At $z < 0$  all $\eta_{nl}(z)$ are the positive and increasing ("attractive") or negative and decreasing ("repulsive") functions in the entire interval $-\infty < z \leq 0$ . The words "attractive" and "repulsive" are associated with the fact that for purely attractive potentials $V_l$ all $\eta_{nl}$ are positive and for purely repulsive potentials they are negative. An interaction of a general type gives both positive and negative eigenvalues.

(5)  With increasing principal quantum number n, the function $\eta_{nl}$ decreases, so that $\eta_{nl}(z) \sqrt{n} \to 0$ .

(6)  All $\eta_{nl}(z)$ are analytic in the entire $z$-plane except the cut along the real axis from $z = 0$ to $z = \infty$ .

(7)  Since at $z < 0$ the functions $\eta_{nl}$ are real, the Schwarz reflection principle is valid for them in the entire complex plane:

$$\eta_{nl}(z^*) = \eta_{nl}^*(z) . \tag{5.19}$$

(8)  The attractive (repulsive) $\eta_{nl}(z)$ in the upper $z$-halfplane have a positively definite (negatively definite) imaginary part.

(9)  $\eta_{nl}(z) \to 0$  for  $|z| \to \infty$ .

Properties 1–9 are valid for the general-type potential $V$. For a short-range potential satisfying, at some $a > 0$, the condition

$$\int\limits_0^\infty r\, |V(r)|\, e^{r/a}\, dr < \infty \tag{5.20}$$

we may prove stronger limitations for the eigenvalues $\eta_{nl}(z)$ [21, 36, 38].

(10)  The function $\eta_{nl}(z)$ has only one singularity in the $z$-plane, namely, the root branch point at $z = 0$ . In this way, going to a uniformizing variable $k = \sqrt{(2mz/\hbar^2)}$ , we obtain a function which is analytic in the region of analyticity of the Jost function $f_l(k)$ [8], that is, at

$$\mathrm{Im}\; k > -\frac{1}{2a} . \tag{5.21}$$

For finite potentials $(V(r) = 0$ at $r > r_0)$ $\eta_{nl}(k)$ is, as the Jost function, analytic in the entire $k$-plane.

(11)  The following representation is valid for $\eta_{nl}(k)$ in the extreme $k \to 0$ :

$$\eta_{nl}(k) = \eta_{nl}(0) + \sum_{j=0}^\infty a_j^{nl} k^{2j} + i k^{2l+1} \sum_{j=0}^\infty b_j^{nl} k^{2j} \tag{5.22}$$

where $\eta_{nl}(0)$, $\{a_j^{nl}\}$ and $\{b_j^{nl}\}$ are *real* numbers.

The Schwarz reflection principle (5.19) in the $k$-plane may be written as

$$\eta_{nl}(-k^*) = \eta_{nl}^*(k) . \tag{5.23}$$

All of the most important quantities of the scattering theory on energy shell may be expressed in terms of the eigenvalues $\eta_{nl}(k)$ , namely, the Jost function

$$f_l(+k) = \prod_n (1 - \eta_{nl}(k)); \tag{5.24}$$

the scattering phase shift

$$\delta_l(k) = \sum_n \mathrm{arctg}\, \frac{\mathrm{Im}\, \eta_{nl}(k)}{1 - \mathrm{Re}\, \eta_{nl}(k)} \tag{5.25}$$

where $k > 0$, which corresponds to $z = E + i\varepsilon$, that is to the upper rim of the cut in the $z$-plane;
the $S_l$-matrix

$$S_l(k) = f_l(-k)/f_l(k) = \prod_n \frac{1 - \eta_{nl}(-k)}{1 - \eta_{nl}(k)} \tag{5.26}$$

etc.

All the analytic properties of the $S$-matrix follow from the expression (5.26) in a natural way:

(a) $S_l(k) = S_l^{-1}(-k)$,

(b) on the imaginary semiaxis $k$, where $E < 0$ and $\eta_{nl}$ are real, $S_l$ is also real;

(c) for real $k$ it follows from (5.23) that $|S_l(k)| = 1$.

Further, one can easily see from (5.26) that the poles of the $S$-matrix correspond to the values

$$\eta_{nl}(k_0) = 1 \tag{5.27}$$

and from the Schwarz principle (5.23) we obtain also that $\eta_{nl}(-k_0^*) = 1$, i.e. $S_l(k)$ is sure to have another pole at $k = -k_0^*$. At $z < 0$ $(k = i\varkappa)$ eq. (5.27) determines the location of a bound state. From property (4) it follows immediately that for the $n$-th bound state to appear in the $l$-th partial wave it is necessary and sufficient that

$$\eta_{nl}(0) > 1. \tag{5.28}$$

For the complex values of $z = \hbar^2 k^2/2m$ $(k = \pm\alpha - i\beta)$ the equation (5.27) defines the resonance and virtual states. Now, using the properties presented above, we describe the behaviour of $\eta_{nl}(z)$ as $z$ varies from $-\infty$ to 0 and from $0 + i\varepsilon$ to $\infty + i\varepsilon$ (which corresponds to the variations of $k$ from $+i\infty$ to 0 and from 0 to $+\infty$). As the argument varies in such a way, the values $\eta_{nl}$ describe a curve in the $\eta$ plane, which is called the trajectory of the eigenvalue $\eta_{nl}(z)$. As $z$ varies from $-\infty$ to 0, the function $\eta_{nl}(z)$ increases monotonically (we examine the case of attractive interaction) from 0 to $\eta_{nl}(0)$ and remains real. At the point $z = 0$ the trajectory turns to the upper half-plane of the complex plane $\eta$. The angle at which the trajectory turns to the complex plane depends on the angular momentum $l$. Indeed, from (5.22) we obtain

$$\operatorname{tg}\alpha_l = \lim_{k \to 0} \frac{\operatorname{Im}\left(\eta_{nl}(k)\right)}{\operatorname{Re}\left(\eta_{nl}(k)\right)} \sim k^{2l-1}. \tag{5.29}$$

Thus, the $S$-wave trajectory $(l = 0)$ goes to the complex plane at $90°$ $(\alpha_0 = \pi/2)$; at $l \neq 0$ we get $\alpha_l = 0$ and the trajectory touches the axis $\operatorname{Re}\eta$. Here, the larger $l$, the higher is the order of contact. Further, at $z \to \infty + i\varepsilon$ the trajectory again goes to the origin from the upper $\eta$ halfplane. Fig. 1.7

presents a typical trajectory for an attractive potential. In section 3.1.2 we present the trajectories of eigenvalues for several of the most extensively used forms of the potential $V$.

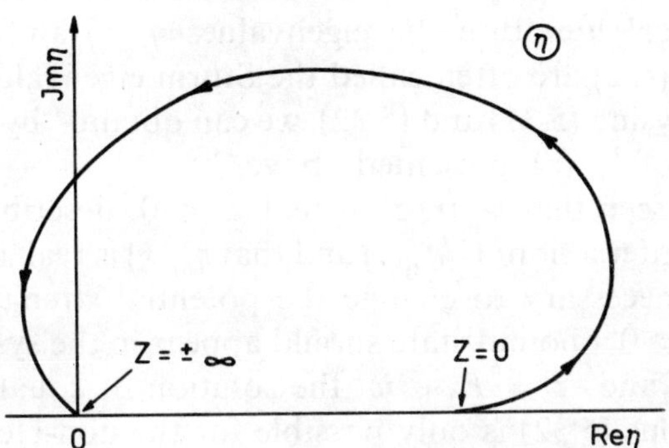


Fig. 1.7
A typical trajectory $\eta(E)$ in $\eta$-plane for an attractive potential.

It should be noted that the $t$-matrix and the interaction $V$ may be series-
-expanded in terms of the Hilbert-Schmidt eigenfunctions $g_{nl}(z)$:

$$t_l(k', k; z) = -\sum_n \frac{\eta_{nl}(z)}{1 - \eta_{nl}(z)} g_{nl}^*(k', z^*) g_{nl}(k, z) \,,$$

$$V_l(k', k) = -\sum_n \eta_{nl}(z)g_{nl}^*(k', z^*) g_{nl}(k, z) \,. \tag{5.30}$$

These formulae were extensively used in solving the Faddeev equations (see e.g. [32, 37]). To understand the physical meaning of the Hilbert-Schmidt problem (5.14) more properly, we shall reformulate it in a differential form.

We shall look for the solution to the equation

$$\left[ H_{0l} + \frac{1}{\eta_{nl}(z)} V_l \right] \psi_{nl}(r, z) = z \, \psi_{nl}(r, z) \tag{5.31}$$

where

$$H_{0l} = -\frac{\hbar^2}{2m} \left[ \frac{d^2}{dr^2} - \frac{l(l+1)}{r^2} \right]$$

and $\psi_{nl}(r, z)$ satisfies the boundary conditions

$$\psi_{nl}(r, z) \to r^{l+1} \qquad r \to 0$$

$$\psi_{nl}(r, z) \to e^{ikr} \qquad r \to \infty, \quad \text{at} \quad \text{Im } k > 0 \,. \tag{5.32}$$

Since we treat $z$ on the physical sheet limited by the upper rim of the cut $(z = E + i\varepsilon)$, $\eta_{nl}(z)$ in (5.31) are the same eigenvalues as in the expression (5.14) and $\psi_{nl}(r, z)$ are the same eigenfunctions in the co-ordinate representation. The boundary-value problem (5.31)–(5.32) is a particular case of the Sturm-Liouville boundary-value problem [32] which has been thoroughly studied in mathematics. In the physical literature the eigenvalues $\eta_{nl}(z)$ and the corresponding eigenfunctions $\psi_{nl}(r, z)$ are often called the Sturm eigenvalues and eigenfunctions. Now, by analysing (5.31) and (5.32), we can obtain, "by fingers" practically all the properties of $\eta_{nl}(z)$, presented above.

Thus, it can be seen that $\psi_{nl}(r, z)$ at real $z < 0$ describes a bound state in a system with the interaction $V_l/\eta_{nl}(z)$ and that $\eta_{nl}(x)$ is real, i.e. $\eta_{nl}(z)$ shows how many times it is necessary to change the potential strength in order that at a given energy $z < 0$ a bound state should appear in the system. It is clear that for the complex $z$ and $z = E + i\varepsilon$ the solution of equation (5.31) with the boundary conditions (5.32) is only possible for the non-Hermitian interaction $V_l/\eta_{nl}(z)$, i.e. $\eta_{nl}(z)$ is to be complex.

Now we shall analyse the Hilbert-Schmidt expansion in the case of the three-particle Lippmann-Schwinger equation.

### 1.5.2. The Three Particle Case

As shown above, in the case of two particles the Hilbert-Schmidt problem for the Lippmann-Schwinger equation is completely equivalent to the Sturm-Liouville problem for the Schrödinger equation. In this case the Lippmann-Schwinger equation is of Fredholm type, i.e. it has a unique solution at both $z < 0$ and $z = E + i\varepsilon$ and it is equivalent to the Schödinger equation together with the corresponding boundary conditions. In the three particles case the Lippmann--Schwinger equation

$$T(z) = V_{123} + G_0(z)\, V_{123}\, T(z) \tag{5.33}$$

(where $V_{123} = V_1 + V_2 + V_3$, $G_0(z) = (z - H_0)^{-1}$) has no longer a unique solution and is equivalent to the three-particle Schrödinger equation only at energies below the lowest two-particle threshold (taking into account the exchange processes). At higher energies the kernel of eq. (5.33) becomes non--compact [21, 30] and the equation becomes a non-Fredholm type equation, that is, the standard theory of integral equations cannot be applied. As discussed in Section 1.4.2, the Faddeev equations [29] are Fredholm-type equations for a three-particle system, so the Hilbert-Schmidt theory may be applied to them. Thus, we may formulate the problem of Hilbert-Schmidt type for a three-particle system in two ways, namely, for the Schrödinger equation in a differential form (or, correspondingly, for the Lippmann-Schwinger equation in an integral form) and for the Faddeev equations. We shall examine these two approaches successively. For simplicity we shall consider a system of three identical spinless particles of mass $m$. The relevant Schrödinger equations is

$$[H_0 + 1/\mu_n(z)\, V_{123}(x, y)]\psi_n(x, y) = z\, \psi_n(x, y) \tag{5.34}$$

where $x$ and $y$ are the usual Jacobi co-ordinates:

$$x = \frac{1}{\sqrt{2}}(r_2 - r_3); \qquad y = \frac{1}{\sqrt{6}}(r_2 + r_3 - 2r_1)$$

$$H_0 = -\frac{\hbar^2}{2m}(\Delta_x + \Delta_y). \tag{5.35}$$

We impose the following boundary conditions on the function $\psi_n(x, y)$ [37] at the origin

$$\psi_n(x, y) = \text{const} \quad \text{for} \quad x \to 0, \ y \to 0 \tag{5.36}$$

and at infinity:

$$\psi_n(x, y) = C_{1n}\frac{e^{iq\varrho}}{\varrho^{5/2}} + C_{2n}\hat{S}\Phi_d(x)\frac{e^{i\sqrt{3/2}\,ty}}{y} \tag{5.37}$$

where $q = \sqrt{2mz/\hbar^2}$, $t = \sqrt{(4m/3\hbar^2)(z - \varepsilon_d)}$, $\varrho^2 = x^2 + y^2$, $\varepsilon_d$ and $\Phi_d$ are respectively the energy eigenvalue (the characteristic energy) and the eigenfunction of two-particle bound state $(\varepsilon_d < 0)$; $\hat{S}$ is the symmetrization operator. At $\varepsilon_d < E < 0$ only the second term remains in the right-hand side of (5.37). The choice of the asymptotic behaviour in the form of (5.37) corresponds to the choice of the solutions containing only outgoing waves at infinity. It is evident that at $z < \varepsilon_d$ and $\mu_n(z) = 1$ the energy eigenvalues (characteristic energies) and the eigenfunctions of three-particle bound states are obtainable from (5.34)–(5.37). At $\varepsilon_d < z < 0$ there arises a channel of decay to a "deuteron", i.e. a bound two-particle subsystem, and a free particle which corresponds to a two-particle root cut in the $z$ plane from the point $z = \varepsilon_d$ to $z = +\infty$. At $z = 0$ a channel of decay to three free particles opens, which corresponds to a three-particle logarithmic cut [30] in the $z$-plane from the point $z = 0$ to $z = +\infty$. Therefore, the energies above the two-particle threshold are understood to be $z = E + i\varepsilon$, i.e. at the upper rim of the cuts.

For finding eigenvalues $\mu_n(z)$ and eigenfunctions $\psi_n$ we can also formulate a variational principle [37]. These functions may be used to expand a three--particle $T$-matrix:

$$T(z) = V_{123} + \sum_m \frac{V_{123}\,|\psi_m\rangle\,\langle\psi_m|\,V_{123}}{1 - \mu_m(z)} + T'(z) \tag{5.38}$$

where $T'(z)$ does not contain pole singularities in $z$.

From (5.38) it can be seen that poles of the three-particle $T$-matrix are defined by the condition:

$$\mu_n(z_0) = 1 \ . \tag{5.39}$$

At $z_0 < \varepsilon_d$ the equation (5.39) determines the three-particle bound states on the physical sheet and at $\text{Re } z_0 > \varepsilon_d$ the three-particle resonances on non-physical sheets. However, as we shall see below, the eigenvalues $\mu_n(z)$, do not, generally speaking, coincide with the Hilbert-Schmidt eigenvalues $\xi_n(z)$ for the Faddeev equations. They $(\mu_n$ and $\xi_n)$ coincide only at $z = z_0$ i.e. when $\mu_n(z_0) = \xi_n(z_0) = 1$, which is in the poles of the three-particle scattering amplitude. In the general case this is clear without further reasoning, because the kernels of the three-particle Lippmann-Schwinger and Faddeev equations are very different.

Now we shall examine the Hilbert-Schmidt problem for the Faddeev equations and, for the sake of simplicity, we shall confine ourselves to the same case of three identical spinless particles of mass $m$ and $s$-wave separable two-particle interactions, i.e. $V(p, p') = -\lambda g(p)g(p')$. The homogeneous Faddeev equation for such a system is

$$\psi(k, p) =$$

$$= \frac{2m}{\xi_n \hbar^2} \int \frac{t_0\left(k, |p/2 + p'|; z - \dfrac{3\hbar^2 p^2}{4m}\right)}{\left(p^2 + pp' + p'^2 - \dfrac{mz}{\hbar^2}\right)} \psi(|p + p'_2|, p') \, \mathrm{d}p' \tag{5.40}$$

where $k = \frac{1}{2}(k_2 - k_3)$; $p = \frac{1}{3}(k_2 + k_3 - 2k_1)$ and $k_j$ is the momentum of the $j$-th particle.

After making the partial-wave expansion, substitution of $t$-matrix in separable form, and symmetrization of the kernel, we may obtain, from eq. (5.40) the equation [30, 32]

$$\varphi_{nl}(p, z) = \frac{1}{\xi_{nl}(z)} \int_0^\infty K_{nl}(p, p'; z)\varphi_{nl}(p', z)p'^2 \, dp' \tag{5.41}$$

where the kernel $K_{nl}(p, p'; z)$ is the partial wave projection of the initial three-dimensional kernel of eq. (5.40) (for details see Ref. [30]) and is of Hilbert-Schmidt type. Here, the angular momentum of the two-particle subsystem is assumed to be zero; $l$ is the relative angular momentum of an incomming particle in the c.m.s. system. Equation (5.41) defines the eigenvalues $\xi_{nl}(z)$ and the eigenfunctions $\varphi_{nl}$ of the Hilbert-Schmidt problem in the three-body case.

From (5.40) it can be seen that $\xi = 1$ corresponds, as in the two-body case, to the physical solutions for the homogeneous equation, i.e. to the states with a purely outgoing wave in the asymptotic region. At $z < \varepsilon_d$ (where $\varepsilon_d$ is the energy of the lowest two-particle threshold) this corresponds to bound states. Therefore, $\xi_{nl}(z)$ at $z < \varepsilon_d$ are real, $\xi_{nl}$ and $\varphi_{nl}$ satisfy the Schwarz principle:

$$\xi_{nl}(z^*) = \xi_{nl}^*(z) \text{ and } \varphi_{nl}(p, z^*) = \varphi_{nl}^*(p, z) \tag{5.42}$$

The Hilbert-Schmidt eigenvalues $\xi_{nl}$ have the following singularities in the $z$-plane [32, 38]:

(i) at the two-particle threshold $z = \varepsilon_d$, $\xi_{nl}$ has a root branch point

$$\xi_{nl}(z) \underset{z \to \varepsilon d}{\sim} C_1(\varepsilon_d - z)^{l+1/2} + C_2, \tag{5.43}$$

H    (ii) at the three-particle threshold $z = 0$, $\xi_{nl}$ has a logarithmic branch point

$$\xi_{nl}(z) \underset{z \to 0}{\sim} f(z) + C z^{l+2} \ln(-z). \tag{5.44}$$

The three-particle $T$-matrix may be presented, as in case of the two-body problem, in the form of the following expansion [37]:

$$T\left(k_1, k_2, k_3; k_1', k_2', k_3'; z\right) =$$

$$= \sum_n \frac{\xi_n(z)}{\xi_n(z) - 1} g_n(k_1, k_2, k_3; k_1', k_2', k_3'; z) \tag{5.45}$$

where $g_n$ is expressed in terms of eigenfunctions $\varphi_{nl}$. It is important that a denominator $(\xi_n(z) - 1)$ appears in eq. (5.45). In this way, the equation

$$\xi_{nl}(z_0) = 1 \tag{5.46}$$

actually defines the poles of the three-particle $T$-matrix, i.e. the bound states on the physical sheet of $z$ and the resonance states on the non-physical sheets of $z$. Examples of numerical studies of the three-particle Hilbert-Schmidt eigenvalues for particular physical systems may be found in a number of works (see, e.g. [40–42]) and will be discussed in detail in Chapter 3.

## References

1. HARDY G. H., Divergent Series, 2nd ed., Clarendon Press, 1956
2. ZEL'DOVICH Ya. B., ZhETF (in Russian) *39* (1960) 776 [Sov. Phys. JETP *12* (1961) 542].
3. BATEMAN H., ERDÉLYI A., Higher Transcedental Funstions, Vol. II, Mc Graw-Hill, N. Y.–Toronto–London 1955.
4. BERGGREN T., Nucl. Phys. *A109* (1968), 265.
5. GYARMATI B., VERTSE T., Nucl. Phys. *A160*(1971) 523.
6. SHNOLL E. E., Theor. Math. Phys. (in Russian) 8 (1971) 140.

7. BAKER G. A., Adv. in Theor. Phys. 1, ed. Brueckner, Academic Press, N. Y.–London 1965; BAKER G. A., Jr., GRAVES-MORRIS P., Padé Approximants, Encyclopedia of Mathematics and its Applications, V. 13, 14, G. C. Rota Ed., Addison-Wesley Publ. Co., 1981, London-Amsterdam; BASDEVANT J. L., Fortschritte der Physik *20* (1972) 283.

8. NEWTON R. G., Scattering Theory of Waves and Particles, 2$^{nd}$ ed., Springer, Berlin, Heidelberg 1982.

9. NUTTAL J., J. Math. Anal. Appl., *31* (1970) 147.

10. BAKER G. A., GAMMEL J. L., J. Math. Appl. *2* (1961) 21

11. ZINN-JUSTIN J., Phys. Rep. *10* (1971) 33.

12. WALL H. S., Analytic Theory of Continued Fractions; van Nostrand, Princeton, New Jersey 1948.

13. SCHLESSINGER L., Phys. Rev. *167* (1968) 1411.

14. LUKE Y. L., Mathematical Functions and their Approximations, Academic Press, N. Y.–San Francisco–London 1975.

15. HARTT K., Phys. Rev. *C22* (1980) 1377; *C23* (1981) 2 399; *C29* (1984) 695.

16. KRASNOPOL'SKY V. M., KUKULIN V. I., HORÁČEK J., Czech. J. Phys. *B35* (1985) 805.

17. KRASNOPOL'SKY V. M., CHLEBNIKOV S. Yu. in: Theory of Quantum Systems With Strong Interaction, Kalinin Univ., 1983, p. 80 (in Russian).

18. BAKER G. A., GAMMEL J. L., The Padé Approximant in Theoretical Physics, Academic Press, N. Y. 1970.

19. JOHNSON R. C., New Theory of Padé Approximants, Preprint, Durham University 1971.

20. LOEFFEL J. J., MARTIN A., SIMON B., WIGHTMAN A. S., Phys. Lett. *30B* (1969) 656.

21. SMITHIES V, Integral Equations, Cambridge, New York. 1958.

22. KLOET W. M., TJON J. A., Ann. Phys. *79* (1973) 407.

23. WHITTAKER E. T., WATSON G. N., A Course of Modern Analysis, Cambridge University Press, 1927.

24. DIENES P., The Taylor Series, Dover, New York 1957.

25. PIŠÚT J., ECHaYa [Sov. J. Part. Nucl. *9* (1978) 246]. (in Russian) *9* (1978) No. 3, 602.

26. YNDURAIN F. J., CERN Preprint TH 1372, 1971, YNDURAIN F., Ann. Phys. *75* (1973) 171.

27. FROISSART M., Recherche Cooperative sur Programme, Prog. 25, Vol. 9, C.N.R.S., Strasbour 1969.

28. WIENER N., Cybernetics or Control and Communication in the Animal and the Machine. The technology press, N. Y.-Paris 1948.

29. FADDEEV L. D., ZhETF (in Russian) *39* (1960) 1459 [Sov. Phys. JETP *12* (1961) 1014]; DAN USSR (ser. fiz.) (in Russian) *6* (1961) 384.

30. SCHMID E., ZIEGELMANN H., The Quantum Mechanical Three-Body Problem, Pergamon Press, Braunschweig 1974.

31. BAZ A. I., ZEL'DOVICH Ya. B., PERELOMOV A. M., Scattering, Reactions and Decay in Nonrelativistic Quantum Mechanics, Nauka, Moscow, 2nd ed. 1971 (in Russian). [English translation of lst ed.: Israel Program for Scientific Translations, Jerusalem, 1966].

32. SITENKO A. G., The Lectures in the Scattering Theory (in Russian), Kiev 1971. [English translation: Pergamon Press, Oxford, 1971].

33. FULLER R. C., Phys. Rev. *C3* (1971) 1 042.

34. GLÖCKLE W., HEISS W. D., Nucl. Phys. *A122* (1968) 343.

35. RIESZ F., SZENT-NAGY B., Leçons d'analyse fonctionelle, Budapest 1953.

36. SASAKAWA T., Nucl. Phys. *A160* (1971) 321.

37. BADALYAN A. M., SIMONOV Yu. A., The Variational Method for the 3-body Resonances (Preprint series):
I. The Asymptotic Behaviour of the Resonance Wave Function, Preprint ITEP No. 966, Moscow 1972;

II. The Equation for the Resonance Wave Function, Preprint ITEP No. 952, Moscow 1972; Yad. Fiz. (in Russian) *17* (1973) 441 [Sov. J. Nucl. Phys. *17* (1973) 225].

38. WEINBERG S., Phys. Rev. *131* (1963) 440.
39. MERKURIEV S. P., FADDEEV L. D., Quantum-Mechanical Scattering Theory for Few Body Systems (in Russian) Nauka, Moscow 1985.
40. NARODETSKI I. M., GALPERN E. S., LYAKHOVITSKY V. N., Yad. Fiz. (in Russian) *16* (1972) 707 [Sov. J. Nucl. Phys. *16* (1973) 395].
41. BELYAEV V. B., MOLLER K., Preprint JINR E4-9601, Dubna 1976.
42. HARMS E., Phys. Rev. *D9* (1974) 993.
43. EVGRAFOV M. A., Analytic Functions (in Russian), Nauka, Moscow 1968. [English translation: Saunders, Philadelphia, 1966].
44. BELYAEV V. B., Lectures on Few Body Scattering Theory (in Russian), Energoizdat, Moscow 1985.
45. GOLDBERGER M. L., WATSON K. M., Collision Theory, Chap. 3, 8, John Wiley, N. Y.–London–Sydney 1964.
46. BAKER G. A. Jr., GAMMEL J. L., WILLS J. G., J. Math. Anal. Appl. *2* (1961) 405.

# Chapter 2

# General Concepts of the Theory of Resonance States and Processes

In this chapter we shall briefly treat some parts of the theory of resonance phenomena, such as the methods for analysing experimental data to find resonance states and their parameters, the relationships between the resonances and poles of the $S$-matrix in the scattering theory, the Kapur-Peierls approach, the description of the Gamow and Siegert states, etc. Detailed description of these items may be found in a number of monographs (see, e.g., [1–3]) and articles [4–8]. Here we shall expound this material as uniformly as possible to introduce a certain language and to lay the foundations for understanding and active mastering of the material of the subsequent sections where we shall discuss concrete realizations and applications of the general aspects described here.

## 2.1. Resonances in Scattering Theory

Any quantum-mechanical system can be experimentally found to be either in a bound stationary state, or in a state of scattering, i.e. in a continuous-spectrum state (where the fragments of the system are scattered by each another). In other words, we may experimentally determine the discrete or the continuous spectrum of the system. The resonance (quasi-stationary) states have a definite lifetime, that is, they decay after some time. In scattering experiments we cannot observe them directly, instead, we see the states of the continuous spectrum into which they decay, i.e. the products of their decay. Thus, in the scattering experiments, the resonances manifest themselves not directly, but via the characteristics of the scattering states, namely, the $S$-matrix, the scattering amplitude, the phase shifts, etc. Such a viewpoint is assumed as a basis of the description of resonances in scattering theory. There exists, however, an alternative point of view which relates the resonances to unstable particles and is more widely used in field theory, (see for example, [33]). However, this approach requires a formalism quite different from that developed in this book.

Finally, there is a unifying point of view where the resonances are treated on nearly the same basis as the true bound states and enter the expansions of wave functions and scattering amplitudes in the same way as the bound states. This approach is widely discussed in this and subsequent chapters.

### 2.1.1. *Parametrization of S-matrix in and outside the Vicinity of a Resonance*

In Section 1.5 we gave the expressions for the Jost function (1.5.24) and for the $S$-matrix (1.5.26) in terms of the Hilbert-Schmidt eigenvalues. Here, using these expressions and the properties of the eigenvalues, we obtain a number of formulae characterizing the behaviour of various physical quantities near resonance. For simplicity, we shall limit ourselves here to the case of a short-range interaction potential. Modification of the general formulae necessary when making allowance for long-range interaction is discussed partly in the subsequent chapters (in particular, Chapter 5). As shown in Chapter 1, the location of the poles of $S$-matrix $S_l(k)$ (the zeros of the Jost function $f_l(k)$) is determined by the equality

$$\eta_{nl}(k_0) = 1 . \tag{1.1}$$

The solution of the Schrödinger equation

$$\left\{\frac{\hbar^2}{2m}\left[-\frac{d^2}{dr^2} + \frac{l(l+1)}{r^2}\right] + V(r) - E\right\}\psi_l(k, r) = 0 \tag{1.2}$$

regular at the origin exhibits asymptotic behaviour [1]

$$\psi_l(k, r) \xrightarrow[r\to\infty]{} [f_l(k)h_l^{(-)}(kr) - f_l(-k)h_l^{(+)}(kr)] \tag{1.3}$$

where $h_l^{(\pm)}(k\,r)$ are the spherical Riccati-Hankel functions (see [1]).
At a pole of the $S$-matrix (at $k = k_0$), $f_l(k_0) = 0$ and

$$\Phi_l(k_0, r) = \lim_{k\to k_0} \psi_l(k, r) \tag{1.4}$$

that is, the asymptotic part of the wave function contains only an outgoing wave at $\mathrm{Re}\, k_0 > 0$ and ingoing wave at $\mathrm{Re}\, k_0 < 0$ . Let us examine the various possible locations of the $S$-matrix pole.

(i) Let point $k_0$ be located in the upper $k$ halfplane, i.e. $k_0 = \alpha + i\beta$ ; $\beta > 0$ . The Hamiltonian eigenvalue corresponding to the function (1.4) and equal to $z_0 = (\hbar^2/2m)k_0^2$ lies in this case on the first energy sheet[1]) and is complex. This,

---

[1]) The first (physical) energy sheet corresponds to the upper $k$-halfplane ($\mathrm{Im}\, k \geq 0$), and the second (non-physical sheet) to the lower one. The passage from the first to the second $z$ sheet is through a cut going along the positive real semiaxis of energies ($0 \leq z < \infty$). The upper rim of the cut ($z = E + i\varepsilon$) corresponds to the positive real semiaxis $k$, and the lower rim ($z = E - i\varepsilon$) to the negative semiaxis $k$.

however, contradicts hermiticity of the Hamiltonian which cannot have complex eigenvalues on the first sheet. Therefore, $\alpha = 0$, $k_0 = i\beta$, and

$$z_0 = E = -\frac{\hbar^2}{2m}\beta^2 < 0 .$$

Thus, in the upper $k$-halfplane the $S$-matrix poles for the Hermitian Hamiltonian may only be located on the imaginary semiaxis. They correspond to the states with negative energy, that is, to the bound states and, in this case, the asymptotic behaviour of the wave function (1.4) is of the form

$$\Phi_l(i\beta,r) \underset{r\to\infty}{\sim} e^{-\beta r} . \tag{1.5}$$

(ii) If the point $k_0$ lies in the lower $k$-halfplane, i.e. $k_0 = \alpha - i\beta$ $(\beta \geq 0)$, the value $z_0 = (\hbar^2/2m)k_0^2$ is located on the second energy sheet and may be complex. If $\alpha > 0$, the corresponding state and pole will be called resonant[2]) and the asymptotic part of the wave function of such a state is of the form

$$\Phi_l(k_0, r) \underset{z\to\infty}{\sim} e^{i\alpha r} e^{\beta r} \tag{1.6}$$

that is, it represents an outgoing wave whose amplitude increases exponentially thereby corresponding to a decaying state. We shall examine the time-dependent function of this state [3]

$$\psi_l(r, t) \sim \Phi_l(k_0, r) \exp\{-iz_0 t/\hbar\} \tag{1.7}$$

where

$$z_0 = \frac{\hbar^2}{2m} k_0^2 = \frac{\hbar^2}{2m}(\alpha^2 - \beta^2) - 2i\frac{\alpha\beta\hbar^2}{2m} = E_R - \frac{i}{2}\Gamma , \tag{1.7a}$$

that is,

$$\psi_l(r, t) \sim \Phi_l(k_0, r) \exp\left\{-\frac{iE_R t}{\hbar}\right\} \exp\left\{-\frac{\Gamma}{2\hbar}t\right\} .$$

In other words, the amplitude of the function decreases exponentially with time, which is associated with the finite lifetime of such a state equal to

$$\tau = \frac{2\hbar}{\Gamma} = \frac{m}{\hbar\alpha\beta} . \tag{1.7b}$$

---

[2]) Further, we shall discuss in more detail the niceties and the meaning of the terminology.

According to the Schwarz reflection principle (see (1.5.23)), the $S$-matrix has also a pole at the point $-k_0^* = -\alpha - i\beta$. The corresponding state will be called the antiresonance state and its asymptotic tail is of the form

$$\Phi_l(-k_0^*, r) \underset{r \to \infty}{\sim} e^{-i\alpha r}\, e^{\beta r} \tag{1.8}$$

which corresponds to an ingoing wave (at $\alpha > 0$) with an increasing amplitude, i.e. to the produced state whose wave function amplitude increases exponentially in time.

In the lower $k$ halfplane there may also be poles lying on the imaginary $k$ axis, i.e. $k_0 = -i\beta$. Such states are called virtual states and their wavefunction asymptotic part is of the form:

$$\Phi_l(k_0, r) \underset{r \to \infty}{\sim} e^{\beta r} \tag{1.9}$$

that is, the wave function increases exponentially without oscillating[3]). To simplify the notation, hereafter the bound states will be called the $b$-states, the virtual (anti-bound) states the $a$-states, and the resonance and anti-resonance states the $r$-states. The letters $a, b,$ and $r$, when used as indices, will show (if in a given place we do not ascribe any other meaning to them) that the labelled quantity corresponds to the virtual, bound, and resonance states, respectively.

Now, let the $S$-matrix have only one $r$-pole lying in the vicinity of the positive real $k$ semiaxis. This corresponds to the fact that only one Hilbert-Schmidt eigenvalue $\eta_m(k)$ will be equal to unity in the vicinity of the real $k$ semiaxis at $k = k_0$. We discriminate this eigenvalue in the expansion for the $S$-matrix (1.5.26) (for the sake of simplicity we consider below the case of $l = 0$ and the index $l$ will be omitted)

$$S(k) = \frac{1 - \eta_m(-k)}{1 - \eta_m(k)} \prod_{n \neq m} \frac{1 - \eta_n(-k)}{1 - \eta_n(k)}. \tag{1.10}$$

Here $k$ is positive and real, which corresponds to $z = E + i\varepsilon$ $(E \geq 0)$. For $\eta_m(k)$ near $k_0$ we write down the most general represenation compatible with the Schwarz reflection principle:

$$\eta_m(k) = 1 + (k - k_0)(k + k_0^*). \tag{1.11}$$

Substituting (1.11) in (1.10) we get:

$$S(k) = \frac{(-k - k_0)(-k + k_0^*)}{(k - k_0)(k + k_0^*)} \prod_{n \neq m} \frac{1 - \eta_n(-k)}{1 - \eta_r(k)}.$$

------

[3]) We do not discuss here the so-called false $S$-matrix poles which appear only for special kinds of the interaction operators (for discussion of the properties of these poles see Ref. [2]).

The relation obtained may be rewritten as follows

$$S(k) = \frac{(-k - k_0)(-k + k_0^*)}{(k - k_0)(k + k_0^*)}\, e^{2i\delta_P}\,. \tag{1.12}$$

Here we have introduced a potential (background) phase shift $\delta_P$ which describes the contribution to the $S$-matrix from all the rest poles with $n \neq m$, that is, the background here presents the contribution to the scattering from all the rest states, except for the discriminated $r$-state.

The expression (1.12) may be rewritten in the form:

$$S(k) = \frac{k^2 - \alpha^2 - \beta^2 - i2\alpha\beta k}{k^2 - \alpha^2 - \beta^2 + i2\alpha\beta k}\, e^{2i\delta_P} = \frac{E - E_R - (i/2)\Gamma}{E - E_R + (i/2)\Gamma}\, e^{2i\delta_P} \tag{1.13}$$

where

$$E_R = \frac{\hbar^2}{2m}(\alpha^2 + \beta^2)\,; \qquad \Gamma = 2\frac{\hbar^2}{m}k\beta\,.$$

This parametrization differs slightly from the parametrization (1.7a) where

$$E_R = \mathrm{Re}\left(\frac{\hbar^2}{2m}k_0^2\right) = \frac{\hbar^2}{2m}(\alpha^2 - \beta^2)$$

and

$$\Gamma = -2\,\mathrm{Im}\left(\frac{\hbar^2}{2m}k_0^2\right) = 2\frac{\hbar^2}{m}\alpha\beta\,.$$

However, as the expansion (1.13) is only valid in the vicinity of $k_0$, we get $k \sim \alpha$ and because, according to the assumption, the pole lies near the real $k$-axis, $\beta$ is small and, therefore, the two parametrizations are practically the same. We note that the parametrization (1.13) correctly reflects the dependence of the resonance width on energy $\Gamma \sim k \sim \sqrt{E}\,(\Gamma_l \sim k^{2l+1}$ for $l \neq 0)$ in case of short-range interaction.

From (1.13) we can easily obtain the expression for the scattering phase shift:

$$\delta = \frac{1}{2}\arg S = \delta_P - \mathrm{arctg}\,\frac{\Gamma/2}{E - E_R} = \delta_P + \delta_R \tag{1.14}$$

where

$$\delta_R = -\,\mathrm{arctg}\,\frac{\Gamma/2}{E - E_R}$$

is the resonance phase shift.

In accordance with the above mentioned assumption of remoteness of the rest $S$-matrix singularities we may consider the "potential" phase shift $\delta_P$ to vary slowly in the vicinity of $E = E_R$ . Hence, at energies near $E_R$ we get

$$\left. \frac{d\delta}{dE} \right|_{E \simeq E_R} \sim \frac{d\delta_R}{dE} \sim \frac{1}{\Gamma} . \tag{1.15}$$

Thus, from (1.15) it follows that the smaller the resonance width $\Gamma$ the stronger is the change in the scattering phase shift at energies $E \sim E_R$ . At $E = E_R$ , $\delta_R$ goes through $\pi/2$ and $\delta_R(E_R - \Delta) - \delta_R(E_R + \Delta) \approx \pi$ at $\Delta \sim \Gamma/2$. This means that at small $\Gamma$ there is a sharp jump of the scattering phase shift by nearly $\pi$ in a very narrow interval of energies. This jump gives rise to respective sharp bumps in the cross section which are usually called resonances. That is, in the standard terminology the resonances are taken to mean $r$-states which lie in the 4-th quadrant of the $k$-plane near the real $k$-axis, that is, the $r$-states with a small width $\Gamma \ll E_R$ . However, at low energies the near $a$-states also lead to resonance phenomena. Fig. 6.4 (of Chapter 6) illustrates the behaviour of phases in case of relatively narrow $2^+$ and rather broad $4^+$ resonances in the $^4\mathrm{He} + {}^4\mathrm{He}$ nuclear system and Fig. 6.1. shows phase shifts in case of a near $a$-state (singlet deuteron) in the $^1S_0$ n-p scattering.

From the previous discussion it follows that, although the scattering phase shift in the presence of a resonance exhibits the characteristic resonance behaviour, there does not exist in the *general case* unambiguous relationships between the resonance parameters $(E_R$ and $\Gamma)$ and the corresponding behaviour of the phase shift. The condition

$$\delta(E_R) = \frac{\pi}{2}$$

which is sometimes used, is only applicable in the particular case $\delta_P = 0$ . For the narrow resonances, $E_R$ may be approximately determined simply by using the location of the phase jump; for the broad ones, the phase behaves sufficiently smoothly, so such a determination is not possible. Some criteria used usually in the literature to determine the resonance parameters are discussed below.

From the $S$-matrix parametrization of the partial scattering amplitude $g(k)$ follows that near the resonance

$$g(k) = \frac{1}{2ik} \left[ S(k) - 1 \right] = \frac{1}{2ik} \left[ S(k) - e^{2i\delta_P} + e^{2i\delta_P} - 1 \right] =$$

$$= \frac{1}{2ik} \left[ S(k) - e^{2i\delta_P} \right] + \frac{1}{2ik} \left[ e^{2i\delta_P} - 1 \right] = g_R(k) + g_P(k)$$

$$\tag{1.16}$$

where

$$g_P(k) = \frac{1}{2ik} \left[ e^{2i\delta_P} - 1 \right] \tag{1.16a}$$

is the "potential" (background) part of the scattering amplitude conditioned by all the singularities of the $S$-matrix (except for the discriminated resonance pole) and

$$g_R(k) = -\frac{1}{2k} \frac{\Gamma}{E - E_R + (i/2)\,\Gamma}\, e^{2i\delta_P} \tag{1.17}$$

is the resonance part of the amplitude. A similar parametrization may be written for the $t$-matrix:

$$t(k) = k\,g(k) = -\frac{1}{2} \frac{\Gamma}{E_R - E + (i/2)\,\Gamma}\, e^{2i\delta_P} + t_P(k)\,. \tag{1.18}$$

These formulae may be used to seek for resonances in processing experimental data. This may be made in different ways. One of the popular methods is to parametrize the partial-wave scattering amplitude via the superposition of the Breit-Wigner resonance terms and the smooth background part (of the (1.16a) type) approximated by a smooth function of energy (see, for example, [45] where the standard programs to make such a fitting are presented). One of the most widely used methods for analysing the resonance amplitude is the construction of the Argand diagram [9]. To construct this diagram, we introduce the variable

$$x = \frac{E_R - E}{\Gamma/2}\,. \tag{1.19}$$

For simplicity we neglect the "potential" phase $\delta_P$ for the time being. Then, the $t$-matrix (corresponding to the pure resonance term in eq. (1.18)) in the variable $x$ is of the form:

$$t(k) = \frac{1}{x - i}\,, \tag{1.20}$$

$$\mathrm{Re}\,t(k) = \frac{x}{x^2 + 1}\,; \qquad \mathrm{Im}\,t(k) = \frac{1}{x^2 + 1}\,. \tag{1.21}$$

From (1.21) it follows that $t(k)$ as a function of $k$ in the complex $t$-plane forms a circle of radius $1/2$ described by the equation

$$\left( \mathrm{Im}\,t - \frac{1}{2} \right)^2 + (\mathrm{Re}\,t)^2 = \frac{1}{4}\,. \tag{1.22}$$

It should be noted that the property of the Argand diagram is the common feature of pure elastic scattering (i.e. without absorbtion). This circle exhibits the following properties:

(i) the centre of the circle is located at the point i/2;

(ii) when energy increases, the points $t(E)$ moves on the circle in the counter-clockwise direction;

(iii) the point corresponding to $E = E_R$, i.e. to the resonance maximum of the amplitude, where the resonance phase $\delta_R = \pi/2$ corresponds to the top of the circle;

(iv) the rate of variations of $t$ with energy is

$$\left| \frac{dt}{dx} \right| \sim \frac{1}{1 + x^2}. \tag{1.23}$$

As can be seen from (1.23) a maximum rate is reached at $x = 0$, i.e. at $E = E_R$. Here the density of points on the circle is the lowest and increases strongly towards the base of the circle. An Argand diagram is shown in Fig. 2.1. The numerals inside the circle are the values of the parameter $x$.

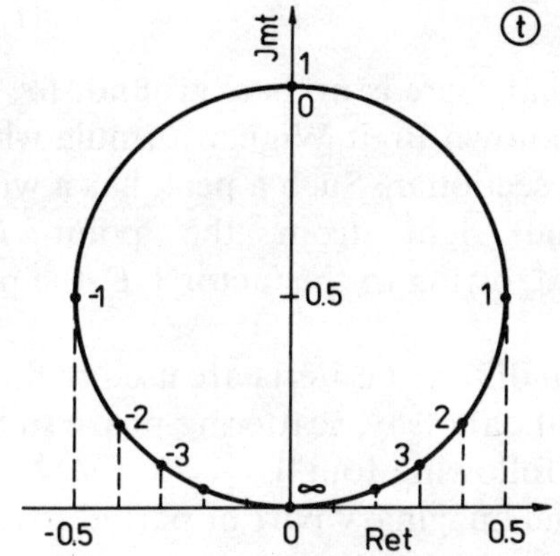


Fig. 2.1
An Argand diagram for pure elastic scattering.

In the real situations the background phase shift is usually non-zero, $\delta_P \neq 0$, and thus $t_P \neq 0$. The more general expression for the $t$-matrix is of the form:

$$t = a + A \frac{e^{2i\delta_P}}{x - i} \tag{1.24}$$

where $A$ is a real parameter $(0 \leq A \leq 1)$ allowing for absorbtion processes in the system. In this case the Argand diagram changes. Instead of a circle in the $t$-plane there appears now a loop or a circle of radius $< 1/2$ with the base at the point $a$ (the so-called Argand loop). Points on the loop as functions of energy are located counterclockwise. The location of the resonance at $E = E_R$ corresponds to a point opposite to $a$.

We shall examine now the form of the partial wave cross section of elastic scattering in the vicinity of resonance:

$$\sigma_l(E) \sim \sin^2 \delta_l(E) = \sin^2 (\delta_R + \delta_p) =$$
$$= \sin^2 \delta_R \cos^2 \delta_P + \cos^2 \delta_R \sin^2 \delta_P +$$
$$+ 2 \sin \delta_R \cos \delta_R \sin \delta_P \cos \delta_P . \qquad (1.25)$$

From (1.14) it follows that

$$\sin \delta_R(E) = \frac{\Gamma/2}{[(E - E_R)^2 + (\Gamma/2)^2]^{1/2}} ,$$

$$\cos \delta_R(E) = \frac{E - E_R}{[(E - E_R)^2 + (\Gamma/2)^2]^{1/2}} , \qquad (1.26)$$

whence it is seen that near $E = E_R$ only the first term survives in (1.25):

$$\sigma_l(E) \sim \sin^2 \delta_R \cos^2 \delta_P = \frac{(\Gamma/2)^2}{(E - E_R)^2 + (\Gamma/2)^2} \cos^2 \delta_P . \qquad (1.27)$$

Under the assumption that there is no background $(\delta_P = 0)$, the expression (1.27) becomes the well-known Breit-Wigner formula which describes the resonance peaks in the cross section $\sigma_l$. Such a peak has a width $\Gamma$ and a maximum shifted slightly to the right from the point $E_R$ because in fact $\sigma_l(E) \sim (1/E) \sin^2 \delta_l$ and, owing to the factor $1/E$, the peak moves to the right from the point $E_R$.

In the literature several different criteria are used to find the resonance energy in analysing experimental data (say, scattering phase shifts). The most frequently used criteria are the following four[4]):

(i) the maximum of the imaginary part of partial scattering amplitude $t_l(E)$, i.e. max Im $t_l(E)$;

(ii) the maximum of the modulus of the amplitude itself, i.e. max $|t_l(E)|$ ;

(iii) the maximum of the velocity (of motion of the point $t_l(E)$ in the Argand diagram, i.e. max vel $t_l(E)$ (see (1.23)).

(iv) the minimum of the conventional inelasticity parameter $v_l$[5]) where

$$t_l(E) = (v_l e^{2i\delta_l} - 1)/2i .$$

---

[4]) Any of the criteria listed below is not sufficient by itself for absolutely reliable identification of resonances. In practical applications these criteria are used in combination with other probable assumptions.

[5]) To avoid confusion with the Hilbert-Schmidt eigenvalues we have denoted the inelasticity parameter by $v_l$.

Now, we shall examine the criteria separately. We neglect the background phase $\delta_P$ and adopt $t$-matrix parametrization of the form

$$t_l(E) = \frac{A}{x - i} \qquad (1.20a)$$

where $0 \leq A \leq 1$ and the case $A = 1$ corresponds to pure elastic scattering. Now, we readily find for the first criterion:

$$\operatorname{Im} t_l(E) = \frac{A}{x^2 + 1},$$

i.e. $\operatorname{Im} t_l(E)$ shows a maximum at $x = 0$, i.e. at $E = E_R$.
    For the criterion (ii) we find:

$$|t_l(E)| = \frac{A}{(x^2 + 1)^{1/2}}$$

which also has a maximum at $x = 0$.
    For the criterion (iii) we find:

$$\operatorname{Re} t_l(E) = \frac{Ax}{x^2 + 1},$$

$$\operatorname{Im} t_l(E) = \frac{A}{x^2 + 1}.$$

that is, in the Argand diagram there is a circle centred at a point $(0, A/2)$ with a radius of $A/2$. Further, we may write

$$t_l(E) = \left[\frac{A}{(x^2 + 1)^{1/2}}\right] \exp\left\{ i \arctan (1/x)\right\}$$

so that on the Argand diagram the motion velocity of the point $t_l(E)$ is

$$\frac{\mathrm{d}}{\mathrm{d}E} \arg t_l(E) = \frac{2}{\Gamma(x^2 + 1)},$$

i.e. shows a maximum at $x = 0$ and $E = E_R$.
    Finally, for the criterion (iv) we find

$$v_l = \mathrm{e}^{-2i\delta_l} [2i\, t_l + 1]$$

whence

$$|v_l|^2 = [2it_l(E) + 1]\,[-2it_l^*(E) + 1] =$$

$$= 4\,|t_l(E)|^2 - 4\operatorname{Im} t_l(E) + 1 = 1 + \left[\frac{4A(A - 1)}{x^2 + 1}\right]$$

and

$$\frac{d|v_l|^2}{dE} = \frac{16A(A-1)}{\Gamma} \frac{x}{(x^2+1)^2}$$

equals zero at $x = 0$ . At this point,

$$\frac{d^2|v_l|^2}{dE^2} = -\frac{32A(A-1)}{\Gamma^2} > 0 ,$$

i.e. $|v_l|$ has a minimum at $x = 0$ , i.e. at $E = E_R$ .

Thus, all four criteria lead to the same result. However, this conclusion is based on the special (i.e. Breit-Wigner) form of the resonance amplitude (1.20a). In the general case of multichannel scattering all four criteria also yield the same results for narrow resonances, but in the case of broader resonances, such as the known $\Delta_{33}$ -resonance in the $\pi N$-system, different criteria may yield different results [34]. A very detailed comparison between the results of all four criteria using simple model examples may be found in Ref. [35].

The study [35] has shown that the first two criteria (i) and (ii), work well (that is, give very accurate predictions) for resonances lying both below and above the inelastic threshold. The third criterion, i.e. the criterion of maximum velocity, works less well (except for the case of very narrow resonances). However, this criterion is the most clear indicator of the presence of the resonance proper though it fails to show the exact resonance location. And finally, the fourth criterion, min $|v_l|^2$, works in many cases only slightly less well than the first two, but is very bad for small inelasticies.

It should be added that for multichannel resonances the trajectories in the Argand diagram are, as a rule, of a very complicated character [35] and the derivation of resonance parameters from them is not an easy task (see discussion in Chapter 6).

It should be noted in passing that a vast literature is devoted to the problem of deriving the resonance parameters from experimental data and to the relevant questions (see, for example [9, 36, 46] and the references therein). We shall touch upon the subject partly in Chapter 6 of this book; for details, however, we refer the reader to the handbook and to the reviews cited above because the main aim of this book is to present the methods of theoretical *calculations* of the resonance states in terms of some microscopic models.

Concluding the discussion of the parametrization of two-particle amplitudes near resonance we note that however ingenious and elaborate a theory relating the resonance phenomena to the $S$-matrix poles may be, it is still far from being applicable to all cases where the ressonance or, generally, long-lived states manifest themselves. In nature there exist systems which cannot be treated in such a simple way. In such systems the near $S$-matrix poles need not correspond to a seemingly resonant behaviour of cross sections; on the contrary, some

$S$-matrix poles do not necessarily lead to a resonant behaviour of cross sections [10]. As a rule, such a situation is surely observed in systems with more than two particles. We shall discuss this correspondence in more detail.

### 2.1.2. *Resonances without S-matrix Poles*

In this book we have accepted the assumption (which is shared by the overwhelming majority of physicists) that the $S$-matrix poles on nonphysical energy sheets near the real axis correspond to all (or nearly all) resonance states observed. In this section we would like to stress, however, that this viewpoint is only a hypothesis, although a highly probable one. It is well known that the correspondence between the resonances and the $S$-matrix poles is by no means so precisely determined as the correspondence between the bound states and the $S$-matrix poles on the real axis of the *first* energy sheet. In order that we may put into correspondence the observed resonances and the $S$-matrix poles on the nonphysical sheets, it is necessary to satisfy at least several conditions. First, the interaction potential has to be sufficiently "good", so that the corresponding $S$-matrix may be continued to the nonphysical sheet (that is, to the lower $k$ half-plane). For example in the Friedrichs model with non-analytic potential the $S$-matrix cannot be, in general, analytically continued through the cut [37]. Secondly, the $S$-matrix poles distant from the real axis do not lead to the observed effects. Thirdly, on the contrary, we may construct potentials which display resonance behaviour of the cross sections (and of phase shifts) and for which the $S$-matrix is analytic at $\mathrm{Im}\ k < 0$, and *has no* resonance poles. We shall below examine such an example following Ref. [38].

In the case of a pole $E_R = E_R - i\Gamma/2$ near the real axis, the partial $S$-matrix near $E_R$ is of the form (1.13). The corresponding resonance phase shift $\delta_R$ leading to the Breit-Wigner formula for the cross section is given by the second term in (1.14). Now we shall try to approximate the function $\delta(E)$ ("potential" phase $\delta_P$ is considered to be zero) in such a way that in the interval $|E - E_R| < \Delta(\Delta \gg \Gamma)$, i.e. where the given resonance pole manifests itself, the approximated phase shift $\tilde{\delta}(E)$ may have differed arbitrarily little from the Breit-Wigner form but at least the respective $S$-matrix $\tilde{S} = \exp\{2i\tilde{\delta}(E)\}$ would not have poles at $E = E_R - i\Gamma/2$ ; in Ref. [38] such an approximation is given in the explicit form

$$\tilde{\delta}(E) = Z_m(E)\, \delta_n(E)\, F(E, C$$

where $\delta_n(E)$ is the polynomial of degree $2n$-1 approximating $\delta_R(E)$ in the interval $|E - E_R| < \Delta$ with any given precision. The function $Z_m(E)$ ensures a regular asymptotic behaviour at $E \to 0$ and in the interval $|E - E_R| < \Delta$ it can be made arbitrary close to unity. It is of the form

$$Z_m(E) = \left(\frac{E}{E_R}\right)^{1/2} \sum_{k=0}^{m} \frac{(2k - 1)!!}{(2k)!!} \left(\frac{E_R - E}{E_R}\right)^k .$$

Finally, the function

$$F(E, C) = \exp\left\{-C \ln^2\left[1 + \frac{E - E_R}{E_R + b}\right]\right\} , \qquad b > 0 , \qquad C > 0$$

exhibits regular asymptotic behaviour at $E \to \infty$ , namely, it decreases more rapidly than any inverse degree of $E$. Choosing $C$ appropriately, we can also make it close to unity in the interval examined. Thus, the approximation $\tilde{\delta}_R$ allows to present the resonance phase shift $\delta_R$ in the interval $|E - a| < \Delta$ (for an arbitrary width $\Gamma$) with any given precision, but the analytically-continued $\tilde{\delta}(z)$ has *no* singularities in the complex plane $z$, except for the kinematic cut $[0, +\infty)$ (owing to the function $Z_m$) and the dynamic cut $(-\infty , -b]$ , and is equal to zero in the branch point $-b$. The corresponding $S$-matrix $\exp\{2i\tilde{\delta}(z)\}$ has no singularities (with the exception of the cuts) on any of the sheets of the Riemann surface either and approaches 1 as $|z| \to \infty$ . Furthermore, for such a function $\tilde{\delta}(E)$, the inverse scattering problem is unambiguously solvable because the rapid decrease of $\tilde{\delta}(E)$ at $E \to \infty$ ensures that the Gelfand-Levitan equation is of Fredholm type. This means that there exists a local potential giving a purely resonance phase shift, but the corresponding $S$-matrix has not a single resonance pole. It would be of interest to understand what is the form of the potential leading to such a resonance behaviour without poles and why such a picture is never observed for the conventional potentials in atomic and nuclear physics.

### 2.1.3. Behaviour of Scattering Wave Function near Resonance

Within the framework of scattering theory the resonance phenomena, despite their non-stationary nature, may be described by means of *time-independent* scattering formalism. A scattering state corresponding to an energy $E$ is a super-position of a wave incident onto the centre and a wave outgoing from the centre (see, for example, (1.3)); in the elastic scattering case the moduli of the amplitudes of these waves are the same, which corresponds to the conservation of the particle number, namely, the flux of incoming particles is equal to the flux of outgoing particles.

The presence of resonances leads to a number of qualitative features in the scattering characteristics, many of which we have treated in the previous sections. Here we shall discuss how the resonances manifest themselves in the scattering wave function.

We shall examine the scattering of a particle by a potential $V(r)$ which is of the form of an attractive well separated from the external region by a repulsive potential or a centrifugal barrier (see Fig. 2.2).

If we change the energy $E$ of the incident particle, we can see that the amplitude of the scattering wave function $\psi_E$ describing the behaviour of the particle in the potential $V(r)$, increases strongly at some energies $E \simeq E_R$ in the inner region $0 \le r \le R$ (see also fig. 7.1). Such a phenomenon is observed in some, generally, sufficiently narrow energy interval having a width $\Gamma$. This

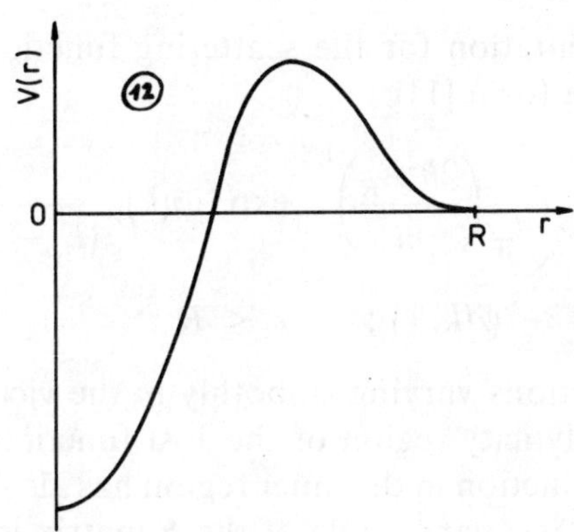



Fig. 2.2

An attractive potential with a centrifugal barrier.

corresponds to a specific interference of incident and reflected waves when, in the inner region behind the barrier, a standing wave is formed the node of which coincides approximately with the boundary of the barrier. The particle does not traverse the barrier immediately but stays for some time in the inner region where the particles accumulate.

As we know [3], an increase in the amplitude of the wave function in the inner region at certain energies, which was described above, is associated with the fact that the function $\psi^{(+)}(k, r)$ continued to a complex energy plane $z$ has a near pole at $z = z_0 = E_R - i\Gamma/2$. The wave function of the continuous spectrum may be written as follows (for simplicity we assume that $l = 0$):

$$\psi^{(+)}(k, r) = \begin{cases} A(k)\chi_k^{(0)}(r) & \text{for } r \le R \\ \sqrt{\dfrac{2}{\pi}} \sin(kr + \delta) & \text{for } r > R \end{cases} \tag{1.28}$$

where $\chi_k^{(0)}(r)$ is the Schrödinger equation solution regular at the origin which may be normalized to unity in the inner region. In the external region $(r > R)$ the function (1.28) varies sharply near the complex resonance energy $z = z_0$ because the $S$-matrix and, hence, $\delta(k)$ go to infinity. In the inner region $\chi_k^{(0)}(r)$ is a regular function of $k$ and the function $\psi^{(+)}$ may have a pole only because of the polar behaviour of $A(k)$. Indeed, it is not difficult to show [3] that $|A(k)|^2$ is of the form:

$$|A(k)|^2 = \frac{1}{\pi} \sqrt{\frac{2\hbar^2 E_R}{m}} \frac{\Gamma/2}{(E - E_R)^2 + \Gamma^2/4} . \tag{1.29}$$

A more general representation for the scattering function in the inner region near resonance is of the form [11]:

$$\psi^{(+)}(k, r) = \frac{1}{\sqrt{\pi}} \left( \frac{2\hbar^2 E_R}{m} \right)^{1/4} \exp\{i\varphi(k)\} \frac{\sqrt{\Gamma/2}}{(E - E_R) + i\Gamma/2} \chi_k^{(0)}(r) +$$

$$+ \tilde{\psi}(k, r) ; \quad r \leq R \tag{1.30}$$

where $\varphi$ and $\tilde{\psi}$ are functions varying smoothly in the vicinity of the resonance and analytic in the analyticity region of the Jost function.

Thus, the scattering function in the inner region has also a pole in the complex energy plane where the resonance pole of the $S$-matrix is located.

It is clear that it makes sense to speak of resonance phenomena only if the increase in the amplitude of the wave function is sufficiently sharp, that is, the width $\Gamma$ is not large (i.e. $\Gamma/E_R < 1$). In the language of poles this corresponds to a small distance of the resonance pole from the physical region $(E > 0)$. A small width $\Gamma$ or a long lifetime correspond to the fact that the particle remains for a long time behind the barrier and produces a state closely resembling a stationary bound state which, however, after a considerable time $\tau \sim \hbar/\Gamma$ decays to the state of a continuous spectrum. This time may be considerable not only on the nuclear but also on our usual scale; let us remember the long lifetimes of many radioactive isotopes. It is clear that many properties of such a narrow resonance are much closer to the properties of a bound state than to those of the scattering states. From this point of view the description of resonance phenomena in terms of scattering theory does not seem to be completely satisfactory and obvious. It would be much more natural to introduce a set of resonant wave functions reflecting the genetic relationships between the resonance and bound states and to reformulate the theory in the language of these functions. The next sections of this chapter are devoted to an examination of such possibilities.

## 2.2. Quasi-Stationary (Gamow or Siegert) States

As we have already mentioned, the narrow resonances (the long-lived decaying states with a small width $\Gamma$ and, accordingly, with a long lifetimes $\tau$ (1.7b)) show properties very similar to those of bound states. The scattering wave function near such a state is localized in the inner region of interaction (behind the

barrier) and has a large amplitude there; under the barrier it decays exponentially, and only outside the barrier at large distances does it reach the asymptotic
regime inherent to a scattering wave function. The amplitude of the asymptotic
part is very small (not so far from the barrier) in comparison with the amplitude
of the function in the inner region. That is, the whole function, except for its far
asymptotic part, closely resembles the function of a bound state (see also fig. 7.1).

Although all the diverse resonance phenomena (see Section 2.1) may be described in terms of scattering functions, in many cases it is more convenient to use for
this purpose a set of wave functions corresponding to some quasi-stationary states
which in the extreme case of large lifetimes, i.e. when the barrier becomes infinitely
high, turn into usual bound states. The Gamow states corresponding to complex-
-energy poles of $S$-matrix prove to be such a convenient expansion basis. These
wave functions were first introduced by Gamow at the outset of the development
of quantum mechanics in the alpha decay theory [12]. Later, they were used by
Siegert when deriving the dispersion formulae for nuclear reactions [13], by
Humblet and Rosenfeld [14] in nuclear reaction theory, and by many other
authors. In Section 2.1 we have shown that the Gamow wave functions (1.4) may
be obtained by analytic continuation of a general solution of the scattering
problem (1.3) in energy to an $S$-matrix pole. They have asymptotic parts containing *only an outgoing* wave and, therefore, they are solutions for the Schrödinger
equation (1.2) with the boundary conditions

$$\Phi_{ln}(0) = 0 \; ;$$

$$\frac{\mathrm{d}\Phi_{ln}(R)}{\mathrm{d}r} = ik_n\Phi_{ln}(R) \; ; \qquad E_n = \frac{\hbar^2 k_n^2}{2m} \; . \tag{2.1}$$

It can be seen that at $E_n < 0$ the conditions (2.1) define the bound-state wave
functions which are a special case of the Gamow states (1.4). It is clear that the
condition (2.1) excludes the stationary states with positive energy[6]. It is, however, satisfied by the complex eigenvalues $E_n$ and the associated eigenfunctions
$\Phi_{nl}$ which correspond to $S$-matrix poles on the second (nonphysical) energy
sheet. It may be noted that the complex eigenvalues appear in our boundary-
-value problem because the boundary conditions which define the corresponding functional space of solutions are no longer „Hermitian".

The presence of only outgoing waves in the asymptotic part of these functions
means that there are no incoming particles and there occur only outgoing
particles, so that the corresponding state is decaying and its wave function
decreases with time, as given by eq. (1.7). At the same time, at great distances
the amplitude of such a function increases exponentially with $r$ (see (1.6), (1.8)).
Physical interpretation of this fact is very simple, namely, at given time $t_0$, the

---

[6] This is true for local real interactions. In case of complex or nonlocal potentials such states are
possible. For more details see Section 5.1.6.

particles emitted from under the barrier at time $t = t_0 - r/v$ (where $v$ is the velocity of particles) are found at a given distance $r$. But at time $t$, earlier than $t_0$, the amplitude of the wave function at the centre, i.e. $\Phi(0, t)$, was larger by a factor of $\exp(r/v\tau)$ ($\tau$ is the lifetime) and the number of emitted particles was correspondingly also higher. Furthermore, it can easily be shown [3] that

$$\Phi(r, t_0) = \Phi(0, t_0 - r/v) .$$

Thus, the Gamow functions are an idealization corresponding to pure decaying states. In a real physical situation, such a state could not exist from zero time and must have been produced as a result of some process in the past. When using the Gamow functions, we lose the information about this prehistory. For a sufficiently large lifetime, however, the manner of resonance state production does not affect the dynamics of the state decay, so the use of the Gamow functions is fully justified.

Although the physical meaning of such functions is clear, their applications give rise to many problems of mathematical character, the majority of which arise because the Gamow functions are not orthonormalized in the conventional sense. For example, the conventional normalization integral

$$\int_0^\infty |\Phi_{ln}(r)|^2 \, dr$$

diverges in the upper limit because of the presence of an increasing exponential function in the asymptotic part of $\Phi_{ln}$. Also, it is not at all easy to formulate the completeness conditions for such a basis. Fairly voluminous literature is devoted to these problems (see, for example, [17, 21, 25]).

Owing to such difficulties, the Gamow functions are seldom used in *concrete* calculations. However, these difficulties may be overcome in different ways. Some of them will be considered in the subsequent sections of this chapter, and the others will be discussed in Chapters 3 to 5.

### 2.2.1. *Relationships between the Gamow Functions and the Green Function*

As we know from scattering theory [1, 2], the poles of the Green function coincide with the $S$-matrix poles[7]). We shall now find the relation between the residues of the Green function at these poles and the Gamow functions.

For the sake of simplicity we shall discuss the case $l = 0$ and assume that the interaction potential is truncated, i.e.

$$V(r) = 0 \qquad \text{for} \quad r > R_0 . \tag{2.2}$$

---

[7]) Here we do not take into consideration the problem of the so-called false poles of the $S$-matrix [2, 3].

We shall find a Green function $G^{(+)}(r, r'; k)$ for the Schrödinger equation

$$\frac{d^2}{dr^2} \Phi(r) + [k^2 - v(r)]\Phi(r) = 0 \tag{2.3}$$

$$v(r) = \frac{2m}{\hbar^2} V(r) ,$$

with the boundary conditions corresponding to outgoing waves:

$$\Phi(0) = 0 , \tag{2.4}$$

$$\frac{d}{dr} \Phi - ik\Phi = 0 , \qquad r \geq R \geq R_0 .$$

For such a Green function we may write the equation [15]:

$$\frac{\partial^2}{\partial r^2} G^{(+)}(r, r'; k) + [k^2 - v(r)] \, G^{(+)}(r, r'; k) = \delta(r - r') . \tag{2.5}$$

The boundary conditions for the Green function at points $r = 0$ and $r = R$ are

$$G^{(+)}(0, r'; k) = 0 ; \tag{2.6}$$

$$\left( \frac{\partial}{\partial r} G^{(+)}(r, r'; k) - ikG^{(+)}(r, r'; k) \right)_{r=R} = 0 , \tag{2.7}$$

respectively.

It is evident that near the pole $E_n = \hbar^2 k_n^2/2m$ (in the general case $E_n$ is complex) we may write the representation

$$G^{(+)}(r, r'; k) = \frac{A_n(r, r')}{k^2 - k_n^2} + g(r, r'; k) \tag{2.8}$$

where $A_n(r, r')$ is the residue at the pole $k^2 = k_n^2$; $g(r, r'; k)$ is a regular function for $k = k_n$. After substituting (2.8) in (2.5) and rearranging the terms somewhat, we obtain

$$\frac{1}{k^2 - k_n^2} \left\{ \frac{\partial^2}{\partial r^2} A_n(r, r') + [k^2 - v(r)] A_n(r, r') \right\} +$$

$$+ \left\{ \frac{\partial^2}{\partial r^2} g(r, r'; k) + [k^2 - v(r)]g(r, r';k) - \delta(r - r') \right\} = 0 .$$

$$\tag{2.9}$$

Adding the term $k_n^2 A_n(r, r')/(k^2 - k_n^2)$ to, and then subtracting it from, (2.9) we obtain after going over to the limit $k \to k_n$:

$$\frac{\partial^2}{\partial r^2} A_n(r, r') - \left[ k_n^2 - v(r) \right] A_n(r, r') = 0 \tag{2.10}$$

and

$$\frac{\partial^2}{\partial r^2} g(r, r'; k_n) + \left[ k_n^2 - v(r) \right] g(r, r'; k_n) + A_n(r, r') = \delta(r - r') . \tag{2.11}$$

Further, substituting (2.8) in the boundary conditions (2.6) and (2.7) and going over also to the limit $k \to k_n$ , we get:

$$A_n(0, r') = 0 , \tag{2.12}$$

$$g(0, r'; k_n) = 0 \tag{2.13}$$

and

$$\left[ \frac{\partial}{\partial r} A_n(r, r') - ik_n A_n(r, r') \right]_{r=R} = 0 , \tag{2.14}$$

$$\left[ \frac{\partial}{\partial r} g(r, r'; k_n) - ik_n g(r, r'; k_n) \right]_{r=R} = \frac{i}{2k_n} A_n(R, r') . \tag{2.15}$$

We see now that equation (2.10) for function $A_n(r, r')$ and the boundary conditions (2.12) and (2.14) coincide respectively with equation (2.3) and the boundary conditions (2.4) for the function $\Phi(r)$ .From this it follows that these functions are proportional, i.e.

$$A_n(r, r') = \Phi_n(r) B(r') . \tag{2.16}$$

Now we shall multiply equation (2.3) by $g(r, r'; k)$, and equation (2.11) by $\Phi_n(r)$, integrate them term-by-term between 0 and $\infty$ , and subtract the first from the second. The result is

$$\left[ \Phi_n(r) g'(r, r'; k_n) - g(r, r'; k_n) \Phi_n'(r) \right]_0^R +$$

$$+ \int_0^R \Phi_n(r) A_n(r, r') \, dr = \int_0^R \Phi_n(r) \delta(r - r') \, dr . \tag{2.17}$$

We use in (2.17) the boundary conditions (2.4), (2.12), (2.13) and (2.15) and also the expression (2.16). As a result, we obtain the following expression for $B(r')$:

$$B(r') = \frac{\Phi_n(r')}{\int_0^R \Phi_n^2(r)\, dr + \frac{i}{2k_n} \Phi_n^2(R)}.$$

Then we finally find the expression for the residue of the Green function:

$$A_n(r, r') = \frac{\Phi_n(r)\, \Phi_n(r')}{\int_0^R \Phi_n^2(r)\, dr + \frac{i}{2k_n} \Phi_n^2(R)}. \tag{2.18}$$

It should be recalled that the residue of the Green function at a bound-state pole equals [2]:

$$A_B(r, r') = \Phi_B(r)\, \Phi_B(r')$$

where $\Phi_B$ is the bound state wave function. As the Gamow functions are described by the same equation with the same boundary conditions as the bound-state function, it is natural to set the denominator in (2.18) to unity, i.e. to *define* the normalization condition for the Gamow functions as follows:

$$\int_0^R \Phi_n^2(r)\, dr + \frac{i}{2k_n} \Phi_n^2(R) = 1 . \tag{2.19}$$

Although the parameter $R$ enters the normalization condition, this condition does not actually depend on $R$ at a value of $R$ greater than the interaction radius $R_0$. The parameter $R$ appears in (2.19) from the boundary conditions (2.4) and (2.15). Therefore, the character of the dependence of the normalization condition on $R$ is the same as that of the dependence of the solution on the selection of the boundary (i.e. on the matching radius).

For a bound state we may use the limit $R \to \infty$ in (2.19). In this case $\Phi_n(R) \to 0$ and (2.19) turns out to be the conventional normalization condition. For higher partial waves we may carry out analogous calculations. The only difference is that the function $kL_l(kR)$ appears in (2.4) instead of $ik$. Here

$$L_l(kR) = \tilde{h}_l^{(+)\prime}(kR)/\tilde{h}_l^{(+)}(kR)$$

is the logarithmic derivative of the Riccati-Hankel function; $\hat{h}_l^{(+)}(x) = h_l^{(+)}(x)/x$; $h_l^{(+)}$ is the spherical Hankel function [1]. Then the normalization condition for the Gamow function in an arbitrary partial wave $l$ is of the form:

$$\int_0^R \Phi_{ln}^2(r)\, \mathrm{d}r + \Phi_{ln}^2(R)\left[\frac{\mathrm{d}}{\mathrm{d}k^2}\,(kL_l(kR))\right]_{k=k_n} = 1 \,. \tag{2.20}$$

Thus, we have obtained the relation of the residue of the Green function to the Gamow functions normalized by the conditions (2.19) and (2.20). This residue is

$$A_n(r, r') = \Phi_{ln}(r)\,\Phi_{ln}(r') \,. \tag{2.21}$$

An important circumstance should be mentioned here: as in the case of bound states, it is $\Phi_{ln}^2$, rather than $|\Phi_{ln}|^2$ that enters the normalization conditions (2.19) and (2.20), although the Gamow function is complex. This fact is of fundamental importance which will be discussed below.

Let us examine now the properties of $\Phi_{ln}(kr)$ as a function of $k$. For this purpose we shall discuss the solution for the Schrödinger equation (1.2), $\psi_l(k, r)$, regular at the origin, satisfying the condition [16]

$$\lim_{r\to 0} (2l + 1)!!\, r^{-l-1}\psi_l(k, r) = 1 \,. \tag{2.22}$$

We also introduce two solutions $f_l^{(\pm)}(k, r)$, nonregular at the origin, whose asymptotic behaviour takes the form of ingoing and outgoing waves (the Jost solutions) [16], satisfying the boundary condition

$$\lim_{r\to\infty} e^{\mp ikr} f_l^{(\pm)}(k, r) = i^{\pm l} \,. \tag{2.23}$$

The solution $\psi_l(k, r)$ may be written [16] as a linear combination of the Jost solutions $f_l^{(\pm)}(k, r)$:

$$\psi_l(k, r) = \frac{i}{2}\left[f_l(k)\,f_l^{(-)}(k, r) - f_l(-k)\,f_l^{(+)}(k, r)\right] \,. \tag{2.24}$$

Here $f_l(k)$ is the Jost function which may be defined as follows

$$f_l(k) = \lim_{r\to 0} \frac{(-kr)^l}{(2l-1)!!}\, f_l^{(-)}(k, r) \,. \tag{2.25}$$

Then the scattering function $\psi_l^{(+)}(k, r)$ may be written as

$$\psi_l^{(+)}(k, r) = \frac{i}{2}\left[f_l^{(-)}(k, r) - S_l(k)\,f_l^{(+)}(k, r)\right] \tag{2.26}$$

where

$$S_l(k) = \frac{f_l(-k)}{f_l(k)} . \tag{2.27}$$

In order to obtain the solution for the Schrödinger equation, regular at the origin, with a pure outgouing wave in the asymptotic region, it is necessary to analytically continue the function (2.24) to the $S$-matrix pole $k_n$ making allowance for the equality

$$f_l(k_n) = 0 . \tag{2.28}$$

As a result, we obtain

$$\psi_l(k_n, r) = \underset{k \to k_n}{\mathrm{cont}}\ \psi_l(k, r) = \hat{\Phi}_{ln}(r) = \frac{-i f_l(-k_n)}{2} f_l^{(+)}(k_n, r) \tag{2.29}$$

i.e. $\hat{\Phi}_{ln}(r)$ is proportional to the Gamow function defined by the residue of the Green function (2.21)

$$\Phi_{ln}(r)\Phi_{ln}(r') = \lim_{E \to E_n} (E - E_n)\, G_l^{(+)}(r, r', k) . \tag{2.30}$$

Furthermore, the Green function may be expressed in terms of the solutions for the Schrödinger equation $\psi_l$ and $f_l^{(+)}$ [1, 17]:

$$G_l^{(+)}(r, r'; k) = -\frac{2m}{\hbar^2 k f_l(k)}\, \psi_l(k, r_<) f_l^{(+)}(k, r_>) . \tag{2.31}$$

Substituting (2.31) and (2.29) in the expression (2.30), we get

$$\lim_{E \to E_n} (E - E_n)\, G_l^{(+)}(r, r'; k) = i\frac{f_l(-k_n)}{\dot{f}_l(k_n)} f_l^{(+)}(k_n, r_<) f_l^{(+)}(k_n, r_>) \tag{2.32}$$

where

$$\dot{f}_l(k_n) = \left[\frac{\mathrm{d}}{\mathrm{d}k} f_l(k)\right]_{k=k_n} \tag{2.33}$$

From (2.32) we obtain finally the important relation between the Gamow function and the Jost solution

$$\Phi_{ln}(r) = \left[i\frac{f_l(-k_n)}{\dot{f}_l(k_n)}\right]^{1/2} f_l^{(+)}(k_n, r) . \tag{2.34}$$

The asymptotic form of $\Phi_{ln}(r)$ is

$$\Phi_{ln}(r) \underset{r \to \infty}{\longrightarrow} \cdots h_l^{(+)}(k_n r) \underset{r \to \infty}{\longrightarrow} \cdots e^{ik_n r} . \tag{2.34a}$$

The Gamow function $\Phi_{ln}$ (2.34), as well as $\hat{\Phi}_{ln}$ (2.29), exhibits an asymptotic behaviour taking the form of a pure outgoing wave. However, the asymptotics is expressed in terms of the Green function residue at a pole, analogously with the bound state function. Also, as the attractive part of the interaction $V(r)$ gets deeper, the Gamow function (2.34) corresponding to the resonance state becomes a bound-state function when the correspong $S$-matrix pole transforms from the resonance pole to an appropriate bound-state pole.

Using the behaviour of the Jost function $f_l(k)$ [1] and the expression (2.34), we may write an analytic dependence of $\Phi_{ln}(r)$ on $k_n$ in the extreme $k_n \to 0$; that is, when the corresponding state "sits down" on the threshold [18]:

$$\Phi_{ln}(r) \underset{k_n \to 0}{=} \begin{cases} \sqrt{k_n}\,\varphi_{0n}(r) & \text{if } l = 0 , \\ \varphi_{ln}(r) & \text{if } l \neq 0 \end{cases} \tag{2.35}$$

where $\varphi_{0n}$ and $\varphi_{ln}$ are regular functions of $k_n$ for $k_n \to 0$ .

### 2.2.2. Normalization and Orthogonality of the Gamow Functions

As we mentioned in Section 2.1, the resonance poles are located in the $k$-plane by pairs symmetric with respect to the negative imaginary semi-axis $k$, so that, if $k_n$ is an $S$-matrix pole, then $-k_n^*$ is also a pole.

We denote the Gamow function corresponding to the pole $-k_n^*$ by $\tilde{\Phi}_{ln}(r)$ and $\tilde{k}_n \equiv -k_n^*$, i.e.

$$\tilde{\Phi}_{ln}(r) = \Phi_{ln}(\tilde{k}_n, r) . \tag{2.36}$$

The function $\tilde{\Phi}_{ln}$, as $\Phi_{ln}$, satisfies the same Schrödinger equation (1.2) but with a boundary condition corresponding to a pure *ingoing* wave (1.8). There is a simple relation between $\tilde{\Phi}_{ln}$ and $\Phi_{ln}$:

$$\tilde{\Phi}_{ln}(r) = \Phi_{ln}^*(r) \tag{2.37}$$

i.e.

$$\tilde{\Phi}_{ln}^*(r) = \Phi_{ln}(r) .$$

The functions $\tilde{\Phi}$ and $\Phi$ are sometimes called in-time conjugate functions, because the substitution of $-k_n^*$ for $k_n$ leads to a change in the sign of $\Gamma$ which, in turn, converts the function (1.7), decreasing in time, into an increasing function.

As mentioned above, the Gamow functions $\Phi_{ln}$ cannot be normalized in the usual way because the integral

$$\int_0^\infty |\Phi_{ln}|^2 \, \mathrm{d}r = \int_0^\infty \Phi_{ln}^* \Phi_{ln} \, \mathrm{d}r = \int_0^\infty \tilde{\Phi}_{ln} \Phi_{ln} \, \mathrm{d}r$$

diverges in the upper limit $\left(\text{at } k_n = \alpha - i\beta\right)$.

On the other hand, the Gamow function enters the Green function residue without complex conjugation (see (2.21)), which can be written in the following way:

$$\Phi_{ln}(r)\Phi_{ln}(r') = \tilde{\Phi}_{ln}^*(r)\Phi_{ln}(r') \, .$$

This suggests that the functions $\tilde{\Phi}$ and $\Phi$ may form a *biorthogonal* system and that a normalization integral of the form

$$\int_0^\infty \tilde{\Phi}_{ln}^*(r)\Phi_{ln}(r) \, \mathrm{d}r = \int_0^\infty \Phi_{ln}^2(r) \, \mathrm{d}r$$

is expedient to examine. The integrand in such an integral increases, but oscillates. As shown in Section 1.1, integrals of this type, even if they do not exist in the convetional sense, may be regularized by means of some regularization procedure. Zel'dovich [19] has proposed the following obvious procedure for regularizing such integrals (see Chapter 1):

$$I = \lim_{\varepsilon \to 0} \int_0^\infty \mathrm{e}^{-\varepsilon r^2} \Phi_{ln}^2(r) \, \mathrm{d}r \, . \tag{2.38}$$

On the basis of such a definition we shall show that $\Phi_{ln}$ and $\tilde{\Phi}_{ln}$ can form a biorthogonal system. We shall write the Schrödinger equation for $\Phi_{ln}$ and multiply it on the left by $\Phi_{ln'} = \tilde{\Phi}_{ln'}^*$, and the Schrödinger equation for $\Phi_{ln'}^*$ by $\Phi_{ln}$. After that, we subtract the equations from one another and integrate the resultant expression with the weight $\mathrm{e}^{-\varepsilon r^2}$ from 0 to $\infty$, obtaining

$$\int_0^\infty \mathrm{e}^{-\varepsilon r^2} \frac{\mathrm{d}}{\mathrm{d}r} \left[\Phi_{ln'}(r) \, \Phi_{ln}'(r) - \Phi_{ln'}'(r) \, \Phi_{ln}(r)\right] \mathrm{d}r =$$

$$= (k_{n'}^2 - k_n^2) \int_0^\infty \mathrm{e}^{-\varepsilon r^2} \tilde{\Phi}_{ln'}^*(r) \, \Phi_{ln}(r) \, \mathrm{d}r \, .$$

After integrating by parts and using the boundary conditions (2.1) on the left-hand side of the equality, we find

$$2\varepsilon \int_0^\infty r\, e^{-\varepsilon r^2}\left(\Phi_{ln'}\Phi_{ln}' - \Phi_{ln'}'\Phi_{ln}\right)\, dr \; .$$

The asymptotics of the integrand is of the form $r\, e^{-\varepsilon r^2}\, e^{i(k_n + k_{n'})r}$ . According to (1.1.16), such an integral vanishes at $\varepsilon \to 0$. In this way we find:

$$\left(k_{n'}^2 - k_n^2\right)\lim_{\varepsilon \to 0}\int_0^\infty e^{-\varepsilon r^2}\tilde{\Phi}_{ln'}^*(r)\,\Phi_{ln}(r)\, dr = 0 \; .$$

If we define a scalar (i.e. inner) product of the Gamow functions by means of such a regularized integral, then

$$\left(\tilde{\Phi}_{ln'} \mid \Phi_{ln}\right) = 0 \qquad \text{for} \quad k_n^2 \neq k_{n'}^2$$

and also

$$\left(\tilde{\Phi}_{ln}\mid\Phi_{ln}\right) = 1 \; . \tag{2.39}$$

The quantity $\left(\tilde{\Phi}_{ln}\mid\Phi_{ln}\right)$ is meant to be the expression (2.20) found earlier. For the orthogonality condition we may write a formula similar to (2.20) and taking the form

$$\int_0^R \Phi_n(r)\Phi_{n'}(r)\, dr + i\,\frac{\Phi_n(R)\,\Phi_{n'}(R)}{k_n + k_{n'}} = 0 \tag{2.40}$$

for the $S$-wave.

The matrix elements containing the Gamow functions are defined in this approach also with the help of the regularization:

$$\left(\tilde{\Phi}_{ln} \mid O(r) \mid \Phi_{ln'}\right) = \lim_{\varepsilon \to 0}\int_0^\infty e^{-\varepsilon r^2}\tilde{\Phi}_{ln}^*(r)\, O(r)\,\Phi_{ln'}(r)\, dr \tag{2.41}$$

where $O(r)$ is an operator.

It should be noted that such a regularization is convenient in the theoretical aspect, but in practice proves often to be unstable and tedious [20] because, when making numerical calculations, $\varepsilon$ has to be taken sufficiently small, thereby necessitating calculations of integrals with oscillating integrands over a large interval and to within a high accuracy to ensure the regularization of the integral due to the oscillations of the integrand.

Moreover, in Chapter 1 we have shown that the Zel'dovich regularization may only be operative for the Gamow functions which have $\mathrm{Im}\ k_n < \mathrm{Re}\ k_n$. Nevertheless, the results obtained here are correct for any of the Gamow functions. The orthonormality of the set of the Gamow functions $\tilde{\Phi}$ and $\Phi$ may be proved with the help of analytic continuation in $k$ [17] without using the Zel'dovich regularization. The idea of the proof is simple. We assume that $k_n$ and $k'_n$ are located in the upper half-plane. Then, all the integrals appear to be well defined since the integrands decrease exponentially. After that, we calculate the integrals and continue the result in $k$ to the lower half-plane. It may easily be verified that such a procedure yields the same result as the Zel'dovich regularization, but without limitations on the relation between $\mathrm{Re}\ k_n$ and $\mathrm{Im}\ k_n$ [17]. It is clear that the matrix elements with the Gamow states may also be defined in this way. In Chapter 5 we shall show that the orthonormality of the Gamow functions may easily be proved by means of analytic continuation in the coupling constant and that it follows directly from the orthonormality of bound-state functions.

### 2.2.3. Series Expansions in Terms of Gamow States

In actual problems we often want to series-expand the studied object (wave function, Green function, off-shell scattering amplitude, etc.) in a discrete set of functions. This procedure must meet certain requirements: the basis used must be complete, have a sufficiently simple form, etc. In different problems different expansion bases are used, namely, the harmonic oscillator functions, the hyperspherical functions, the Wigner and Eisenbud [7], Kapur and Peierls [8], Hilbert and Schmidt bases, etc. However, the Gamow function basis has recently been used more and more frequently. This basis has a number of advantages over those mentioned above, of which we shall mention only three main ones.

(i) This basis is generated by the Hamiltonian of the examined problem proper. This distinguishes it advantageously from the bases of the harmonic oscillator functions which are generated by the simplest Hamiltonians, often having little in common with the Hamiltonian of the considered problem. The remote relevance of the basis to the problem to be solved leads to a slow convergence of such expansions.

(ii) *The basis does not depend on energy,* which distinguishes it from the Kapur-Peierls and Hilbert-Schmidt types of bases.

(iii) The use of the Gamow functions as a basis makes it possible to factorize the dependence on energy (on $k$).

A great number of works are devoted to studying Gamow functions and the expansions on their basis (see, for example [15, 17, 21]). In a majority of them the Mittag-Leffler theorem [22] for the expansion of an analytic function in terms of its singularities is used for obtaining all kinds of expansions. We will not go into the details of these problems (see also Chapter 12.1 in the book by Newton [2]), but we shall present a few useful relations with a minimum of comment. We wish to remind readers that, as everywhere in this section, the interaction $V(r)$ is assumed to be finite-range; it is highly probable that the results discussed may also be generalized for the potentials of a more general type. However, this question has not been studied in sufficient detail yet [23].

For the Green function $G_l^{(+)}(r, r'; k)$, we may write the following expansion:

$$G_l^{(+)}(r, r'; k) = \sum_{n=1}^{N} \frac{\Phi_{ln}(r)\tilde{\Phi}_{ln}^*(r')}{k_{ln}(k - k_{ln})} +$$

$$+ \frac{1}{2} \sum_{n=N+1}^{\infty} \left[ \frac{\Phi_{ln}(r)\tilde{\Phi}_{ln}^*(r')}{k_{ln}(k - k_{ln})} - \frac{\tilde{\Phi}_{ln}(r)\Phi_{ln}^*(r')}{k_{ln}^*(k + k_{ln}^*)} \right] ,$$

$$r, r' < R_0 \tag{2.42}$$

where $N$ is the number of $a$- and $b$-states. Thus, the first term in (2.42) corresponds to the contribution of the bound and virtual states to the expansion, and the second to that of the resonance and antiresonance states.

Using (2.24), (2.27), and (2.31), we can write for the scattering function $\psi_l^{(+)}(k, r)$:

$$\psi_l^{(+)}(k, r) = A_l(k)\, G_l^{(+)}(r, R; k) . \tag{2.43}$$

Substituting the formula (2.42) for the Green function in (2.43), we obtain the expansion of the scattering function in terms of the Gamow functions (for $r < R$):

$$\psi^{(+)}(k, r) = A_l(k) \left\{ \sum_{n=1}^{N} \frac{\tilde{\Phi}_{ln}^*(r)\Phi_{ln}(R)}{k_{ln}(k - k_{ln})} + \right.$$

$$\left. + \frac{1}{2} \sum_{n=N+1}^{\infty} \left\{ \frac{\tilde{\Phi}_{ln}^*(r)\Phi_{ln}(R)}{k_{ln}(k - k_{ln})} - \frac{\Phi_{ln}^*(r)\tilde{\Phi}_{ln}(R)}{k_{ln}^*(k + k_{ln}^*)} \right] \right\} \tag{2.44}$$

where $A_l(k)$ is expressed merely through the Jost function $f_l(k)$ and the Jost solution $f_l^{(+)}(k, R)$ .

Finally, we shall present the forms of expansions for $t$ and $S$-matrices. We define the $t$-matrix by the expression:

$$t = V + VG^{(+)}V$$

or in the momentum representation after the expansion in partial waves:

$$t_l(p, q; k) = - \frac{1}{pq} \left\{ \int\limits_0^\infty dr\, u_l(pr)\, V(r)\, u_l(qr) + \right.$$

$$\left. + \int\limits_0^\infty \int\limits_0^\infty dr\, dr'\, u_l(pr)\, V(r)\, G_l^{(+)}(r, r'; k)\, V(r')\, u_l(qr') \right\}$$

where $u_l(z) = (1/2\pi z)^{1/2} J_{l+1/2}(z)$ is the Riccatti–Bessel function. Using the expansion for the Green function, we obtain for $t_l$ [23]:

$$t_l(p, q; k) = - \frac{1}{pq} \left\{ \langle u_l(p)|\, V\, |u_l(q)\rangle + \right.$$

$$+ \sum_{n=1}^N \frac{\langle u_l(p)|\, V\, |\Phi_{ln}\rangle \langle \tilde{\Phi}_{ln}|\, V\, |u_l(q)\rangle}{k_{nl}(k - k_{nl})} +$$

$$+ \frac{1}{2} \sum_{n=N+1}^\infty \left[ \frac{\langle u_l(p)|\, V\, |\Phi_{ln}\rangle \langle \tilde{\Phi}_{ln}|\, V\, |u_l(q)\rangle}{k_{ln}(k - k_{ln})} - \right.$$

$$\left.\left. - \frac{\langle u_l(p)|\, V\, |\tilde{\Phi}_{ln}\rangle \langle \Phi_{ln}|\, V\, |u_l(q)\rangle}{k_{ln}^*(k + k_{ln}^*)} \right] \right\} \qquad (2.45)$$

where

$$\langle u_l(p)|\, V\, |u_l(q)\rangle = \int\limits_0^R dr\, u_l(pr)\, V(r)\, u_l(qr)$$

$$\langle u_l(k)|\, V\, |\Phi_{ln}\rangle = \langle \tilde{\Phi}_{ln}|\, V\, |u_l(k)\rangle = \langle \Phi_{ln}|\, V\, |u_l(k)\rangle^* =$$

$$= \langle u_l(k)|\, V\, |\tilde{\Phi}_{ln}^*\rangle = \int\limits_0^R dr\, u_l(kr)\, V(r)\, \Phi_{ln}(r) .$$

A general feature of the expansions (2.42), (2.44), (2.45) and of other expansions of this type which use the Gamow basis is the factorization of the energy (or $k$) dependence because the expansions are of the form:

$$Q(x, k) = \sum_n \varphi_n(x)\, \psi_n(k)$$

where $\psi_n(k)$ have a very simple form. Such a property is a great advantage of the basis, especially if the expanded quantities are subsequently integrated over energy. This property follows from the fact that the Gamow functions themselves do not depend on energy because they are eigenfunctions of the initial Hamiltonian.

Using a known relation between $t$ and $S$-matrices, we obtain the expression for the partial wave $S$-matrix:

$$
S_l(k) = 1 - \frac{2i}{k} \left\{ \langle u_l(k)| \, V \, |u_l(k)\rangle + \right.
$$
$$
+ \sum_{n=1}^{N} \frac{(\langle u_l(k)| \, V \, |\Phi_{ln}\rangle)^2}{k_{ln}(k - k_{ln})} +
$$
$$
\left. + \frac{1}{2} \sum_{n=N+1}^{\infty} \left[ \frac{(\langle u_l(k)| \, V \, |\Phi_{ln}\rangle)^2}{k_{ln}(k - k_{ln})} - \frac{(\langle u_l(k)| \, V \, |\Phi_{ln}\rangle^*)^2}{k_{ln}^*(k + k_{ln}^*)} \right] \right\}.
$$

$$(2.46)$$

A lot of numerical examples of such expansions may be found in [24] where the convergence of the expansions of this type has been studied in detail.

In conclusion we shall say a few words about the completeness of the Gamow basis. It is known [2] that the functions of stationary states in the scattering theory (bound states + scattering states) form a complete set, that is, any solution for the Schrödinger equation with a "good" potential at real energy may be expanded in these functions on the interval $0 \le r < \infty$. The above-mentioned expansions in the Gamow functions allow one to speak about the "completeness" of the Gamow basis in some limited sense. Thus, from (2.42) and (2.43) it can be seen that the scattering function and the Green function may be expanded in the Gamow functions but in the limited interval $0 \le r < R$, where $R$ is arbitrary (if only $V(r) = 0$ for $r > R$), but *finite*. The completeness condition for the Gamow functions, which is the same as the completeness condition in the scattering theory [2], was not achieved. Moreover, in [24] it has been shown that the Gamow basis is overcomplete even in the interval $(0, R)$, i.e.

$$
\frac{1}{2} \sum_{n=1}^{\infty} \langle r \mid \Phi_{ln}\rangle \langle \tilde{\Phi}_{ln} \mid r'\rangle = \delta(r - r')
$$

$$(2.47)$$

and that the Gamow functions may be, to a high degree, linearly dependent:

$$
\sum_{n=1}^{\infty} \frac{\langle r \mid \Phi_{ln}\rangle \langle \tilde{\Phi}_{ln} \mid r'\rangle}{k_{ln}} = 0 .
$$

$$(2.48)$$

Nevertheless, in practical applications the above mentioned shortcomings caused by the permanent limitation on the number of expansion terms will not lead to any troubles with convergence [24]. The properties of the Gamow basis are being intensively studied [23, 24].

### 2.3. The Kapur-Peierls Formalism and the Basis of Eigenfunctions for the Operator Green Function

In the preceding sections we have examined the Gamow states in connection with the residues of the Green function at complex poles $k = k_n$. Here we shall consider the relationships between the Gamow functions $\Phi_{ln}$ and another widely used family of functions $\varphi_s$ defined as eigenvalues of the operator Green function $G(k)$ for arbitrary fixed $k$ [6]. We shall also show that in case of potential scattering the functions $\varphi_s$ coincide with the Kapur-Peierls resonance functions [8]. For the sake of simplicity we choose the system of units where $\hbar = 2m = 1$. Then, the energy $E = k^2$.

Let us formulate more precisely the problem of the decay of a state produced by a particle in a force field $V(r)$. For this purpose we introduce two channels, namely, the "outer" channel of scattering states where a particle exhibits a continuous energy spectrum with the origin at a point $E = 0$ and the "inner" channel containing a discrete set of bound states which decay due to coupling with the outer channel.

The total Hamiltonian of the system $H$ is a Hermitian operator independent of particle energy. This Hamiltonian may be presented, with the help of projection operators (for more details see Chapter 4), as a sum of four operators:

$$H = H_{ii} + H_{io} + H_{oi} + H_{oo}$$

where, for example, $H_{io}$ is the Hamiltonian part responsible for the transition from the outer to the inner channel, and so on. The resolvent $R(z) = (z - H)^{-1}$ may also be divided into four corresponding parts. The part corresponding to the inner channel will be called the Green function $G(k)$:

$$G(k) = R_{ii}(k^2)$$

that is, the Green function $G(k)$ is the projection of the resolvent $R(z)$ on the states of the inner channel. Thus, for example, in the case of scattering by a potential containing a barrier we may always select a sphere of radius $R_c$ containing the whole potential[8]). Then the domain inside the sphere will be the inner channel, and the region outside it will be the outer channel. In this case, the ordinary Green function $G(r, r'; k)$ is the Green function $G(k)$, but the two vectors $r$ and $r'$ must not go out beyond the sphere boundaries. It is clear that such a Green function contains both the direct effects generated by the potential and the indirect effects associated with the particle emission through the sphere. Therefore, it would be incorrect to assert that $G(k)$ (i.e. $R_{ii}(k^2)$) corresponds to the Hamiltonian $H_{ii}$ which does not describe the particle runaway to beyond the sphere. $G(k)$ is a resolvent of the effective Hamiltonian $H_{eff}(k)$ which, despite the

---

[8]) Here, as in the preceding sections, we consider cut-off interactions $V(r)$.

fact that it operates in the inner channel (it contains $H_{ii}$), contains a part describing the particle ejection to beyond the sphere boundary and return to the inside again. The effective Hamiltonian of such a kind depends on energy and is non-Hermitian. It may be found using the operator equation.

$$k^2 - H_{\text{eff}}(k) = [G(k)]^{-1}$$

and is a single-valued function of $k$. As an example of such an effective Hamiltonian we may consider the Hamiltonian of the optical model for particle scattering by a composite target.

As the effective Hamiltonian depends on energy, its eigenvalues also depend on $k$:

$$H_{\text{eff}}(k) \, |\varphi_s(k)\rangle = e_s(k) \, |\varphi_s(k)\rangle \,. \tag{3.1}$$

Equation (3.1) with the boundary conditions

$$\varphi_s(k, 0) = 0 \,, \tag{3.2}$$

$$\frac{\mathrm{d}}{\mathrm{d}r} \, \varphi_s(k, r)|_{r=R_0} = ik\varphi_s(k, R_0) \tag{3.3}$$

defines the so-called Kapur-Peierls functions $\varphi_s(k, r)$.

For potential scattering, equation (3.1) is of the form (for the $S$-wave):

$$\left[ -\frac{\mathrm{d}^2}{\mathrm{d}r^2} + v(r) \right] \varphi_s(r) = e_s(k) \, \varphi_s(r) \,. \tag{3.1a}$$

This equation is similar to the Schrödinger equation, but in the general case we get $e_s \neq k^2$. In this example the effective Hamiltonian does not depend explicitly on $k$. The dependence of the eigenfunctions and eigenvalues on $k$ results from the dependence of the boundary conditions (3.2) and (3.3) on $k$.

Since $H_{\text{eff}}$ in the general case depends *on energy*, and the boundary condition (3.3) *on $k$*, so that the eigenfunctions and eigenvalues of the effective Hamiltonian depend on $k$, the energy is *not* an eigenvalue in (3.1) and the Kapur-Peierls eigenvalues $e_s(k)$ have no direct physical meaning. Nevertheless, the theory of resonance states [8] may be constructed on the basis of eigenfunctions $\varphi_s$ and eigenvalues $e_s$.

The eigenfunctions of $H_{\text{eff}}$ are also eigenfunctions of the operator $G(k)$, which follows from the definition of the Green function given above. The left and right eigenfunctions differ from one another:

$$G(k) \, |\varphi_r(k)\rangle = [k^2 - e_r(k)]^{-1} \, |\varphi_r(k)\rangle \,, \tag{3.4}$$

$$\langle \tilde{\varphi}_s(k)| \, G(k) = [k^2 - \tilde{e}_s(k)]^{-1} \, \langle \tilde{\varphi}_s(k)| \,. \tag{3.5}$$

However, it may be shown [6] that a simple relationship exists between them:

$$\tilde{\varphi}_s(k, r) = \varphi_s^*(k, r) = \varphi_s(-k^*, r) \tag{3.6}$$

and

$$\tilde{e}_s(k) = e_s(k) = [e_s(-k^*)]^* .$$

The matrix element $\langle \tilde{\varphi}_s(k) \mid G(k) \mid \varphi_r(k) \rangle$ may be calculated in two ways by acting with the operator $G(k)$ to the left and to the right. By comparing the results, we obtain

$$\langle \tilde{\varphi}_s(k) \mid \varphi_r(k) \rangle = \delta_{r,s} {}^9) \qquad \text{if} \quad e_r \neq e_s . \tag{3.7}$$

Let us stress that this scalar product and all of the similar products which we shall meet below are defined *only in the space of the inner channel*.

Hence, the Kapur-Peierls functions constitute a biorthogonal set of vectors, which is complete in the subspace of the inner channel [25, 26]. However, they depend parametrically and in a relatively complicated way on $k$. Moreover, in the general case $\tilde{\varphi}_r(k)$ is *not orthogonal* to the function $\varphi_s(k')$ taken at *another* energy.

Using the completeness of the Kapur–Peierls functions, we may expand the Green function $G(k)$ in these functions:

$$G(k) = \sum_s \frac{|\varphi_s\rangle \langle \tilde{\varphi}_s|}{k^2 - e_s(k)} . \tag{3.8}$$

This formula is not very suitable for practical use because of a complicated dependence of $\varphi_s$ and $e_s$ on $k$, but it permits one to determine the singularities of $G(k)$ and to find also the relation between the functions $\varphi_s$ and the Gamow functions $\Phi_{ln}$. The Green function will have a pole at the point $k = k_n$ if at some $s = s(n)$ the following equation is valid:

$$k_n^2 = e_s(k_n) . \tag{3.9}$$

The latter equation establishes the relation between the energy of the Gamow state $E_n = k_n^2$ (at this energy the Green function and the $S$-matrix have a pole) and the Kapur-Peierls eigenvalue $e_s(k)$. Thus, to find the resonance energy $E_n$ in the Kapur-Peierls formalism, it is necessary to determine the eigenvalue $e_s(k)$ and then to solve the equation (3.9).

Let us examine now the relationship of $\varphi_s$ to the Gamow functions $\Phi_{ln}$. To this end, we shall find a residue of the Green function $G(k)$ (3.8) at the point $E = E_n$:

$$\lim_{k \to k_n} (k^2 - k_n^2) G(k) = \frac{|\tilde{\varphi}_s(k_n)\rangle \langle \varphi_s(k_n)|}{1 - e_s'(k_n)} \tag{3.10}$$

---

<sup></sup>⁹) Since $k$ in the function $|\varphi_s(k)\rangle$ is a parameter, such notation should not give rise to any confusion.

where $e_s'(k_n)$ is a derivative of $e_s$ with respect to energy:

$$e_s'(k_n) = \frac{1}{2k_n} \frac{de_s(k)}{dk}\bigg|_{k=k_n} . \tag{3.11}$$

Comparing (3.10) with (2.21), we get:

$$\Phi_n = [1 - e'(k_n)]^{-1/2}\varphi_s(k_n) \tag{3.12}$$

To normalize the Gamow functions, we obtain the following expression from (3.12) and (3.7):

$$Z_n = \langle \tilde{\Phi}_n \mid \Phi_n \rangle = [1 - e_s'(k_n)]^{-1} . \tag{3.13}$$

It should be noted that the scalar product in (3.13) is defined, as in all the formulae of the present section, only in the subspace of the inner channel, i.e.

$$Z_n = \int_0^{R_c} \Phi_n^2 \, dr .$$

Now we shall find the relationship between the formula (3.13) and Gamow function normalization (2.19). For simplicity, we shall consider the case of $s$-wave scattering. We shall introduce the notation $q^2 = e_s(k)$ for an arbitrary and fixed $k$. The general solution for equation (3.1a) may be written as

$$\varphi_s(k, r) = Af^{(-)}(q, r) + Bf^{(+)}(q, r) .$$

From the boundary condition (3.2) it follows that

$$A/B = -f(q)/f(-q)$$

where $f(\pm q) \equiv f^{(\pm)}(q, 0)$ and the second boundary condition (3.3) determines the discrete set of admissible values of $q = Q_n$

$$ik = \frac{\varphi_s'(R_c)}{\varphi_s(R_c)} = -iq \frac{A\,e^{-iqR_c} - B\,e^{iqR_c}}{A\,e^{-iqR_c} + B\,e^{iqR_c}}$$

or

$$-\frac{k}{q} = \frac{f(q)\,e^{-iqR_c} + f(-q)\,e^{iqR_c}}{f(q)\,e^{-iqR_c} - f(-q)\,e^{iR_c}} \tag{3.14}$$

where $R_c$ is the channel radius. Since the eigenvalues of equation (3.1) are determined by the poles of the Green function (3.9), the equality $k^2 = q^2$ must be fulfilled in eq. (3.14). In this case two possibilities are offered, namely, $q = -k$ and $q = k$. It can easily be verified that in both cases eq. (3.14) leads to the equation $f(k_n) = 0$ the solutions of which will give poles of $G(k)$ for the values $k = k_n = \pm Q_n$. From this it can be seen that the Kapur-Peirls condition $e_s(k_n) = k_n^2$ (3.9) is equivalent, in the case of potential scattering, to the condition $f(k_n) = 0$.

From (3.14) we may easily find $dk/dq$ by differentiating the two parts with respect to $q$ and taking into account that $f(q)\big|_{q=Q_n} = 0$. We get:

$$-\frac{dk}{dq}\bigg|_{q=Q_n} = \left[1 + 2q\,\frac{f'(q)}{f(-q)}\,e^{2iqR_c}\right]_{q=Q_n}.$$

This leads to the following expression for $e_s'$ (3.11):

$$e_s' = \frac{Q_n}{k_n}\left(\frac{dq}{dk}\right)_{k=Q_n} = \left(1 + 2Q_n\,\frac{f'(Q_n)}{f(-Q_n)}\,e^{2iQ_nR_c}\right)^{-1}. \qquad (3.15)$$

Now we may obtain a formula for the normalization constant $Z_n$ (3.13):

$$Z_n = \left[1 - e_s'(k_n)\right]^{-1} = 1 - \frac{f(-k_n)}{2k_n\,f'(k_n)}\,e^{2ik_nR_c} \qquad (3.16)$$

whence

$$\int_0^{R_c} \Phi_n^2(r)\,dr = 1 - \frac{f(-k_n)}{2k_n\,f'(k_n)}\,e^{2ik_nR_c}. \qquad (3.17)$$

We take into consideration that the channel radius $R_c$ is larger than the interaction radius; hence $\Phi_n(R)$ at $R = R_c$ is already in the asymptotic region. Recalling the asymptotic form of (2.34a), we may rewrite eq. (3.17) as

$$\int_0^{R_c} \Phi_n^2(r)\,dr = 1 - \frac{i}{2k_n}\,\Phi_n^2(R_c)$$

which coincides with (2.19). This is another method for obtaining the Gamow function normalization condition. As can be seen, this result is not dependent on the location of the poles $k_n$. This confirms once again the conclusion that the range of application of formulae (2.19), (2.20), (2.39)–(2.41) and others is much wider than that of the Zel'dovich regularization procedure.

In conclusion we shall again emphasize that in practical uses the Kapur-Peierls basis is not very suitable because, as mentioned above, $\varphi_s(k, r)$ and $e_s(k)$ are complicated functions of $k$ which have branch points in $k$ [6]. Therefore, if we want to write, for example, an expansion of type (2.42) for $G(r, r'; k)$, but with the help of $\varphi_s$ instead of $\Phi_n$, then the expansion will be of a much more complicated form than (3.8), the dependence of $G(r, r'; k)$ on $k$ cannot be factorized, and the simple character of this dependence will be hidden by the complicated analytic structure of $\varphi_s$ and $e_s(k)$. Nevertheless, the Kapur-Peierls basis is widely used in theoretical constructions because of its discreteness and clear relevance to the scattering problem. Besides, the Kapur-Peierls approach, as regards the physical meaning and mathematical realization, is very similar to the Feshbach projection method [44] which will be described in Chapter 4 below. The projection method, with its usable modifications found by many authors (see Chapters 4 and 7), is extremely widely used in the concrete investigations of the resonance processes in molecules, atoms, and nuclei.

### 2.4  Resonances in Two-Level System. A Model Example

Following Ref. [6], we shall consider here an exactly solvable model of resonance, namely, the two-level system, decaying to the continuum, which allows us to illustrate the general formalism described above.

Let an electron in a one-dimensional space $(-\infty < x < \infty)$ be able to move in two channels, inner and outer. In the inner channel the electron may be in the state $|\psi\rangle = \alpha |a\rangle + \beta |b\rangle$ ($\alpha$ and $\beta$ are real numbers) which is produced by a linear combination of localized basis orbitals $|a\rangle$ and $|b\rangle$ the unperturbed energies of which are real, positive, and equal to $e_a$ and $e_b$, respectively. The unperturbed Hamiltonian of the inner channel is

$$H_i^{(0)} = |a\rangle\, e_a\, \langle a| + |b\rangle\, e_b\, \langle b| .$$

(4.1)

The resolvent corresponding to $H_i^{(0)}$ is

$$R_{ii}^{(0)}(z) = \frac{|a\rangle\,\langle a|}{z - e_a} + \frac{|b\rangle\,\langle b|}{z - e_b} .$$

(4.2)

Further, we assume that in the outer channel the electron moves freely in the entire space $-\infty < x < \infty$. Then the unperturbed Hamiltonian of the outer channel may evidently be written as

$$H_o^{(0)} = -\frac{d^2}{dx^2}$$

(4.1a)

and its resolvent

$$\langle x|\, R_{oo}^{(0)}\, |x'\rangle = \frac{1}{2ik}\, e^{ik\,|x - x'|}$$

(4.2a)

is a single-valued function of $k = z^{1/2}$, where $z$ is the complex energy parameter. Everywhere in this section we also use the system of units where $\hbar^2/2m = 1$.

The coupling between the inner and the outer channels is realized by the Hamiltonian $H_c$. Owing to this Hamiltonian, an electron in the state $|\xi\rangle$ of the inner channel may fly out to the continuum of the outer channel (the state $|\xi\rangle$ decays). Conversely, an electron from the outer channel reaching the point $x = 0$ (the interaction is considered to be point-like and localized at point $x = 0$, that is $H_c \sim \delta(x)$) is transferred, with the help of the Hamiltonian $H_c$, to the inner channel, to the state $|\xi\rangle$, where it may remain for some time (formation of the resonance state $|\xi\rangle$). We shall denote by $Q$ the operator which transfers the electron from the outer to the inner channel. Then $H_c$ will be of the form:

$$H_c = \lambda |\xi\rangle \, \hat{Q} \, \delta(x) + \lambda \, \delta(x) \, \hat{Q}^{(+)} \langle\xi| \tag{4.3}$$

where $\lambda$ is the channel coupling constant. It is clear how $H_c$ acts on an arbitrary function of the outer channel $\psi(x)$:

$$H_c \, \psi(x) = \lambda \, \psi(0) \, |\xi\rangle \, .$$

For the total Hamiltonian

$$H = H_i^{(0)} + H_o^{(0)} + H_c$$

we may write the resolvent

$$R(z) = (z - H)^{-1}$$

which is a solution for the equation

$$R(z) = R^{(0)}(z) + R^{(0)}(z) \, H_c R(z) \, .$$

This equation may be solved in a simple way. For example, after simple calculations we obtain the following expression for the matrix element of $R(z)$ between the states of the outer channel:

$$\langle x| \, R_{oo}(z) \, |x'\rangle = \frac{e^{ik \, |x-x'|}}{2ik} + \frac{e^{ik \, |x|}\lambda^2 F(k^2) \, e^{ik \, |x'|}}{2ik[1 - (\lambda^2/2ik) \, F(k^2)] \, 2ik} \tag{4.4}$$

where

$$F(k^2) = \langle\xi| \, R_i^{(0)} \, |\xi\rangle = \frac{\alpha^2}{k^2 - e_a} + \frac{\beta^2}{k^2 - e_b} \, . \tag{4.5}$$

Further, we shall be interested in $R_{ii}(z)$, i.e. the projection of the resolvent $R(z)$ on the inner channel (we shall denote it by $G(k)$). It is of the form

$$G(k) = R_{ii}^{(0)}(k^2) + \lambda^2 \, R_{ii}^{(0)}(k^2) \frac{|\xi\rangle \, \langle\xi|}{2ik - \lambda^2 F(k^2)} R_{ii}^{(0)}(k^2) \tag{4.6}$$

or, in more detail,

$$\langle a| G(k) |a\rangle = \frac{1}{k^2 - e_a} + \frac{\alpha^2\lambda^2}{(k^2 - e_a)^2 \left[2ik - \lambda^2 F(k^2)\right]},$$

$$\langle a| G(k) |b\rangle = \langle b| G(k) |a\rangle =$$

$$= \frac{\lambda^2\alpha\beta}{(k^2 - e_a)(k^2 - e_b)\left[2ik - \lambda^2 F(k^2)\right]},$$

$$\langle b| G(k) |b\rangle = \frac{1}{k^2 - e_b} + \frac{\beta^2\lambda^2}{(k^2 - e_b)^2 \left[2ik - \lambda^2 F(k^2)\right]}. \tag{4.7}$$

Now, if we want to study resonances in such a system, we have to examine the singularities of the Green function $G(k)$ (4.7). As a function of $k$ it possesses only poles. If the states of the inner channel are nondegenerate $(e_a \neq e_b)$, then, as can be easily verified, $G(k)$ has *no* poles for *unperturbed* energies $e_a$ and $e_b$ at finite $\lambda$ (to verify this, it is sufficient to substitute (4.5) in (4.7)). The poles $k_n$ of $G(k)$ are evidently determined by the equality

$$2ik_n = \lambda^2 F(k^2) = \frac{\lambda^2\alpha^2}{k_n^2 - e_a} + \frac{\lambda^2\beta^2}{k_n^2 - e_b}. \tag{4.8}$$

The resonance wave function may be determined as

$$|\Phi_n\rangle = N_n \left(\frac{\alpha |a\rangle}{E_n - e_a} + \frac{\beta |b\rangle}{E_n - e_b}\right) \tag{4.9}$$

where $E_n = k_n^2$ and

$$N_n = \left[\frac{i}{\lambda^2 k_n} - F'(E_n)\right]^{-1/2}; \qquad F'(E) = \frac{dF}{dE}. \tag{4.10}$$

The normalization $Z_n$ of the resonance wave function may be defined in terms of $F(E)$:

$$Z_n = \langle \tilde{\Phi}_n | \Phi_n\rangle = -N_n^2 F'(E_n) = \left(1 - \frac{F(E_n)}{2E_n F'(E_n)}\right)^{-1}. \tag{4.11}$$

Equation (4.8) is an algebraic equation of the fifth order, it has five roots, i.e. $G(k)$ has five poles and, consequently, there are in the system five resonance states and five corresponding resonance wave functions.

The physical meaning of these singularities (and of the states corresponding to them) becomes clear in the limit of long lifetimes, that is, at small $\lambda$, (i.e. when the coupling between channels is weak). Here we may expand all the quantities in powers of $\lambda$. It is clear that at small $\lambda$ some of the solutions will group near the unperturbed state $|a\rangle$. These will be the states into which the state $|a\rangle$ turns after imposing the perturbation. Some of the levels will group near the level corresponding to the state $|b\rangle$. Indeed, it turns out [6] that two solutions are associated with the level $|a\rangle$ the energies and functions of which we shall denote by $E_{\pm a}$ and $|\Phi_{\pm a}\rangle$, respectively. Expansion in terms of $\lambda$ will yield the following expression for the resonance energies:

$$E_{\pm a} = e_a \pm \frac{\lambda^2 \alpha^2}{2i\sqrt{e_a}} + \frac{\lambda^4 \alpha^2}{4e_a}\left[\frac{\alpha^2}{2e_a} - \frac{\beta^2}{(e_a - e_b)}\right] + \dots \qquad (4.12)$$

and for the wave functions:

$$|\Phi_{\pm a}\rangle = |a\rangle \pm \frac{\lambda^2 \alpha^2}{2i\sqrt{e_a}\,4e_a}|a\rangle \pm \frac{\lambda^2 \alpha\beta}{2i\sqrt{e_a}(e_a - e_b)}|b\rangle + \dots \qquad (4.13)$$

In eq. (4.12) the second term is the resonance width in the first order perturbation theory

$$E_{\pm a} = e_a \pm \frac{i}{2}\Gamma\,; \qquad \Gamma = \frac{\lambda^2 \alpha^2}{\sqrt{e_a}}\,,$$

the third term, proportional in the given model to $\lambda^4$, is the level shift caused by the coupling to both the continuum and the second level $|b\rangle$. In more realistic models the energy shift is not necessarily of higher order in coupling constant than the width. Rather, as a rule, they are of the same order.

The two functions $\Phi_{\pm a}$ are linearly independent. However, they are related to each other by the expression

$$\langle a\,|\,\Phi_a\rangle^* = \langle a\,|\,\Phi_{-a}\rangle.$$

A detailed analysis of the location of the two singularities of $G(k)$ in the $k$ plane shows that one of them is located in the fourth quadrant, and the second is symmetric with respect to the vertical axis in the third quadrant. That is, one of these states is resonant (decaying) and the second antiresonant (to be produced). The expansion of $Z_n$ in $\lambda$ is of the form

$$Z_n = \langle \tilde\Phi_a\,|\,\Phi_a\rangle = 1 + i\frac{\lambda^2 \alpha^2}{4e_a\sqrt{e_a}} + \dots \cong 1 + \frac{i}{2}\left(\frac{\mathrm{Im}\,E_a}{\mathrm{Re}\,E_a}\right). \, (4.14)$$

The second pair of solutions associated with the state $|b\rangle$ is quite the same as that discussed above.

The fifth solution for (4.8) is, however, of an entirely different character. If we use again the expansion in $\lambda$ we obtain:

$$k_0 = \frac{i\lambda^2}{2}\left(\frac{\alpha^2}{e_a} + \frac{\beta^2}{e_b}\right)\left(1 + \frac{\alpha^2\lambda^2}{e_a^2} + \frac{\beta^2\lambda^2}{e_b^2} + O(\lambda^4)\right) \tag{4.15}$$

and, correspondingly, for the wave function

$$|\Phi_0\rangle = \frac{\lambda^2}{\sqrt{2}}\left(\frac{\alpha^2}{e_a} + \frac{\beta^2}{e_b}\right)^{1/2}\left(\frac{\alpha\,|a\rangle}{e_a} + \frac{\beta\,|b\rangle}{e_b}\right) + O(\lambda^4) . \tag{4.16}$$

From (4.15) it can be seen that the energy of this state $E_0 = k_0^2 < 0$ is very small, i.e. we obtain a near-threshold bound state. To understand how a bound state arises in the system we shall examine the resolvent (4.4) in the outer channel. It corresponds to free motion of an electron plus scattering by the "potential" localized at the origin. We may introduce a $T$-matrix describing the electron scattering in the following way:

$$R_{oo}(z) = R_{oo}^{(0)}(z) + R_{oo}^{(0)}(z)T(z)\,R_{oo}^{(0)}(z) \tag{4.17}$$

i.e. the $T$-matrix is of the form

$$\langle x|\,T(k^2)\,|x'\rangle = \frac{\lambda^2\,F(k^2)}{1 - \lambda^2\,F(k^2)/(2ik)}\,\delta(x)\,\delta(x') . \tag{4.18}$$

Such a $T$-matrix corresponds to an effective energy-dependent potential of the form

$$V_\lambda(k, x) = \lambda\,F(k^2)\,\delta(x) .$$

At small $k^2$, because $e_a$ and $e_b$ are positive, $F(k^2)$ may be negative (see (4.5)), i.e. at small $k^2$ this effective potential is attractive and may produce a bound state in the outer channel.

In the outer channel the wave function of this state is of the conventional form

$$\psi_0(x) = A_0\,e^{-\gamma_0|x|} \tag{4.19}$$

where

$$\gamma_0 = |k_0|$$

and

$$A_0 = \frac{-F(-\gamma_0^2)}{2\gamma_0[1/\gamma_0\lambda^2 - F'(-\gamma_0^2)]} . \tag{4.20}$$

In terms of the discussed model we may easily construct, on the basis of the Mittag-Leffler theory, the expansion of the Green function in the resonance functions [6]:

$$G(k) = \sum_{n=-2}^{2} \frac{|\Phi_n\rangle \langle \tilde{\Phi}_n|}{2k_n(k - k_n)} .$$

(4.21)

Comparing (4.21) with the asymptotic form of $G(k)$, we may obtain the relation [6]

$$\frac{1}{2} \sum_{n=-2}^{2} |\Phi_n\rangle \langle \tilde{\Phi}_n| = I$$

(4.22)

where $I$ is the unit matrix. The last relation reflects overfilling of the Gamow functions basis.


## 2.5. Resonance States in the $R$-matrix Theory


The $R$-matrix theory developed by Wigner and Eisenbud [7, 27] is very widely used to describe resonance phenomena in nuclear and atomic physics. The theory is especially used to parametrize experimental data in the fields of nuclear spectroscopy and nuclear reactions, in particular to infer the parameters (energies and widths) of nuclear levels from the experimental scattering and reaction cross sections. Furthermore, the $R$-matrix functions forming the basis of the theory and known also as the Wigner-Eisenbud basis are widely used in various theories of nuclear reactions and in structural calculations. The details of the $R$ matrix method as applied to the theory of nuclear reactions are excellently explained in the review [28]. For completeness we shall briefly discuss a scheme of this approach as applied to describing the resonance states in the simple case of the one-channel potential $s$-wave scattering.

In the $R$-matrix approach, as in the Kapur-Peierls approach, the entire configurational space is divided by a sphere of radius $R_c$ into an inner channel containing the whole potential (here the potential is also considered to be truncated) and an outer channel including all the space outside the sphere.

In the $R$-matrix theory the basis functions (the Wigner-Eisenbud functions $u_n(r)$) are defined as the solutions for the Schrödinger equation

$$-\frac{\hbar^2}{2m} \frac{d^2}{dr^2} u_n(r) + (V(r) - E) u_n(r) = 0 ,$$

(5.1)

with the boundary conditions

$$u_n(0) = 0 \, , \tag{5.2}$$

$$\left. \frac{\mathrm{d} u_n(r)}{\mathrm{d}r} \right|_{r=R_c} = b \tag{5.3}$$

where $m$ is the reduced particle mass.

Thus, the basis consists of the functions $u_n$ satisfying the Schrödinger equation (5.1) at such definite real energies $E_n$ that the derivative of the function at the channel boundary $(r = R_c)$ is equal to a constant $b$ independent of energy. In the majority of cases one sets $b = 0$. We shall show that $u_n(r)$ form an orthonormalized system. To do this, we multiply the equation for $u_n$ on the left by $u_m$, and the equation for $u_m$ by $u_n$. Then we subtract one equation from the other and integrate over $r$:

$$-\frac{\hbar^2}{2m} \int_0^{R_c} \left( u_m \frac{\mathrm{d}^2}{\mathrm{d}r^2} u_n - u_n \frac{\mathrm{d}^2}{\mathrm{d}r^2} u_m \right) \mathrm{d}r + (E_m - E_n) \int_0^{R_c} u_m(r) u_n(r) \, \mathrm{d}r = 0 \, .$$

Integrating the first term by parts and taking into account (5.2), we obtain:

$$-\frac{\hbar^2}{2m} \left( u_m \frac{\mathrm{d}u_n}{\mathrm{d}r} - u_n \frac{\mathrm{d}u_m}{\mathrm{d}r} \right) \Bigg|_{r=R_c} + (E_m - E_n) \int_0^{R_c} u_m(r) \, u_n(r) \, \mathrm{d}r = 0 \, .$$

$$\tag{5.4}$$

On imposing the boundary condition (5.3) with $b = 0$ we obtain immediately at $E_n \neq E_m$,

$$\int_0^{R_c} u_m(r) \, u_n(r) \, \mathrm{d}r = \delta_{nm} \, , \tag{5.5}$$

i.e. the basis is orthonormalized. Moreover, the solution for the boundary problem forms a complete set. Therefore, the solution to equation (5.1) corresponding to an *arbitrary* energy $E$ may be series-expanded in eigenfunctions $u_n$ [28] in the inner region.

$$u_E(r) = \sum_n A_n u_n(r) \tag{5.6}$$

where

$$A_n = \int_0^{R_c} u_n(r) \, u_E(r) \, \mathrm{d}r \, .$$

To determine the expansion coefficients $A_n$ we may use the relation (5.4). Let $u_n$ be an eigenfunction and $u_m = u_E$. Then, from eq. (5.4) we obtain:

$$\frac{\hbar^2}{2m} u_n(R_c) \left(\frac{du_E}{dr}\right)\bigg|_{r=R_c} + (E - E_n) \int_0^{R_c} u_E(r) u_n(r) \, dr = 0 \, ,$$

which gives immediately:

$$A_n = \frac{\hbar^2}{2m} \frac{u_n(R_c)}{E_n - E} \left(\frac{du_E}{dr}\right)\bigg|_{r=R_c} \tag{5.7}$$

Substituting this in eq. (5.6), we get

$$u_E(r) = G(r, R_c) \left(R_c \frac{du_E}{dr}\right)\bigg|_{r=R_c} \tag{5.8}$$

where

$$G(r, R_c) = \frac{\hbar^2}{2mR_c} \sum_n \frac{u_n(r) u_n(R_c)}{E_n - E} \tag{5.9}$$

is the Green function relating the value of the wave function in the inner region to the value of its derivative at the boundary.

The $R$-function is defined in the following way [28]:

$$R = G(R_c, R_c) = \sum_n \frac{\gamma_n^2}{E_n - E} \tag{5.10}$$

where the quantity $\gamma_n^2$ is called the reduced width of the $n$-th level and $\gamma_n$ is the amplitude of the reduced width

$$\gamma_n = \left(\frac{\hbar^2}{2mR_c}\right)^{1/2} u_n(R_c) \, .$$

According to (5.8), the $R$-function is equal to the inverse logarithmic derivative of the function $u_E$ at $r = R_c$ divided by $R_c$:

$$R(E) = \frac{u_E(R_c)}{R_c \left(\dfrac{du_E}{dr}\right)_{r=R_c}} \tag{5.11}$$

One can easily show [28] that the phase shift may be written in terms of the $R$-function in the following way:

$$\delta(E) = \text{arctg} \left[\frac{RP}{1 - RS}\right] + \delta_H \tag{5.12}$$

where $S$ and $P$ are some quantities which may be expressed, as we know, in terms of the value of the asymptotic part of the wave function and its logarithmic derivative at $r = R_c$ [28].

The second term $\delta_H$ in (5.12) is the so-called phase shift of the scattering by a hard sphere. The first term is the resonance contribution. If the energy $E$ is close to one of the levels $E_n$, the $R$-function may be written as

$$R \sim \frac{\gamma_n^2}{E_n - E} . \tag{5.13}$$

Substituting (5.13) in (5.12), we obtain:

$$\delta = \operatorname{arctg}\left(\frac{\frac{1}{2}\Gamma_n}{E_n + \Delta_n - E}\right) + \delta_H . \tag{5.14}$$

Here, the level width

$$\Gamma_n = 2\gamma_n^2 P \tag{5.15}$$

determines the rate of phase shift change when the energy $E$ runs through the resonance value $E_R = E_n + \Delta_n$, while the lifetime of the particle in the state $E_n$ of the inner channel is $\tau_n \sim \Gamma_n^{-1}$. The level shift

$$\Delta_n = -\gamma_n^2 S \tag{5.16}$$

defines the displacement of the level $E_n$ of the inner channel after the coupling to the outer channel is switched on, i.e. gives the shift of the resonance energy $E_R$ from the unperturbed $R$ matrix level $E_n$.

Equation (5.14) may be used to infer the values $E_R$ and $\Gamma$ from the experimental phase; further, it may be made more exact by introducing a two-level approximation, that is, by assuming that two $R$-matrix levels contribute to the resonance part of the phase [29], etc.

Now we shall discuss some shortcomings of, and difficulties with, the $R$-matrix approach. The $R$-matrix states form an infinite discrete spectrum. Those for which $E_n < 0$ may be unambiguously correlated with the true bound states of the system [30], and the functions $u_n$ are very similar to the corresponding functions of the bound states [31]. For the states with positive energy $E_n > 0$ lying below the edge of the potential barrier, a correspondence may be established with the subbarrier (sufficiently narrow) resonances. In this case the functions $u_n$ are also similar to the inner parts of the wave functions of the corresponding resonances. Broader, above-barrier resonances, especially the resonances at high energy, having a diffraction character, cannot be correlated with definite $R$-matrix states [30]. This fact also manifests itself in that the parametrizations of the type of (5.14) work well only for narrow resonances. For broad resonances, the quantities $\Gamma_n$ determined from the $R$-matrix parametrization are very sensitive to the theory parameters, that is, to the parameters $b$ and $R_c$ and also to the approximation type (one-level or multilevel) used. Because of the completeness of the basis applied, the description of the amplitudes of different processes and parameters of the resonance states should not depend, in principle, on theoretical parameters $b$ and $R_c$. However, in practice, owing to the fact that the convergence of $R$-matrix expansions is often slow, one has to limit oneself to a small number of the expansion terms. Therefore, the practical results of the application of the $R$-matrix scheme depend often on the parameters $b$ and $R_c$. The presence of the theory parameters (such as the channel radius $R_c$) entering explicitly into all the results may also be regarded as some shortcomings of the $R$-matrix approach. Nevertheless, frequent attempts to improve the classical $R$-matrix approach, either by introducing a more flexible procedure of matching the functions of the inner and outer channels or through a radical replacement of the Wigner-Eisenbud basis [32] by a "natural" basis, have not led as yet to the construction of a complete and practically suitable scheme similar to the $R$-matrix one. Concrete calculational schemes for describing the resonance states on the basis of the $R$-matrix formalism are discussed in Refs. [30, 39, 40–43].

Chapter 4 deals with the projecting operator formalism which is very similar in spirit to the $R$-matrix scheme. Here, as in the $R$-matrix approach, the entire space is divided into "inner", $\mathcal{H}_P$ and "outer", $\mathcal{H}_Q$, subspaces; this division is, however, not made in the configurational space $R$, but in the Hilbert space $\mathcal{H}$. In many cases, especially in the theory of many-particle resonances, the use of the methods in the Hilbert space is preferable.

In the first two chapters (and, partially, in Chapter 3) we have described the main elements used as building blocks to construct the "edifices" of most of the works in resonance state theory. Such works are vividly exemplified by the study (Minelly et al. [47]) devoted to finding the analytically continued $S$-matrix $S(k)$ (in terms of potential scattering by a cut-off potential) in the complex $k$-plane and determining the $S$-matrix resonance poles. The authors [47] proceed from the known relationship between the $S$-matrix and the Green function in the Kapur-Peierls formalism (see Section 3 of this Chapter) and construct the Taylor-series expansion of the Green function in the vicinity of some points located on a certain grid in the $k$-plane, with the solution of the Schrödinger equation at the point having been given beforehand. After that, using the Taylor series as an example, the authors use the Hadamard method (see Section 3.1 of Chapter 1) to find the poles of the analytically continued $S$-matrix. Similar examples may have been presented for many other works (see the subsequent chapters). Such illustrations are most probably indicative of the validity of the method used here.

*References*


1. TAYLOR J. R., Scattering Theory. The Quantum Theory of Nonrelativistic Collisions, John Wiley, N. Y. 1972.
2. NEWTON R. G., Scattering Theory of Waves and Particles, Springer, New York, Heidelberg, London 1982.
3. BAZ I. A., ZEL'DOVICH Ya. B., PERELOMOV A. M., Scattering, Reactions and Decay in Nonrelativistic Quantum Mechanics, Nauka, Moscow, 2nd ed. 1971 (in Russian). [English translation of lst ed.: Israel Program for Scientific Translations, Jerusalem, 1966].
4. PEIERLS R. E., Proc. Roy. Soc. *A253* (1959) 16.
5. MORE R. M., Phys. Rev. A. *3* (1971) 1217; *4* (1971) 1789.
6. MORE R. M., GERJOY E., Phys. Rev. *A7* (1973) 1288.
7. WIGNER E. P., EISENBUD L., Phys. Rev. *72* (1947) 29.
8. KAPUR P. L., PEIERLS R. E., Proc. Roy. Soc. *A166* (1938) 277.
9. NICHITIU F., Analiza de faza în fizica interactiilor nucleare, Editura Academici RSR, Bucuresti 1980.
10. BECKER S. F., SIEGEL S. G., Ann. Phys. *52* (1969) 479.
11. BAZ A. I., MERKURJEV S. D., Theor. Math. Phys. (in Russian) *27* (1976) 1; *31* (1977) 48.
12. GAMOW G., Z. Phys. *51* (1928) 204.
13. SIEGERT A. J. F., Phys. Rev. *56* (1939) 750.
14. HUMBLET J., ROSENFELD L., Nucl. Phys. *26* (1961) 529.
15. GARCIA-CALDERON G., PEIERLS R., Nucl. Phys. *A265* (1976) 443.
16. de ALFARO V., REGGE T., Potential Scattering, North-Holland, Amsterdam 1965.
17. ROMO W. J., Nucl. Phys. *A116* (1968) 618.
18. KUKULIN V. I., KRASNOPOL'SKY V. M., MISELHI M., Yad. Fiz. (in Russian) *29* (1979) 818. [Sov. J. Nucl. Phys. 29 (1979) 421].
19. ZEL'DOVICH Ya. B., ZhETF (in Russian) *39* (1960) 776 [Sov. Phys. JETP *12* (1961) 542].
20. BANG J., ZIMANY I., Nucl. Phys. *A139* (1969) 534:
COKER W. R., Phys. Rev. *C7* (1973) 2426.
21. BERGREN T., Nucl. Phys. *A109* (1968) 265.
22. BOAS R. R., Entire functions, Academic Press, New York 1954,
CARATHEODORY C., Funktionen-theorie I, Birkhäuser, Basel 1950.
23. ROMO W. J., Preprint, Ottawa University 1978; Nucl. Phys. *A302* (1978) 61.
24. BANG J., GAREEV A. F., GIZZATKULOV M. H., GONCHAROV S. A., Preprint, E4-113 77, Dubna 1978; BANG J., ERSHOV S. N., GAREEV F. A., KAZACHA G. S., Nucl. Phys. *A339* (1980) 89.
25. PEIERLS R., Proc. Camb. Philos. Soc. *44* (1948) 242.
26. FONDA L. et al., J. Math. Phys. *7* (1966) 1643.
27. WIGNER E. P., J. Am. Phys. Soc. *17* (1949) 99.
28. LANE A. M., THOMAS R. G., Rev. Mod. Phys. *30* (1958) 257.
29. CHIEN W. S., BROZIN R. E., Preprint University Minnesota, 1974.
30. TAKEUCHI K., MOLDAUER P. A., Phys. Rev. *2C* (1970) 920.
31. MAHAUX C., WEIDENMÜLLER H. A., Phys. Rev. *170* (1968) 847.
32. SERDOBOLSKY V. I., ZhETF (in Russian) *36* (1959) 1903 [Sov. Phys. JETP *9* (1959) 1354].
33. MATHEWS P. T., SALAM A., Phys. Rev. *112* (1958) 283, *115* (1959) 1079;
ZWANZIGER D., Phys. Rev. *131* (1963) 2818;
HAVLÍČEK M., EXNER P., Czech. J. Phys. *B23* (1973) 594.
34. NOGOVA A., PIŠÚT J., Nucl. Phys. *B61* (1973) 444;
SPEARMAN T. D., Phys. Rev. *D10* (1974) 1660.
35. GAUTHIER L., KAMAL A. N., Ann. Phys. *88* (1974) 193; *90* (1975) 515.
36. DALITZ R. H., Resonance: Its Description, Criteria and Significance. Lecture Notes in Physics *211* (1984) 1.

37. DAVIES E. B., Lett. Math. Phys. *1* (1975) 31.
38. CALUCCI G., GHIRARDI C., Phys. Rev. *169* (1968) 1339.
39. BURKE P. G., TAYLOR K. T., J. Phys. *B8* (1975) 2620.
40. PHILPOTT R. J., Fizika 1977 *9* Suppl. 3, 36.
41. AHMAD S. S., BARRET R. F., ROBSON B. A., Nucl. Phys. *A257* (1976), 378.
42. PHILPOTT R. J., GEORGE J., Nucl. Phys. *A233* (1974) 164.
43. PURCELL J. E., MEDER M. R., Phys. Rev. C *16* (1977) 76.
44. FESHBACH H., Ann. Phys. (N. Y.) *5* (1958) 357.
45. TENNYSON J., NOBLE C. J., Comput. Phys. Commun *33* (1984) 421.
46. BADALYAN A. M., KOK L. P., SIMONOV Yu. A., POLIKARPOV M. I., Phys. Rep. *82* (1982) 177.
47. MINELLY T. A., VITTURI A. V., ZARDI F., Nuovo Cim. *14A* (1973) 827.

# Chapter 3

# Theory of Resonance States Based on the Hilbert-Schmidt Expansion

In this chapter, when discussing the theory of resonance states on the basis of the Hilbert-Schmidt method, we shall build upon the material explained in Section 1.5. Therefore, before studying the material of the present chapter, the reader is advised to look through the above-mentioned section.

## 3.1. Two-particle Resonances

The theory of resonance states in a two-particle system may be constructed in many ways. However, the theory based on the Hilbert-Schmidt expansion has many advantages over most other methods. In terms of this theory many important results of potential scattering theory are obtainable in a natural way (see Section 2.1), and the theory makes it possible to obtain a normalizable function of the resonance state; further, on this basis we may construct a correct perturbation theory for such states, and so on.

### 3.1.1. Perturbation Theory for Resonance States.
### Construction of a Normalized Wave Function

As shown in Section 1.5.1, the Hilbert-Schmidt eigenvalues $\eta_{nl}(z)$ and eigenfunctions $\psi_{nl}(z)$ may be defined either as the eigenvalues and eigenfunctions of the Lippmann-Schwinger equation kernel or as solutions for a generalized Schrödinger equation (1.5.31) with boundary conditions (1.5.32). The $S$-matrix poles, i.e. the $a$, $b$ and $r$-states, are determined by the condition (1.5.27):

$$\eta_{nl}(k_0) = 1 . \tag{1.1}$$

where the type of state is determined by the location of $k_0$ in the complex $k$-plane (see Section 2.1), the energy of the state is $z_0 = \hbar^2 k_0^2/2m$.

In Chapter 2 we discussed a number of approaches in the theory of resonance states. Out of them, the most "natural" approach is based on the Gamow functions (see Section 2.2). These functions were shown to be very suitable in many respects; however, because of the unusual asymptotic behaviour of the wave functions of $a$ and $r$-states (cf. eqs. (2.1.6) and (2.1.9)) they are not quadratically integrable and the use of these functions for evaluating the normalization integrals and the matrix elements of operators involves certain difficulties. In particular, it is necessary to introduce a regularization procedure (see Section 1.1) or an analytic continuation (see Chapter 5). Moreover, the complex values of energy are not observable in experiments where we always deal with real values of energy $E$.

The Hilbert-Schmidt method makes it just possible to juxtapose a resonance state with a normalizable function at a "real" value of energy $z = E + i0$.

Let us examine a generalized Schrödinger equation (1.5.31) for such a complex value $z = z_0$ when $\eta_{nl}(z_0) = 1$, i.e. at the $S$-matrix resonance pole. At this energy, equation (1.5.31) will coincide with the ordinary Schrödinger equation (2.1.2), and the eigenfunction of the Sturm-Liouville problem $\Psi_{nl}$ with the Gamow function $\Phi_{nl}$. If we add a positive imaginary part $i\delta$ to $\eta_{nl}$, the generalized potential $v/\eta_{nl}$ becomes complex and its imaginary part becomes negative (absorptive). With increasing $\delta$, the $S$-matrix pole moves upwards from the 4-th quadrant of the $k$-plane [1] (see Section 5.1.6). At some $\delta$ the pole reaches the positive semi-axis $k$, i.e. the upper rim of the cut in the $E$ plane. At this positive real energy $E_0$ (and, correspondingly, at complex $\eta_{nl}(E_0)$) equation (1.5.31) with the boundary conditions (1.5.32) will have a normalizable solution $\psi_{nl}(r, E_0)$. This function is often called the resonance function in the Hilbert-Schmidt (Sturm-Liouville) formalism. It is clear that for narrow resonances $\Gamma \ll E_R$, the imaginary part of the eigenvalue $\eta_{nl}(E_0)$ is small; therefore,

$$\mathrm{Re}\ \eta_{nl}(E_0) \sim 1 \quad \text{and} \quad E_0 \sim E_R . \tag{1.2}$$

The condition (1.2) is often used for finding the resonance energy $E_0$. It is also clear that for narrow resonances the Hilbert-Schmidt wave function $\psi_{nl}(r, E_0)$ is very similar to the Gamow function, but, in contrast to the latter, the former may be normalized without redefining the integration procedure though with the weight function $V(r)$:

$$\int_0^\infty \psi_{nl}(r, E_0 + i\varepsilon)\, V(r)\, \psi_{ml}(r, E_0 + i\varepsilon)\, r^2\, \mathrm{d}r = -\,\eta_{nl}(E_0)\, \delta_{mn} . \tag{1.3}$$

Let us stress that, owing to the addition of $+i\varepsilon$ to the energy (energy is given on the first sheet, that is $\mathrm{Im}\ k > 0$), the integral (1.3) always converges even if $V(r)$ is not a short-range potential.

Owing to the normalizability of functions $\psi_{nl}(r, E)$ and to the fact that $n$ runs through a discrete series of values, we may develop for the quasi-stationary states a perturbation theory similar in a definite sense to conventional perturbation theory [2, 3]. We shall demonstrate this by an example of calculating the corrections to the eigenvalues $\eta_{nl}(E)$.

Let us add a small perturbation $\lambda\, V_1(r)$ to the potential $V(r)/\eta_n$ of eq. (1.5.31). The corresponding equation (1.5.14) in the integral form will be (we introduce the function $\varphi_n(r) = r\,\psi_n(r)$ and omit the index $l$)

$$\eta_n(E)\,\varphi_n(r, E) = \int_0^\infty G_0(E; r, r')\left[V(r') + \eta_n(E)\,\lambda\, V_1(r')\right]\varphi_n(r', E)\,dr'$$

$$(1.4)$$

where $G_0(E; r, r')$ is a free Green function.
Series expansion of $\eta_n(k)$ and $\varphi_n$ in $\lambda$ is

$$\eta_n(k) = \eta_n^{(0)}(k) + \lambda\,\eta_n^{(1)}(k) + \cdots$$

$$\varphi_n = \varphi_n^{(0)} + \lambda\,\varphi_n^{(1)} + \cdots .$$

$$(1.5)$$

If the perturbation is small, the series (1.5) converge sufficiently rapidly. At $\lambda \to 0$, $\varphi_n$ and $\eta_n$ turn into $\varphi_n^{(0)}$ and $\eta_n^{(0)}$. We substitute (1.5) in (1.4) and retain in equation (1.4) only the terms of the first order in $\lambda$:

$$\eta_n^{(0)}\varphi_n^{(1)} + \eta_n^{(1)}\varphi_n^{(0)} = \int_0^\infty G_0(E; r, r')\, V(r')\, \varphi_n^{(1)}(r')\,dr' +$$

$$+ \eta_n^{(0)} \int_0^\infty G_0(E; r, r')\, V_1(r')\, \varphi_n^{(0)}(r')\,dr' . \qquad (1.6)$$

No we shall multiply (1.6) by $V(r)\,\varphi_n^{(0)}(r)$, integrate the result over $r$ from 0 to $\infty$. Taking into account, as in (1.6) that $G_0 V\varphi_n^{(0)} = \eta_n^{(0)}\varphi_n^{(0)}$ we find:

$$\eta_n^{(0)} \int_0^\infty \varphi_n^{(1)}(r)\, V(r)\, \varphi_n^{(0)}(r)\,dr + \eta_n^{(1)} \int_0^\infty \varphi_n^{(0)}(r)\, V(r)\, \varphi_n^{(0)}(r)\,dr =$$

$$= \eta_n^{(0)} \int_0^\infty \varphi_n^{(0)}(r')\, V(r')\, \varphi_n^{(1)}(r')\,dr' + (\eta_n^{(0)})^2 \int_0^\infty \varphi_n^{(0)}(r')\, V_1(r')\, \varphi_n^{(0)}(r')\,dr'$$

whence we obtain the first correction for the eigenvalues

$$\eta_n^{(1)}(z) = (\eta_n^{(0)}(z))^2 \int_0^\infty (\varphi_n^{(0)}(r))^2 V_1(r)\,dr \Big/ \int_0^\infty (\varphi_n^{(0)}(r))^2 V(r)\,dr . \qquad (1.7)$$

If the normalization condition (1.3) is satisfied, then

$$\eta_n^{(1)}(z) = -\eta_n^{(0)}(z) \int_0^\infty (\varphi_n^{(0)}(r))^2 V_1(r)\, dr \ . \tag{1.8}$$

Now we shall find the correction to the resonance energy due to the perturbation $\lambda V_1$. The unperturbed solution has a resonance at $z = z_0$, i.e.

$$\eta_n^{(0)}(z_0) = 1 \ . \tag{1.9}$$

On imposing the perturbation, the resonance energy changes:

$$z_0' = z_0 + \lambda z_1 + \lambda^2 z_2 + \cdots \ . \tag{1.10}$$

Using the expansion (1.5) for $\eta_n$ in the resonance condition, we obtain

$$\eta_n^{(0)}(z_0') + \lambda \eta_n^{(1)}(z_0') + \cdots = 1 \ . \tag{1.11}$$

Substituting (1.10) in (1.11), series-expanding $\eta_n(z_0')$ near the point $z_0$, and retaining only the terms of the first order in $\lambda$, we find (taking into account (1.8)):

$$1 + \frac{\partial \eta_n^{(0)}(z)}{\partial z}\bigg|_{z=z_0} \lambda z_1 + \lambda \eta_n^{(1)}(z_0) = 1 \ ,$$

i.e. the first-order correction $z_1$ equals

$$z_1 = -\frac{\eta_n^{(1)}(z_0)}{\dfrac{\partial \eta_n^{(0)}}{\partial z}\bigg|_{z=z_0}} = \frac{\eta_n^{(1)}(z_0)}{(\eta_n^{(0)})^2 \dfrac{\partial}{\partial z}\left(\dfrac{1}{\eta_n^{(0)}}\right)\bigg|_{z=z_0}} = -\frac{\displaystyle\int_0^\infty (\varphi_n^{(0)}(r))^2 V_1(r)\, dr}{\eta_n^{(0)} \dfrac{\partial}{\partial z}\left(\dfrac{1}{\eta_n^{(0)}}\right)\bigg|_{z=z_0}} \ .$$

Expressing $\eta_n^{(0)}$ from the condition (1.9) and using the Hellmann-Feynman theorem for $\partial/\partial z(1/\eta_n^{(0)})$ (see Chapter 5)

$$\frac{\partial}{\partial z}\left(\frac{1}{\eta_n^{(0)}}\right) = \int_0^\infty (\varphi_n^{(0)}(r))^2\, dr \bigg/ \int_0^\infty (\varphi_n^{(0)}(r))^2\, V(r)\, dr \tag{1.12}$$

and the normalization condition for $\varphi_n^{(0)}(r)$, we obtain the final expression for the first-order correction to resonance energy

$$z_1 = \int_0^\infty V_1(r)\, (\varphi_n^{(0)}(r))^2\, dr \bigg/ \int_0^\infty (\varphi_n^{(0)}(r))^2\, dr \ . \tag{1.13}$$

The common and important distinctive feature of the expressions (1.3), (1.8) and (1.13) is that they do not contain complex conjugate functions. Instead of the squared modulus, there appears everywhere merely the squared wave function; thus, the integrals in these expressions converge. This is due to the fact that in the Hilbert-Schmidt theory the functions $\varphi_n$ do not form an orthogonal set in the conventional sense (as is the case, for example, for the functions of discrete and continuous spectra). Instead of this, a biorthogonal set appears consisting of right-hand and left-hand eigenfunctions of the Lippmann-Schwinger equation kernel (see (1.5.14)–(1.5.18)) $g_n$ and $\psi_n$, respectively. A similar situation faced us in Chapter 2 (see (2.2.38), (2.2.39), (2.3.7) and (2.5.5)).

### 3.1.2. Determination of the Hilbert-Schmidt Eigenvalues with the Help of Analytic Continuation in Energy

In this section we present a simple way of finding the Hilbert-Schmidt eigenvalues $\eta_{nl}(z)$ and eigenfunctions $\psi_{nl}(z)$. As noted above, many results of scattering theory may be formulated in terms of $\eta_{nl}$ and $\psi_{nl}$. Moreover, these quantities are often used in the separable expansions of the two-particle $t$-matrix when solving the Faddeev equations [4], summing up divergent Born series [5], in some theories of nuclear reactions [6], etc. To find the eigenvalues and eigenfunctions we may directly solve the generalized Schrödinger equation (1.5.31) with boundary conditions (1.5.32). However, at $\mathrm{Re}\, z > 0$ direct solution is often impractical and leads to a numerical instability, because at $\mathrm{Im}\, \sqrt{z} < 0$ the solution in the outer domain increases strongly and there arise difficulties similar to the difficulties with direct calculations of the Gamow states (see Chapter 5).

As noted in Section 1.5, $\eta_{nl}$ and $\psi_{nl}$ can be determined correctly at real $(E > 0)$ and complex values of energy with the help of analytic continuation from the region of $E < 0$ where the Hilbert-Schmidt problem has a simple and well-defined solution. Here we shall present a method allowing us, by using this definition constructively, to find the Hilbert-Schmidt eigenvalues and eigenfunctions with the help of analytic continuation in energy (or in $k$) from the region of bound states $(E < 0)$. Indeed, at $E < 0$ the equation may be solved very simply because the discussed problem reduces in this case to calculating the bound states in the potential $V/\eta_{nl}$ ($\eta_{nl}$ is real). Thus, in the region of $E < 0$ ($k = i\varkappa$, $\varkappa > 0$) the functions $\psi_{nl}(k, r)$ and $\eta_{nl}(k)$, $k = \sqrt{(2mE/\hbar^2)}$ can be calculated in a standard way.

Furthermore, the eigenvalue $\eta_{nl}(k)$ is known to be an analytic function in the entire analyticity region of the Jost function $f(k)$ (in the entire $k$ plane for the finite $V(r)$, see Section 1.5). Knowing the analytic function $\eta_{nl}(k)$ in some interval, we may in principle determine it in the entire analyticity region with the help of analytic continuation. For carrying out the analytic continuation we use the Padé approximant technique (see Section 1.2). To this end we may use the Padé approximants of type II or type III. If we know the coefficients of the series expansion of $\eta_{nl}(k)$ (1.5.22), they may be used to construct a Padé approximant of type I. However, for an arbitrary potential $V$, the calculations of the coefficients of the series (1.5.22) involve considerable difficulties; it is much simpler to find $\eta_{nl}(E_i)$ at $E_i < 0$ by direct numerical solution of a generalized Schrödinger equation (1.5.31) at these energies $E_i{}^1)$. In this way we obtain the reference sets of values $\{E_i\}$, $\{\eta_{nl}^{(i)}\} \equiv \{\eta_{nl}(E_i)\}$ and $\{\psi_{nl}^{(i)}\} \equiv \{\psi_{nl}(r, E_i)\}$ $(i = 1, 2, ..., p)$, where $p$ is the number indicating at how many points the Schrödinger equation was solved for bound state.

We now turn from the set of eigenvalues $\{E_i\}$ to the sequence of wave numbers $\{\varkappa_i\} \equiv \{-i \sqrt{2mE_i/\hbar^2}\}$ and construct the Padé approximant

$$\eta_{nl}(k) \approx \eta_{nl}^{[N,M]}(k) = P_N^{nl}(\varkappa)/Q_M^{nl}(\varkappa); \qquad N + M + 1 \leq p . \quad (1.14)$$

The coefficients of the polynomials $P_N^{nl}$ and $Q_M^{nl}$ are uniquely determined by the sets $\{\varkappa_i\}$ and $\{\eta_{nl}^{(i)}\}$ . Since $\varkappa$ and $\eta$ are real at $E < 0$ , the coefficients of $P_N$ and $Q_M$ are also real. Therefore, the choice of $\varkappa = -ik$ as a variable in the Padé approximant (1.14) ensures automatic satisfaction of the Schwarz principle:

$$\eta^*(-k^*) = \{P_N(-i(-k^*))/Q_M(-i(-k^*))\}^* =$$
$$= \{P_N(ik^*)/Q_M(ik^*)\}^* = P_N(-ik)/Q_M(-ik) = \eta(k) .$$

Owing to the analyticity of $\eta_{nl}(k)$, the approximate equality in formula (1.14) at $N, M \to \infty$ becomes exact. However, if we series-expand the Padé approximant (1.14) again, then at $l > 0$ the obtained series will not coincide with the series (1.5.22), because some terms with odd powers are absent in the latter series at $l > 0$. Of course, for sufficiently high $N$ and $M$, i.e. for a sufficiently good approximation of the function $\eta_{nl}(k)$ the coefficients of these terms in the series obtained from the Padé approximant (1.14) will be very small; nevertheless, if we prescribe this property of the approximate function in the Padé approximant beforehand, the accuracy of the approximation increases substantially. In order to do this it is sufficient to use a non-standard Padé approximant

$$\eta_{nl}(k) \approx \tilde{\eta}_{nl}^{[N,M]}(k) = \frac{\tilde{P}_L^{nl}(k^2) + \varkappa^{2l+1}\tilde{Q}_{N-L}^{nl}(k^2)}{R_M^{nl}(k^2)} . \qquad (1.15)$$

The series expansion of such a Padé approximant yields the same series as (1.5.22). We note, however, that the approximant (1.14) already gives a high accuracy. For example, all the illustrations presented below were obtained just using (1.14).

It should be stressed again that the Padé approximants, (1.14) and (1.15), which are obtained from the solution of the Schrödinger equation for bound states, usually determine the eigenvalues $\eta_{nl}(k)$ in the entire complex k-plane and

---

$^1)$ Naturally, in practice we proceed conversely: we set the value of $\eta_i$, whereupon we solve the Schrödinger equation with the potential $V/\eta_i$ for bound states, i.e. we find $E_i$ and $\psi(r, E_i)$ .

in a simple analytic form. The accuracy of the Padé extrapolation will decrease as the distance from the location region of the basis points increases. Using formulae (1.5.24)–(1.5.26), we may obtain, in such a simple form, the expres-

Table 3.1

Positions of the bound (b), antibound (a) and resonance (r) $S$-matrix poles for a rectangular well potential, $l = 0$.*)

| $n$ | PA order | 2,2 | 3,3 | 4,4 | 5,5 | 6,6 | 7,7 | 8,8 | Exact |
|---|---|---|---|---|---|---|---|---|---|
| 1b | Im $(kR)$ | 4.29 | 4.29 | 4.295 | 4.273 4 | 4.273 4 | 4.273 4 | 4.273 4 | 4.273 4 |
| 2b | Im $(kR)$ | 0.963 5 | 0.963 5 | 0.963 5 | 0.963 5 | 0.963 5 | 0.963 5 | 0.963 5 | 0.963 5 |
| 2a | Im $(kR)$ | −5.29 | −2.789 | −2.431 | −2.832 | −2.869 | −2.854 | −2.855 | −2.855 |
| 3r ⎰ | Re $(kR)$ | – | – | – | 5.56 | 5.85 | 5.82 | 5.851 | 5.894 |
| ⎱ | Im $(kR)$ | – | – | – | −0.88 | −0.974 | −1.447 | −1.333 | −1.321 |

*) The parameters of the potential are as follows:
$V_0 = 12.25$ MeV, $R = 2$ fm; $2m/\hbar^2 = 0.25$ MeV$^{-1}$ fm$^{-2}$, $k = (2mE/\hbar^2)^{1/2}$ .

sions for the Jost function $f_l(k)$, the $S$-matrix $S_l(k)$, and the scattering phase shifts $\delta_l(k)$ at real $k$ $(k > 0)$.

Now, the determination of the $S$-matrix resonance pole reduces to a standard problem of finding the complex roots of the (polynomial) equation

$$P_N^{nl}(k_0) - Q_M^{nl}(k_0) = 0 .$$ (1.16)

which may be obtained by substituting (1.14) in (1.1). To demonstrate the effectiveness of the Padé parametrization for calculation of the Hilbert-Schmidt eigenvalues and eigenfunctions we shall discuss a very simple case-the calculation of resonance eigenvalues $\eta$ and their trajectories for a rectangular potential well. We shall again use the PA-II technique [7]. For this purpose it is sufficient to calculate just a few bound state energies for various potential depths.

Table 3.1 gives the results of the calculations of the $a$, $b$-, and $r$-states for a rectangular-well potential in comparison with the exact values. In order not to solve equation (1.16) for *every* $n$, we expressed the function $f_l(k)$ according to formula (1.5.24) in terms of the values $\eta_{nl}(k)$ taken to be of the form (1.14) and then we sought simultaneously for *all* the zeros of the Jost function. As can be seen from the table, the location of the $S$-matrix poles is already reproduced sufficiently well at quite modest values of $N$ and $M$.

Fig. 3.1 depicts the trajectories $\eta_{nl}(k)$ in the complex plane at real $-\infty < k < \infty$ for a rectangular-well potential at $n = 1$ and $l = 0, 1, 2, 3, 4$. For comparison, the exact values of $\eta_{10}$ (i.e. at $l = 0$) obtained by direct calculation are also shown (by crosses) in the figure. Figure 3.2 shows trajectories of the Hilbert-Schmidt eigenvalues for the Woods-Saxon potential

$$V(r) = V_0/(1 + \exp((r - R)/a))^{-1}$$

widely used in nuclear physics. For $l > 0$ the trajectories for the Woods-Saxon as well as for the rectangular well potential were obtained for the first time just by the described method [7].

Now, we may express and calculate the scattering phase shifts with the help of formula (1.5.25) in terms of the Padé approximant (1.14). Then for $\tan \delta_l$ we obtain a simple analytic rational expression. It should be emphasized that we have succeeded in obtaining the scattering phase shifts with the help of analytic continuation from only the solution for the bound state problem. This unexpected result is explained by the fact that, besides the values $\{E_i\}$ and $\{\eta_i\}$ in the bound state region we have prescribed, in the approximation for $\eta$, the analytic properties of this function in the entire $k$ plane. Therefore, the approximate

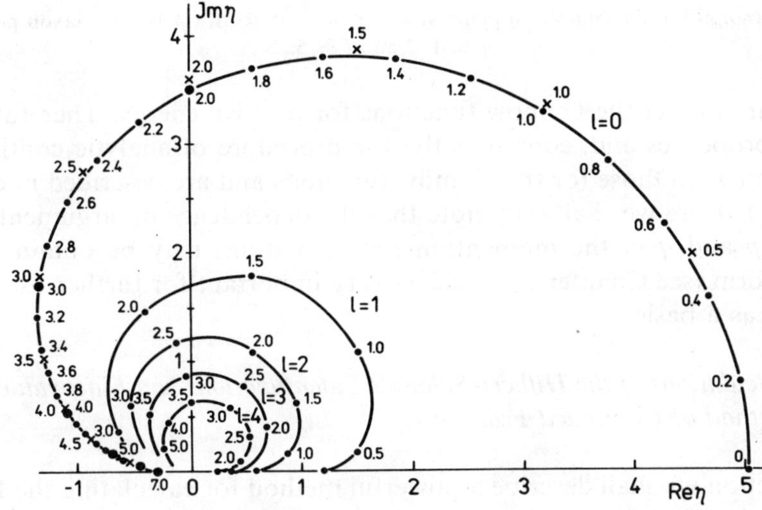

Fig. 3.1

Trajectories $\eta_{nl}(k)$ in the complex $\eta$ plane at real $-\infty < k < \infty$ for a rectangular-well potential; $n = 1$; $l = 0, 1, 2, 3, 4$. The numbers at the curves are the values of $k$ in the units of $\text{fm}^{-1}$. The exact values $\eta_{10}$ are denoted by crosses.

expression (1.14) makes it possible to proceed freely from one part of the $k$-plane to another, thereby relating such quantities as, for example, bound state energies and scattering phase shifts to each other. Analyticity near the threshold is also the basis of the effective range theory [8].

The eigenfunctions $\psi_{nl}(k, r)$ at $E > 0$ may also be obtained by means of analytic continuation of these functions from the region of $E < 0$, where they coincide with the bound state functions in the potential $V/\eta_{nl}$. As shown above,

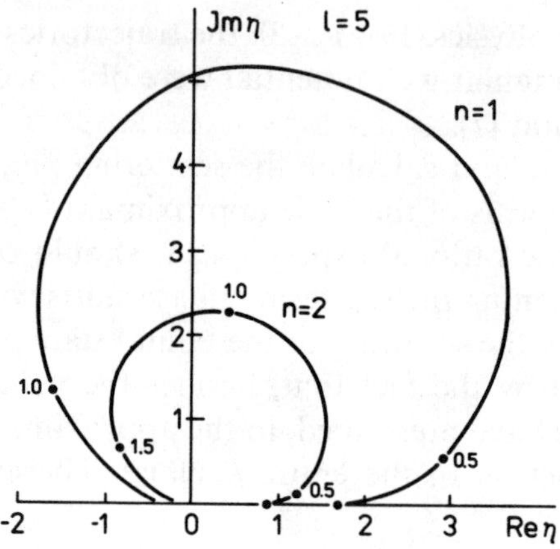



Fig. 3.2

Trajectories $\eta_{nl}(k)$ in the complex $\eta$ plane at $-\infty < k < \infty$ for a Woods-Saxon potential;
$n = 1, 2$ and $l = 5$.

$\psi_{nl}(k, r)$ are in fact the Gamow functions for positive energy. Therefore, their analytic properties and, consequently, the procedure of analytic continuation will be similar to those for the Gamow functions and are described in detail in Section 5.1. Here we shall only note that the dependence on arguments $k$ and $r$ (or on $p$ and $q$ in the momentum representation) may be obtained in an analytic form (see Chapter 5), which is very important for further use of such functions as a basis.

### 3.1.3. Calculations of the Hilbert-Schmidt Eigenfunctions and Eigenvalues by the Method of Continued Fractions

In this section we shall describe a powerful method for calculating the Hilbert--Schmidt eigenfunctions and eigenvalues proposed recently [19, 20] in the context of atomic physics. This method is quite general and has many advantages over the traditional methods. It is iterative and therefore the computer memory requirements are very moderate. Since no matrix inversion is necessary and the method is variational in its essence, very accurate results are easily obtainable. The t-matrix obtained by this method is of the form of a continued fraction in the potential strength parameter $\lambda$ and this is why this method is known as method of continued fractions. The essence of this method is to construct a sequence of operators with increasing null spaces. Contrary to other similar methods, the basis functions of the null space are not given in advance but are generated at each step of the iterations. As a result, very rapid convergence is obtained. There are two forms of this method, the $V$-form and the $G$-form. For simplicity we describe here only the $V$-form. The $G$-form, which is based on an expansion of the free-particle Green function $G_0$, is described in [20].

For lucidity of the presentation we restrict ourselves to the $s$-wave scattering of two nonrelativistic particles interacting through a local or nonlocal potential $U$ which is supposed to be Hermitian. Our task is to calculate the $K$-matrix defined by

$$K = \langle u| \, U \, |\varphi\rangle \tag{1.17}$$

which is related to the scattering phase shift $\delta(k)$ by the standart relation

$$K = -\frac{1}{k} \tan \delta(k) \tag{1.18}$$

and where $\varphi$ is the solution of the Lippmann-Schwinger equation:

$$\varphi = u + \lambda \, G_0 U\varphi \,. \tag{1.19}$$

Here $u$ is an incident plane wave; $U$ is the interaction potential; $G_0$ defines the standing-wave Green function which is given in terms of the Ricatti-Bessel (Ricatti-Neumann) functions $j_l(kr)$ and $n_l(kr)$ by

$$G_0(k; r, r') = \begin{cases} j_l(kr)n_l(kr') \,, & r' > r \,, \\ j_l(kr')n_l(kr) \,, & r' < r \,. \end{cases} \tag{1.20}$$

In eq. (1.19) we have explicitly introduced the potential strength $\lambda$.

Let us suppose that the first Born approximation to $K$, i.e. $\langle u| \, U \, |u\rangle$, is not zero and define a "reduced" potential $U^{(1)}$ by

$$U^{(1)} = U - \frac{U \, |u\rangle \, \langle u| \, U}{\langle u| \, U \, |u\rangle} \,; \tag{1.21}$$

substituting this expression in eq. (1.19), we obtain

$$\varphi = u + \lambda \, G_0 U \, |u\rangle \, \frac{\langle u| \, U \, |\varphi\rangle}{\langle u| \, U \, |u\rangle} + \lambda \, G_0 U^{(1)} \, |\varphi\rangle \,. \tag{1.22}$$

The function $\varphi$ is expressed formally as

$$\varphi = u + \lambda \, (1 - \lambda \, G_0 U^{(1)})^{-1} \, G_0 U \, |u\rangle \, \frac{\langle u| \, U \, |\varphi\rangle}{\langle u| \, U \, |u\rangle} \,, \tag{1.23}$$

where we used the fact that the free particle wave function belongs to the null space of $U^{(1)}$, i.e.

$$U^{(1)}|u\rangle = 0, \qquad \langle u| U^{(1)} = 0 .\tag{1.24}$$

Then, we define two functions $u_1$ and $\varphi_1$ by

$$u_1 = G_0 U u ,\tag{1.25}$$
$$\varphi_1 = (1 - \lambda\, G_0 U^{(1)})^{-1} u_1 .\tag{1.26}$$

In terms of these functions, eq. (1.23) reads

$$\varphi = u + \lambda\varphi_1 \frac{\langle u| U |\varphi\rangle}{\langle u| U |u\rangle} .\tag{1.27}$$

Multiplying by $\langle u| U$ from the left on both sides of eq. (1.27), we obtain after some manipulations:

$$\varphi = u + \varphi_1 \frac{\lambda \langle u| U |u\rangle}{\langle u| U |u\rangle - \lambda \langle u| U |\varphi_1\rangle} .\tag{1.28}$$

The scattering wave function $\varphi$ is expressed in terms of the function $\varphi_1$, defined in eq. (1.26), which in turn satisfies the Lippmann-Schwinger-type equation

$$\varphi_1 = u_1 + \lambda\, G_0 U^{(1)} \varphi_1\tag{1.29}$$

as the starting equation (1.19). Therefore, we may follow the same procedure. By analogy with eqs. (1.21), (1.25), and (1.26), we define $U^{(2)}$, $u_2$ and $\varphi_2$ as

$$U^{(2)} = U^{(1)} - \frac{U^{(1)} |u_1\rangle \langle u_1| U^{(1)}}{\langle u_1| U^{(1)} |u_1\rangle} ,\tag{1.30}$$

$$u_2 = G_0 U^{(1)} u_1 ,\tag{1.31}$$

$$\varphi_2 = (1 - \lambda\, G_0 U^{(2)})^{-1} u_1 .\tag{1.32}$$

Both functions $u$ and $u_1$ now belong to the null space of $U^{(2)}$, i.e.

$$U^{(2)} |u\rangle = U^{(2)} |u_1\rangle = 0 ,$$
$$\langle u| U^{(2)} = \langle u_1| U^{(2)} = 0 .\tag{1.33}$$

Proceeding similarly to the previous step, we obtain the following equations for $\varphi_1$ and $\varphi_2$:

$$\varphi_1 = u_1 + \varphi_2 \frac{\lambda \langle u_1| U^{(1)} |u_1\rangle}{\langle u_1| U^{(1)} |u_1\rangle - \lambda \langle u_1| U^{(1)} |\varphi_2\rangle} ,\tag{1.34}$$

$$\varphi_2 = u_2 + \lambda\, G_0 U^{(2)} \varphi_2 .\tag{1.35}$$

This procedure is repeated and, after some $N$ steps, we have

$$\varphi_N = u_N + \varphi_{N+1} \frac{\lambda \langle u_N| U^{(N)} |u_N\rangle}{\langle u_N| U^{(N)} |u_N\rangle - \lambda \langle u_N| U^{(N)} |\varphi_{N+1}\rangle}, \qquad (1.36)$$

$$u_{N+1} = G_0 U^{(N)} u_N, \qquad (1.37)$$

$$\varphi_{N+1} = u_{N+1} + \lambda G_0 U^{(N+1)} \varphi_{N+1}. \qquad (1.38)$$

By analogy with eq. (1.33), all the functions $u, u_1, ..., u_{N-1}$ belong to the null space of $U^{(N)}$, i.e.

$$U^{(N)} |u_i\rangle = 0, \quad \langle u_i| U^{(N)} = 0 \qquad (1.39)$$

for $i = 0, 1, ..., N - 1$.

Therefore, we expect the operator $U^{(N)}$ to become weaker and weaker with increasing N and, after some M steps, we may terminate this procedure by setting

$$\varphi_M \approx u_M. \qquad (1.40)$$

Our purpose is to calculate the $K$-matrix (1.17) which is in practice much simpler to calculate than the wave function. If we substitute the expression for $\varphi$ (eq. (1.28)) in eq. (1.17), we get

$$K_\lambda = \lambda \langle u| U |\varphi\rangle = \frac{\lambda \langle u| U |u\rangle^2}{\langle u| U |u\rangle - \lambda \langle u| U |\varphi_1\rangle}. \qquad (1.41)$$

Multipyling eq. (1.29) by $\langle u| U$ from the left, we obtain

$$\langle u| U |\varphi_1\rangle = \langle u| U |u_1\rangle + \lambda \langle u| U G_0 U^{(1)} |\varphi_1\rangle =$$

$$= \langle u| U |u_1\rangle + \lambda \langle u_1| U^{(1)} |\varphi_1\rangle \qquad (1.42)$$

where $\langle u_1| = \langle u| U G_0$.

The $K$-matrix is now expressed as

$$K_\lambda = \frac{\lambda \langle u| U |u\rangle^2}{\langle u| U |u\rangle - \lambda \langle u| U |u_1\rangle - \lambda^2 \langle u_1| U^{(1)} |\varphi_1\rangle}. \qquad (1.43)$$

The matrix element $\langle u_1| U^{(1)} |\varphi_1\rangle$ in the denominator of this expression plays the same role for eq. (1.29) as $\langle u| U |\varphi\rangle$ does for eq. (1.19). Therefore, we may write

$$K_\lambda = \frac{\lambda \langle u| U |u\rangle^2}{\langle u| U |u\rangle - \lambda \langle u| U |u_1\rangle - \dfrac{\lambda^2 \langle u_1| U^{(1)} |u_1\rangle^2}{\langle u_1| U^{(1)} |u_1\rangle - \lambda \langle u_1| U^{(1)} |\varphi_2\rangle}}.$$

$$(1.44)$$

By repeating this procedure, we obtain the $K$-matrix in the form of a continued fraction

$$K_\lambda = \cfrac{\lambda\, p_0^2}{p_0 - \lambda p_1 - \cfrac{\lambda^2 p_2^2}{p_2 - \lambda p_3 - \cfrac{\lambda^2 p_4^2}{p_4 - \lambda p_5 - \,\cdots}}} \tag{1.45}$$

where

$$p_{2i} = \langle u_i |\, U^{(i)} \,| u_i \rangle$$

and

$$p_{2i+1} = \langle u_i |\, U^{(i)} \,| u_{i+1} \rangle \,. \tag{1.46}$$

It is therefore necessary to calculate just two matrix elements $p_{2i}$ and $p_{2i+1}$ at each step. Once the coefficients have been calculated for a given energy, the expression (1.45) yields the matrix element of $K$ for all values of $\lambda$. By solving the algebraic equation

$$D(\lambda) = 0 \tag{1.47}$$

where $D(\lambda)$ is the denominator of $K_\lambda$ in (1.45), one obtaines the Hilbert-Schmidt eigenvalues $\lambda_i$ and the eigenfunctions $\psi_i$.

Indeed, for any $\lambda_i$ satisfying eq. (1.47), the function $\varphi_1$ given by (1.29) solves the equation

$$\varphi_1 = \lambda_i G_0 U \varphi_1 \,. \tag{1.48}$$

This can be demonstrated as follows. By substituting eq. (1.21) for $U^{(1)}$ in (1.29), we obtain:

$$\varphi_1 = u_1 + \lambda_i G_0 U \varphi_1 - \lambda_i G_0 U \, |u\rangle \frac{\langle u |\, U \,|\varphi_1 \rangle}{\langle u |\, U \,|u \rangle} \,. \tag{1.49}$$

For any $\lambda_i$ satisfying eq. (1.47), it follows from eq. (1.41) that

$$\langle u |\, U \,|u \rangle = \lambda_i \, \langle u |\, U \,|\varphi_1 \rangle \tag{1.50}$$

and because $G_0 U \,|u\rangle = |u_1\rangle$ , the first and third terms in (1.49) cancel and equation (1.48) follows.

Summarizing, we find that in the method of continued fractions the Hilbert-Schmidt eigenvalues are obtainable by solving the algebraic equation eq. (1.47), which is a very simple task. Simultaneously, the Hilbert-Schmidt eigenfunctions $\psi_i$ (unnormalized) are also obtained as

$$\psi_i = \varphi_1(\lambda = \lambda_i) \,. \tag{1.51}$$

The efficiency of this method is clearly demonstrated by the following example:

$$U(r) = -\frac{2}{\exp(r) - 1} \,. \tag{1.52}$$

For this potential (the Hulthèn potential) the eigenvalues have been known analytically [21] and are at zero energy

$$\eta_n = \frac{1}{\lambda_n} = \frac{2}{n^2}. \tag{1.53}$$

Table 3.2 presents the eigenvalues calculated by solving eq. (1.47). $N$ denotes the number of iterations, i.e. the number of terms in eq. (1.45). The convergence is remarkably rapid. Two iterations are sufficient for making the largest eigenvalue correct up to five significant digits. This applies to a local interaction. It is very important, however, that the method of continued fractions works equally

Table 3.2

| $N$ | $\eta_1$ | $\eta_2$ | $\eta_3$ | $\eta_4$ | $\eta_5$ | $\eta_6$ | $\eta_7$ | $\eta_8$ |
|---|---|---|---|---|---|---|---|---|
| 1 | 1.929 | | | | | | | |
| 2 | 1.995 | 0.418 7 | | | | | | |
| 3 | 2.0 | 0.496 09 | 0.150 44 | | | | | |
| 4 | 2.0 | 0.499 94 | 0.213 17 | 0.066 6 | | | | |
| 5 | 2.0 | 0.5 | 0.221 71 | 0.111 91 | 0.035 58 | | | |
| 6 | 2.0 | 0.5 | 0.222 21 | 0.123 41 | 0.064 89 | 0.018 59 | | |
| 7 | 2.0 | 0.5 | 0.222 22 | 0.124 91 | 0.076 90 | 0.040 10 | 0.011 06 | |
| 8 | 2.0 | 0.5 | 0.222 22 | 0.125 | 0.079 65 | 0.056 01 | 0.026 01 | $6.98 \cdot 10^{-3}$ |
| Exact | 2.0 | 0.5 | 0.222 22 | 0.125 | 0.08 | 0.055 55 | 0.040 81 | 0.031 25 |

well also for non-local interactions [22]. Moreover, our requirement that $U$ should be a Hermitian operator may easily be omitted and the method may be generalized to the case of a non-Hermitian $U$.

In order to calculate the $t$-matrix instead of the $K$-matrix, it is sufficient to change the Green function $G_0$ appropriately.


## 3.2. Resonance in Three-Particle System

The theory of three-particle resonances has been developed much less than the conventional theory of three-particle scattering because of serious mathematical and computational complications accompanying the solution of this problem beginning with its formulation using a mathematically correct definition of the three-particle resonance state and ending with great difficulties in solving multi-dimensional integral equations on non-physical energy sheets. On the other hand, there are considerable difficulties with interpretation of experimental data including the three-particle resonances [9] because, without a complete theory for three-particle resonances, the experimental data themselves do not permit an unambiguous interpretation. The interpretation of experimental data which is often ambiguous, even in the case of two-particle resonances accompanied by high inelasticity (e.g. the problems of dibaryon resonances or of the nucleus-antinucleon near-threshold resonances) becomes still more complicated in the case of a greater number of particles and depends strongly on the models used, on the parametrization methods, etc. Theoretical investigation of three-particle resonances becomes difficult because of insufficient knowledge of the conditions under which they arise and exist and because of the strong sensitivity of their characteristics to slight variation in the two-particle interaction parameters [10, 11].

Nevertheless, the three-particle system remains, for the time being, the only model multiparticle system which allows a successive microscopic examination of resonance states having several two- and three-particle channels of decay. For such an examination to be feasible, it is necessary to solve the exact three-particle dynamical equations, e.g. the Faddeev equations (see Section 1.4). However, to examine the true three-particle resonances (that is, the resonances which can decay into three particles), it is necessary to solve the equations for positive total energy where the equation kernels comprise the moving (logarithmic) singularities which greatly complicate the solution of the corresponding equations. Mathematical methods for solving equations with such kernels have been developed comparatively recently, and are based on integration contour deformation (see the textbook [12] and Chapter 4 of the present book). With the help of this mathematical formalism and the three-particle Hilbert-Schmidt expansion, some interesting results have been obtained [13–15, 17, 23]. The results are tentative however and, for the time being, they can but elucidate little the most important features of the problem (the competition between the two- and three-body channels of decay, the formation of intermediate two-particle resonances etc.). It should be noted that the strict mathematical theory of three--particle resonances treated as the poles of analytically continued three-particle resolvent in the case of the so called dilatation-analytic potential [25] (see also Chap. 7 of the present book) was given in Refs. [26–28] by E. Balslev and other authors [29]. In these works, unfortunately, there was considered only the energy range below three-particle threshold. The important fact, among others, was proved in the papers cited: the resonance poles of S-matrix elements (i.e. the poles of the continued S-matrix) on the Riemann surface at some (not very constraining) conditions, coincide with the poles of three-particle resolvent continued on the non-physical sheet. In the book we do not intend to discuss in details these mathematical problems (see also Chap. 4) and refer the readers to the original works [26–29] and references therein.

### 3.2.1. Three-Particle Hilbert-Schmidt Expansion and a Classification of Three-Particle Resonances

Following Ref. [13], we shall introduce the Hilbert-Schmidt eigenvalues and eigenfunctions for a system of three identical spinles particles, and consider the amplitude $T(f, k; k_0, z_3)$ describing the scattering process $2 \to 3$, that is, the breakup into three free particles; $k_0$ is the relative momentum of a particle and of a bound state in the initial state; $k$ and $f$ are the Jacobi momenta of the final state; $z_3$ is the total three-particle energy in the centre-of-mass system. The Faddeev equation for this amplitude is known to be of the form [12]

$$T(f, k, k_0; z_3) = t_s(f, p_{20}; \bar{z}_2)\varphi_d(p_{10}) +$$

$$+ \int \frac{dk'}{(2\pi)^3} \frac{t_s(f, p, \bar{z}_2)\, T(p_1, k', k_0, z_3)}{\left[ z_3 - \dfrac{\hbar^2}{m}(k^2 + kk' + k'^2) \right]} \tag{2.1}$$

where

$$\bar{z}_2 = z_3 - \frac{3}{4}\frac{\hbar^2 k^2}{m}\,; \qquad p_{10} = k + k_0/2$$

$$p_{20} = k/2 + k_0\,; \qquad p_1 = k + k'/2\,; \qquad p_2 = k/2 + k'\,,$$

$\varphi_d$ is the wave function of the two-particle subsystem ("deuteron"); $m$ is the particle mass. The symmetrized two-particle $t$-matrix $t_s$ is determined by the formula

$$t_s(k', k; z_2) = t(k', k; z_2) + t(-k', k; z_2)\,.$$

Separating the angular variables in (2.1) and assuming that the two-particle potential acts only in the $S$-wave, we obtain

$$T_L(f, k, k_0; z_3) =$$

$$= T^{(0)} + \int_0^\infty \frac{k'\,dk'}{k\pi^2} \int_{|k-k'/2|}^{|k+k'/2|} p\,dp\, \frac{t_0(f, p_2(p), y)\, P_L(y)\, T_L(p, k', k_0; z_3)}{z_3 - \hbar^2/m\,(p^2 + \tfrac{3}{4}k'^2)} \tag{2.2}$$

where

$$y = \frac{p^2 - k^2 - k'^2/4}{kk'}\,; \qquad p_2(p) = \left( p^2 + \frac{3}{4}k'^2 - \frac{3}{4}k^2 \right)^{1/2}\,.$$

$L$ is the total angular momentum; $P_L(y)$ is the Legendre polynomial. The $S$-wave component of the two-particle $t$-matrix $t_0(k', k, z_2)$ is determined by the equation:

$$t_0(k', k; z_2) = V_0(k', k) + \frac{1}{2\pi^2} \int_0^\infty \frac{k''^2 V_0(k', k'')t_0(k', k''; z_2)\,dk''}{z_3 - \hbar^2 k''^2/m} \tag{2.3}$$

where $V_0(k, k')$ is the $S$-wave component of the two-particle potential $V(k, k') = \langle k| V |k'\rangle$ with the normalization condition for the plane waves $\langle k | k'\rangle = (2\pi)^3 \delta(k - k')$. The inhomogeneous term $T^{(0)}$ in (2.2) does not affect the examined quantities; therefore, we shall not define its form more exactly. For the sake of simplicity we shall limit ourselves to a separable two-particle interaction:

$$V_0(k', k) = -\lambda\, g(k')\, g(k) . \tag{2.4}$$

Now, the $t$ matrix is of the form

$$t_0(k', k; z_2) = g(k')\, \tau(\sqrt{z_2})\, g(k) \tag{2.5}$$

where

$$\tau(\sqrt{z_2}) = -\left[\frac{1}{\lambda} + \int\limits_0^\infty \frac{k^2\, dk}{2\pi^2}\, \frac{g^2(k)}{z_2 - \hbar^2 k^2/m}\right]^{-1}; \qquad \mathrm{Im}\, \sqrt{z_2} > 0 .$$

The condition $\mathrm{Im}\, \sqrt{z_2} > 0$ defines the physical sheet of the two-particle $t$-matrix.

After substituting (2.5) in (2.2) we obtain the following representation for $T_L$:

$$T_L(f, k, k_0; z_3) = g(f)\, \tau(\sqrt{\bar{z}_2})\, F_L(k, z_3) . \tag{2.6}$$

The function $F_L(k, z_3)$ satisfies the one-dimensional integral equation

$$F_L(k, z_3) = F_L^{(0)}(k, z_3) + \frac{1}{2\pi^2} \int\limits_0^\infty k'^2\, dk' \times$$

$$\times \left[ 2 \int\limits_{|k-k'/2|}^{k+k'/2} \frac{p\, dp}{kk'}\, \frac{g(p_2)\, P_L(y)\, g(p)\, \tau(\sqrt{\bar{z}_2})}{z_3 - \hbar^2/m(p^2 + \tfrac{3}{4} k'^2)} \right] F_L(k', z_3) . \tag{2.7}$$

If we define the three-particle resonances as poles of the three-particle amplitude $T_L(f, k', k_0; z_3)$ (continued in energy to non-physical sheets) in the complex plane of three-particle energy $z_3$ (see Chapter 4), then from (2.6) we immediately see that the amplitude $T_L$ may have singularities of several types, namely,

(i) the branch points of the two-particle amplitude $\tau(\sqrt{\bar{z}_2})$ located at energies $\bar{z}_2 = \varepsilon_1, \varepsilon_2, \dots$, where $\varepsilon_i$ are the energies of the bound and resonance states in a two-particle subsystem and denote the onsets of the corresponding two-particle cuts;

(ii) the three-particle amplitude $F_L(k, z_3)$ has poles in variable $k$ at an arbitrary energy $z_3$ for the values of $k$ satisfying the condition

$$k = (2/\hbar)\, \left[(m/3)\, (z_3 - \varepsilon_n)\right]^{1/2} .$$

This is a general property of the three-particle amplitude and is independent of the form of the two-particle interaction [12]. The three-particle amplitude contains all singularities of the two-particle amplitude;

(iii) the poles of the function $F_L(k, z_3)$ in total energy $z_3$; at these energy values the homogeneous equation corresponding to (2.7) has a non-trivial solution. These poles of $F_L$ do not depend on kinematic variables.

Since the equation (2.7) is a Fredholm-type integral equation, the poles of $F_L$ form a discrete set $(z_3^{(1)}, z_3^{(2)}, ...)$ in the $z_3$ plane. Some of the poles correspond to three-particle bound states and some are called three-particle resonances [13]. Thus, for finding the three-particle singularities of $T_L$ in $z_3$, it is necessary to solve the equation

$$\xi_n(z_3)\,|F_n(z_3)\rangle = K\,|F_n(z_3)\rangle \tag{2.8}$$

with the kernel $K$ corresponding to the integral operator on the right-hand side of (2.7). The location of the poles of $T_L$ is determined by the equation

$$\xi_n(z_3) = 1 , \tag{2.9}$$

where $\xi_n(z_3)$ is the three-particle Hilbert-Schmidt eigenvalue (see 1.5.46). However, equation (2.9) determines the poles of $T_L$ only on the physical sheet of three-particle energy $z_3$. Only the poles corresponding to bound states exist on this sheet, while the poles corresponding to the resonance are located on non-physical sheets. The analytic structure of the amplitude $F_n(k, z_3)$ is depicted in Fig. 3.3 (the structure of $\xi_n(z_3)$ is the same). There are two kinds of cuts in the $z_3$ plane, namely,

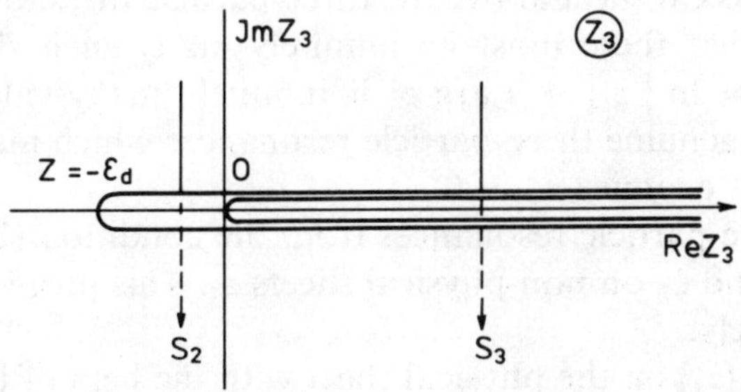


Fig. 3.3

The multisheet structure of the eigenvalue $\xi_n(z_3)$ in the complex $z_3$-plane.

(i) the two-particle root cuts (in the case of three identical particle with separable interaction there is only one such cut) which begin at negative values of $z_3 = \varepsilon_d$, ($\varepsilon_d < 0$ represents the two-particle binding energy) and reach

$z_3 = \infty$ . The cuts correspond to the root branch points in $z_3$ at the two particle thresholds (see 1.5.43):

$$\xi_n(z_3) \underset{z_3 \to +\varepsilon_d}{\sim} C_1(-z_3 + \varepsilon_d)^{1/2} + C_2 ; \tag{2.10}$$

(ii) the three-particle cut which begins at the point $z_3 = 0$, reaches $z_3 = \infty$, and corresponds to a logarithmic branch point at the three-particle threshold (see 1.5.44)

$$\xi_n(z_3) \underset{z_3 \to 0}{\approx} C z^{L+2} \ln (-z_3) + f(z_3) . \tag{2.11}$$

Here, the angular momentum $l$ of the two-particle subsystem is assumed to be zero; $L$ is the relative angular momentum of an incoming particle in the centre-of-mass system and $f(z_3)$ is assumed to be regular near $z_3 = 0$ .

Owing to the multisheet analytic structure of the eigenvalues $\xi_n(z_3)$ in the complex $z_3$ plane, as shown in Fig. 3.3, the analytic continuation of the physical sheet $z_3$ above and below the three-particle threshold leads to *various* non-physical sheets. Since the $z_3$ plane contains at least two cuts, the analytical continuation may be carried out in two ways:

1. The continuation to the non-physical sheet $S_2$ which is realized by continuing the physical sheet from the upper $z_3$ half-plane downwards between the points $z_3 = \varepsilon_d$ and $z_3 = 0$. This non-physical sheet corresponds to so-called "quasi two-particle" resonances which manifest themselves in the three-particle scattering at energies $\varepsilon_d < z_3 < 0$ . Such resonances are possible if the two-particle resonances decay in the field of the third particle or if the two-particle bound state resonates in the field of the third particle (for more details, see the discussion in Chap. 4).

2. The continuation to the non-physical sheets $S_3$ which is carried out by continuing the physical sheet above the three-particle threshold $z_3 = 0$. From (2.11) it follows that there must be infinitely many such sheets because the function $\mathrm{Ln}\,(z) = \ln |z| + i \arg z$ is infinitely many-valued. These sheets correspond to the genuine three-particle resonances which manifest themselves in the scattering at energies $z_3 > 0$ .

To find the three-particle resonances from the condition (2.9) it is necessary to know how to find $\xi_n$ on non-physical sheets $z_3$. This problem may be solved in two ways, namely:

(i) by finding $\xi_n(z_3)$ on the physical sheet with the help of (2.8), it is possible to calculate the complex function $\xi_n(z_3)$ on non-physical sheets [7,15] by means of numerical analytic continuation. In such an approach the main thing is to take into account correctly the analytic properties of the function to be continued. We shall discuss this problem in more detail in the second subsection;

(ii) the second approach is to continue the equation (2.8) proper to the non-physical sheets and to solve it there, thereby finding $\xi_n(z_3)$ directly on these sheets [14, 24]. In both cases the resonance energy may be found from the standard condition

$$\xi_n\!\left(z_3^{(res)}\right) = 1 \,. \tag{2.12}$$

If the resonance is located near the physical region, then on the physical sheet there exists a domain of real energies where the condition

$$\operatorname{Re} \xi_n \cong 1 \,, \qquad |\operatorname{Im} \xi_n| \ll 1 \tag{2.13}$$

is satisfied. Here, the resonance width is estimated according to the formula [17]

$$\Gamma = 2\big[\operatorname{Im} \xi_n(E)/(\mathrm{d}\operatorname{Re}\xi_n(E)/\mathrm{d}\,E)\big]_{E=E_R} \,. \tag{2.13a}$$

Usually, the conditions (2.13) constitute a criterion of the existence of resonance in numerical calculations [13, 17].

As an example of the direct realization of the approach described here one may use the calculations of low-lying three-particle resonance states of the $^6$Li nucleus in the three-body model ($\alpha + n + p$) made in [17]. Two-particle NN and N$\alpha$-interactions were naturally chosen in the separable form; the Coulomb interaction between an $\alpha$-particle and a proton was disregarded for the sake of simplicity. In contrast to the system of three nucleous, the examined system has at least five quasi-discrete states, thereby exhibiting the structure of a three-particle resonance (two lighter particles, a neutron and a proton in the field of a "heavy" $\alpha$-particle) and a three-particle bound state, the $^6$Li ground state [18].

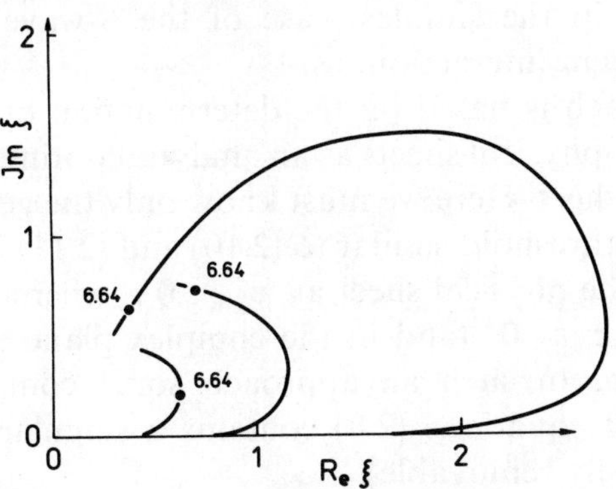


Fig. 3.4

Trajectories $\xi_n(z_3)$ of the $\alpha + p + n$ model of $^6$Li. The first three trajectories with $J^P T = 1^+0$ are shown with points corresponding to an energy 6.64 MeV (denoted by full circles).

From the five quasi-stationary states, three are above the three-particle threshold and can decay into three particles, $^6\mathrm{Li}^* \to \alpha + n + p$. We shall not discuss here the nuclear physics aspects of the calculations [17]. We shall only mention that the calculations reproduced three resonance states out of five.

The trajectories of the eigenvalues $\xi_n$ obtained in [17] are also of interest. Fig. 3.4 shows the trajectories of the first three eigenvalues $\xi_n$ for the channel $J^P T = 1^+ 0$ ($J$ is the total momentum of nucleus in the given state; $P$ is the parity of the state; $T$ is its isospin). The first eigenvalue intersects the unit circle at $\operatorname{Im} \xi_1 = 0$ and goes into the complex plane at $\operatorname{Re} \xi_2 \cong 2$ ; it corresponds to the $^6$Li bound (ground) state. The second eigenvalue demonstrates a typical resonance behaviour; it intersects the unit circle near the real axis $\xi$ at $\operatorname{Re} \xi_2 \cong 1$, $\operatorname{Im} \xi_2 \ll 1$ . The third eigenvalue does not intersect the unit circle and, therefore, does not correspond to the well-expressed resonance states of the system. The figure demonstrates another interesting property of the eigenvalues. The eigenvalues at an energy of 6.64 MeV are denoted by the full circles in the figure. It can be seen that, although at low energies the conditions $|\xi_1| > |\xi_2| > |\xi_3|$ are satisfied, at 6.64 MeV we get $|\xi_1| < |\xi_2|$ . Physically, this corresponds to the well known fact that at higher energies the scattering is already determined by the next resonance states.

### 3.2.2. *The Method of Analytic Continuation in Energy for the Three-Particle Case*

As noted above, for finding the three-particle resonances it is necessary to calculate the Hilbert-Schmidt eigenvalues on non-physical sheets $z_3$. This may be done in two ways. The contour deformation method developed in [14] makes it possible to analytically continue the equation (2.8) to the non-physical sheets $S_2$ and $S_3$. However, the contour deformation technique is rather complicated and requires detailed knowledge of the equation kernel singularities. Therefore, it was realized only in the simplest case of the $S$-wave one-term separable two-particle Yamaguchi interaction.

The second approach is based on the determination of the complex valued function $\xi_n(z)$ on non-physical sheets as an analytic continuation of the function $\xi_n(z)$ on the physical sheet. Here we must know only the general singularities of the solution near the threshold similar to (2.10) and (2.11). In ref. [13], equation (2.8) was solved on the physical sheet at $z_3 < 0$ , whereupon the eigenvalues $\xi_n(z)$ were found at $z > 0$ (and in the complex plane $z$) using a numerical analytic continuation. In such an approach some complications arise also because the kernel of equations (2.8) contains a singularity at $\xi_d < z < 0$ (though relatively easily removable).

Finally, another approach to this problem is possible. This is the same as that explained in item 2 of Section 3.1 for the two-particle Hilbert-Schmidt eigenvalues and is notable for the great simplicity of calculations: the equation (2.8) is only solved *below* the lowest threshold $(z_3 < \varepsilon_d)$ , that is, in the domain of three-particle bound states where sufficiently reliable and convenient methods exist for solving such equations with both non-local and local two-particle interactions. A decisive role is played here by making an accurate allowance for the analytic properties of the eigenvalue $\xi_n(z)$ as a function of $z_3$. As was shown above, these properties are defined by the behaviour of the eigenvalues at the two-particle $(z_3 = \varepsilon_d)$ and three-particle thresholds (2.10) and (2.11), respectively. In accordance with the Cauchy theorem, the function $\xi_n(z)$ may be written as a superposition of the contributions of two singularities:

$$\xi_n(z_3) = f_1(x) + f_2(z_3) \ln(-z_3) z_3^{L+2} \qquad (2.14)$$

where $x = \pm(-z_3 + \varepsilon_d)^{1/2}$, $f_1$ and $f_2$ are the analytic functions of their arguments. For $f_1$ and $f_2$ we may write a representation in the form of the Padé approximants:

$$\xi_n(z_3) \approx \frac{P_N(x)}{Q_M(x)} + z_3^{L+2} \ln(-z_3) \frac{\tilde{P}_{N_1}(z_3)}{\tilde{Q}_{M_1}(z_3)}. \qquad (2.14a)$$

After reducing this expression to the common denominator, we obtain

$$\xi_n(z_3) \approx \xi_n^{[N, M, I]}(z_3) = \frac{\hat{P}_N(x) + z_3^{L+2} \ln(-z_3) \hat{R}_I(x)}{\hat{Q}_M(x)} \qquad (2.15)$$

where $\hat{P}_N$, $\hat{R}_I$ and $\hat{Q}_M$ are respectively the polynomials of degrees $N$, $I$, and $M$ (differing, generally speaking, from $P_N$ and $Q_M)^2$). To find the coefficients of these polynomials (the procedure is the same as that described above for the two-particle case), it is sufficient to calculate $N + M + I + 2$ values of $\xi_n(z_3)$ at $z_3 < \varepsilon_d$, i.e. in the region of three-particle bound states. Furthermore, with the help of analytic expression (2.15), we may continue $\xi_n(z_3)$ to either of the non-physical sheets $S_2$ and $S_3$ because the expression (2.15) makes a sufficiently complete allowance for the analytic structure of $\xi_n(z)$ at both two- and three--particle thresholds and contains the appropriate cuts of $\xi_n(z)$. Also, $\xi_n(z)$ given by (2.15) is real at $z < \varepsilon_d$; therefore, it satisfies the Schwarz principle (see 1.5.42), which is very important for the analytic properties of $\xi_n(z_3)$ to be described correctly. Thus, the expression (2.15) may be used to find $\xi_n(z_3)$ on any sheet of the complex $z_3$ plane (the two-particle sheet is given by the sign of $x$, and the three-particle sheet by the imaginary part of $\ln(-z_3)$). For the resonance energies

$$\xi_{n\,(z_3)}^{[N, M, I]} = 1 \,,$$

i.e. the determination of the three-particle resonance pole location reduces to solving the transcendental equation

$$\hat{P}_N(x) + z_3^{L+2} \ln(-z_3)\, \hat{R}_I(x) - \hat{Q}_M(x) = 0$$

or

$$B_{N'}(x) + z_3^{L+2} \ln(-z_3) \hat{R}_I(x) = 0$$

---

[2]) When going over from (2.14a) to (2.15), we must remember that the terms of even degrees of $x$ yield a polynomial in variable $z$.

where

$$B_{N'} = \hat{P}_N - \hat{Q}_M \quad \text{and} \quad N' = \max\{M, N + 1\}. \tag{2.16}$$

The approach described above was tested using a simple system of three spinless particles with the $S$-wave separable two-particle interaction [13]. The values of $\xi_n$ and $z_3$ at $z_3 < \varepsilon_d$ were taken from Ref. [13], whereupon we constructed the expression (2.15) and, after that, we found $\operatorname{Re}\xi_n(z)$ and $\operatorname{Im}\xi_n(z)$ at $-20\,\text{MeV} < z < 15\,\text{MeV}$. The result obtained was compared with the result of [13] where $\xi_n(z)$ was calculated by means of (2.8) at energies $z < 0$ (but at $z > \varepsilon_d$) and, then, analytically continued numerically to the region $z > 0$.

Fig. 3.5 presents the results of the calculations using formula (2.15) as compared with the results of [13]. We can see good agreement between the two calculations for quite low orders of the Padé approximants. It can also be seen that the contribution of the three-particle cut is very small. It is clear that the calculations by formula (2.15) are much simpler than those made by the direct methods [13–15]. Also, in this approach the contributions of the two-and three-particle channels may conveniently be separated, thereby making it possible to estimate the relative and absolute values of the contributions.

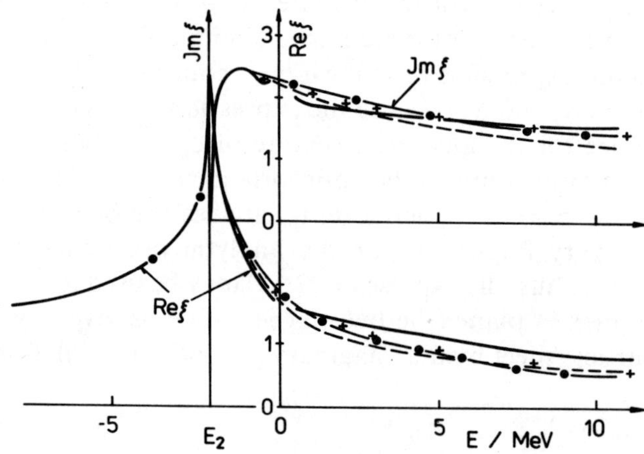



Fig. 3.5

Comparision of the calculation of $\xi_1(z)$ based on (2.15) with the results of [13] (solid line). The dashed line shows the $\eta_1$ of the form $\eta_1 = (a + bx)/(1 + cx)$. The dash-dot line shows the $\eta_1$ of the form $\eta_1 = (a + bx + cx^2)/(1 + dx + ex^2)$ and the crosses present the $\eta_1$ of the form $\eta_1 = (a + bx + cE^2 \ln(-E)[(1 + dx + ex^2)$.

The analytic continuation method described above may also be faced with some of difficulties inherent in all the numerical analytic continuation procedures [16]. We shall discuss them briefly. Since the form (2.15) chosen by us is an "ansatz" or recipe, for analytic continuation and since it allows only for the nearest analytic singularities of the amplitude which are basically of kinematic origin, the numerous other (dynamic) singularities more remote from the physical region may be allowed for only under two conditions, namely,

(a) a sufficiently high power of the approximating polynomials in (2.15);

(b) and, what is most important, a high accuracy corresponding to this power and the necessary volume of the initial data. In other words, the poorer is our knowledge of the exact properties of the function to be continued, which form the basis of an analytic continuation, the higher must be the amount and accuracy of the initial data used to make the continuation. Since the two requirements are in practice always difficult to satisfy and the actual calculations always yield a limited number of initial points within a very moderate accuracy, it is evident that in this case we can obtain only the locations of the nearest (to the physical region and to the region of the thresholds allowed for) singularities with sufficient reliability.

Fortunately, it is these singularities that are as a rule most interesting in practice because the distant (from the physical region or from the thresholds) resonances are very wide and so similar to a smooth continuum that their separation from the latter is unlikely to make sense. Nevertheless, in more "academic" examination and in analysing the convergence of particular numerical calculations of the three-particle resonances the above mentioned arguments should always be borne in mind.

## References


1. KRASNOPOĽSKY V. M., KUKULIN V. I., Phys. Lett. *69A* (1978) 251.
2. ZEL'DOVICH Ya. B., ZhETP (in Russian) *39* (1960) 776; [Sov. Phys. JETP *12* (1961) 542].
3. MANAJENKOV A., Theor. Math. Phys. (in Russian) *12* (1972) 397.
4. BADALYAN A. M., SIMONOV Yu. A., The Variational Method for the 3-Body Resonances:
   I. The Asymptotic Behaviour of the Resonance Wave Function, Preprint ITEP No. 966, Moscow 1972;
   II. The Equation for the Resonance Wave Function, Preprint ITEP No. 952, Moscow 1972.
5. SASAKAWA T., Nucl. Phys. *A160* (1971) 321.
   WEINBERG S., Phys. Rev. *131* (1963) 440.
6. GLÖCKLE W., HÜFNER J., WEIDENMÜLLER H. A., Nucl. Phys. *A90* (1967) 481.
7. KRASNOPOĽSKY V. M., KUKULIN V. I., Phys. Lett. *83A* (1981) 98.
8. NEWTON R. G., Scattering Theory of Waves and Particles, 2$^{nd}$ ed., Springer, Berlin 1982.
9. ALEKSANDROV D. V., GANZA E. A., GLUNKOV Yu. A., NOVATSKY B. G., OGLOBLIN A. A., STEPANOV D. I., Yad. Fiz. (in Russian) *39* (1984) 513 [Sov. J. Nucl. Phys. *39* (1984) 323];
   BRAYSHAW D. D., Phys. Rev. Lett. *39* (1977) 371.
10. BASDEVANT J. L., KREPS R., Phys. Rev. *141* (1965) 1398, 1404.
11. PASQUIER J. Y., PASQUIER R., Proc. I-st International Conference on the Three-Body Problem, Birmingham 1962, p. 461.
12. SCHMID E., ZIEGELMAN H., The Quantum Machanical Three-Body Problem, Pergamon Press and Vieweg, Braunschweig 1974.
13. BELAYEV V. B., MÖLLER K., Preprint of JINR E4-9601, Dubna 1976; Zeitschr. Phys. *279* (1976) 47.

14. MÖLLER K., Preprint ZFK-327, Dresden 1977; Preprint ZFK-351, Dresden 1978; Preprint ZFK-357, Dresden 1978.
15. OFFERMANN R., GLÖCKLE W., Nucl. Phys. *A318* (1979) 138.
16. de ALFARO V., REGGE T., Potential Scattering, North Holland, Amsterdam 1965.
17. MATSUI Y., Phys. Rev. *22C* (1980) 2591.
18. AJZENBERG-SELOVE F., Energy Levels of Light Nuclei A = 5 − 10, Nucl. Phys. *A320* (1979)1.
19. HORÁČEK J., SASAKAWA T., Phys. Rev. *A28* (1983) 2151.
20. HORÁČEK J., SASAKAWA T., Phys. Rev. *A30* (1984) 2274.
21. SITENKO A. G., Lectures in the Scattering Theory (in Russian), Visca Skola, Kiev 1971. [English translation: Pergamon Press, Oxford, 1971].
22. HORÁČEK J., SASAKAWA T., Phys. Rev. *C32* (1985) 70.
23. BELYAEV V. B., MÖLLER K., SIMONOV Yu. A., J. Phys. G: Nucl. Phys. *5* (1979) 1057.
24. ORLOV Yu. V., Pisma v ZhETF *33* (1981) 380 (in Russian). [JETP Lett. 33 (1981) 363].
25. BABBIT D., BALSLEV E., J. Funct. Anal. *18* (1975) 1
    BALSLEV E., COMBES J. M., Comm. Math. Phys. *22* (1971) 280.
26. BALSLEV E., Ann. Inst. Henri Poincare XXXII (1980) 125.
27. BALSLEV E., Comm. Math, Phys. *77* (1980) 173.
28. BALSLEV E., Resonances in Three-Body Scattering Theory, Preprint Series 1981/1982 No. 3., Aarhus.
29. HAGEDORN G. A., Comm. Math. Phys. *65* (1979) 181.

# Chapter 4

# Projection Methods

## 4.1. General Projection Formalism

Projection formalism is a convenient means of studying resonance phenomena arising in scattering and in reactions. It was used for the first time by Feshbach [1] to describe multichannel scattering resonances arising from coupling to closed channels. For this purpose, two mutually orthogonal projectors onto subspaces of open and closed channels are introduced and an energy-dependent effective Hamiltonian is defined in the subspace of the open channels and complex eigenvalues of this Hamiltonian (on the second sheet) determine the location of the resonances. The Feshbach formalism was generalized by Fonda and Newton [2] (see also Chapter 16 of Ref. [3]). After minor modifications projection formalism can also be used to describe potential one-particle resonances [4–7] and, which is of particular importance, many-particle resonances [7].

The problem of calculating resonance energy and width is reduced in the conventional approach to a determination of the Hamiltonian resolvent pole position on nonphysical energy sheets. The projection formalism discussed below in this chapter makes it possible in many cases to simplify the determination of the Hamiltonian spectrum and, in particular, facilitates the search for the $S$-matrix poles (corresponding to the bound and virtual states and to resonances).

We divide the full space $\mathscr{H}$ of our system into two mutually orthogonal subspaces $\mathscr{H} = \mathscr{H}_P + \mathscr{H}_Q$ and introduce two projectors $P$ and $Q$ onto these subspaces:

$$P + Q = 1, \qquad PQ = QP = 0, \qquad P^2 = P, \qquad Q^2 = Q . \quad (1.1)$$

Let the system be described by the Hamiltonian $H$. Our task is to find the complete resolvent (or Green function)

$$G(z) = (z - H)^{-1} . \qquad (1.2)$$

In each of the subspaces $\mathscr{H}_P$ and $\mathscr{H}_Q$ the "projections" of the operator $H$ are defined as

$$H_{PP} = PHP, \qquad H_{QQ} = QHQ$$

and the corresponding resolvents in the subspaces $\mathcal{H}_P$ and $\mathcal{H}_Q$ are

$$G_P = P(z - PHP)^{-1}P ,\qquad\qquad (1.3a)$$

$$G_Q = Q(z - QHQ)^{-1}Q .\qquad\qquad (1.3b)$$

The complete resolvent $G$ may always be expressed formally in terms of $G_P$ or $G_Q$, but in order that such a representation is usable, it is necessary to select $P$ and $Q$ in a special way. We select $\mathcal{H}_P$ to be finite-dimensional, i.e. we construct the projector $P$ from a finite number of square integrable functions[1]):

$$P = \sum_{i=1}^{M} |\varphi_i\rangle \langle\varphi_i| .\qquad\qquad (1.4)$$

It can easily be seen that $P$ is actually a projector $(P^2 = P)$ if the functions $\varphi_i$ are orthogonal and normalized

$$\langle\varphi_i \,|\, \varphi_j\rangle = \delta_{ij} .$$

The resolvent $G_Q$ in $\mathcal{H}_Q$ is expressed formally in terms of the complete resolvent $G$ by the relation (a simple derivation of this formula is given in Section 4.2 below, see expression (2.9))

$$G_Q = G - GP(PGP)^{-1}PG .\qquad\qquad (1.5)$$

The same expression holds also for $G_P$ on replacing $P$ by $Q$, but is not necessary for our purposes. From (1.5) it follows that

$$PG_Q = G_QP = 0 \qquad\qquad (1.6a)$$

hence

$$G_Q = QG_Q = G_QQ .\qquad\qquad (1.6b)$$

Considering these relations, it can readily be verified that $G_Q$ defined by (1.5) is actually the resolvent of the operator $QHQ$ in the subspace $\mathcal{H}_Q$:

$$G_Q(z - QHQ) = G_Q(z - H)Q = (1 - GP(PGP)^{-1}P)Q = Q .$$
$$\qquad\qquad (1.7)$$

It is important to stress that, because in our case $P$ is a finite-dimensional operator, the relation (1.5) yields not only formal but also a constructive

---

[1]) For historical reasons the meaning of the projection operators $P$ and $Q$ introduced by us is opposite to that used in the overwhelming majority of works where projection formalism is used. We hope that these symbols will not make it difficult for the reader accustomed to the standard notation to comprehend the text.

definition of $G_Q$. Let us now reverse the relation (1.5) and express $G$ in terms of $G_Q$. It is clear that, to do this, we need only algebraic operations:

$$G = G_Q + GP(PGP)^{-1}PG .\tag{1.8}$$

Multiplying (1.8) by $G^{-1} = (z - H)$ from the left, from the right, and then from either side we obtain

$$(1 - G^{-1}G_Q) = P(PGP)^{-1}PG ,\tag{1.9a}$$

$$(1 - G_Q G^{-1}) = GP(PGP)^{-1}P\tag{1.9b}$$

and

$$P(G^{-1} - G^{-1}G_Q G^{-1})P = P(PGP)^{-1}P\tag{1.9c}$$

respectively. Substituting relations (1.9a) and (1.9b) into (1.8) and inverting (1.9c) we obtain

$$G = G_Q + (1 - G_Q G^{-1}) \cdot$$
$$\cdot P[P(G^{-1} - G^{-1}G_Q G^{-1})P]^{-1}P(1 - G^{-1}G_Q) .\tag{1.10}$$

Replacing $G^{-1}$ by $(z - H)$ and allowing for (1.6) we arrive at the final expression [7, 8][2])

$$G = G_Q + (1 + G_Q H)D(1 + HG_Q)\tag{1.11}$$

where

$$D = P[P(z - H - HG_Q H)P]^{-1}P .\tag{1.12}$$

Equation (1.11) is a very convenient representation of the complete resolvent. For example, if we make the projector $P$ sufficiently wide, the iterative series for the resolvent $G_Q$ will prove (see Section 4.2) to converge rapidly in the subspace $\mathcal{H}_Q$. Therefore, the expression (1.11) may be used to advantage for approximate calculations of the amplitudes of various processes in many-particle systems. Here, instead of solving the complicated many-particle integral equations, we must only calculate the well-defined matrix elements of the subresolvents of subsystems and invert the complex matrices [9].

In the given case, our task is to find the procedure for deriving the poles of the complete resolvent $G(z)$ from (1.11). If the functions which can approximately describe the resonance states are included in $\mathcal{H}_P$, the Hamiltonian part $QHQ$ will not lead to resonances at energies close to the true resonance energies. This means that the resolvent $G_Q$ does not contain the sought poles. Therefore, as

---

[2]) The same relations were obtained by Newton, and Fonda [2, 3] in terms of the so-called formal resonance theory.

(1.11) is an identity, the sought poles are contained in the operator $D$. This is also evident from the relation

$$D = PGP \tag{1.13}$$

which can be obtained by inverting (1.9c); here, the operator $D$ is given by (1.12). From (1.13) it follows that $D(z)$ contains all the resolvent poles, except those whose residues are orthogonal (in the sense defined below) to all the functions $\varphi_i$ forming the subspace $\mathscr{H}_P$. Since $D$ is a finite-dimensional operator, its poles are defined by the complex zeros in the determinant of the matrix of order $M \times M$:

$$\det \| \langle \varphi_i | z - H - H G_Q(z) H | \varphi_j \rangle \| = 0 . \tag{1.14}$$

The complex roots of (1.14) must be sought on nonphysical sheets of energy surface $z$, i.e. the resolvent $G_Q$ must be continued to nonphysical sheets. It should be noted that by virtue of the relation (1.13) the resonance poles found by solving (1.14) do not depend on the choice of the functions of the initial approximation and coincide with the true poles of the complete resolvent $G$.

The operator $HG_Q(z)H$ arising from the coupling of the initial states $\varphi_i$ with the orthogonal subspace $\mathscr{H}_Q$, i.e. from the coupling of subspaces $H_P$ and $H_Q$, is caled the energy shift operator because in perturbation theory based on such a projection technique [6–8] this operator defines the energy shift of the unperturbed state:

$$\varDelta(z) = PHG_Q(z)HP . \tag{1.14a}$$

The study of the analytic properties of this operator is very useful with a view to understanding the numerous problems of resonance theory, in particular to determining the resonance trajectories for different types of Hamiltonians. Some typical examples of such an analysis are given in the next chapter.

Transcendental equation (1.14) is convenient to use in calculating the location of resonance poles. The simplest way of doing it is to take the one-dimensional projector $P = |\varphi\rangle \langle\varphi|$ where the function $\varphi$ is so selected that it would approximately describe the sought resonance. Then, (1.14) reduces to the following transcendental equation:

$$z_R = \langle\varphi| H |\varphi\rangle + \langle\varphi| H G_Q^{II}(z_R) H |\varphi\rangle = z_0 + \varDelta(z_R) \tag{1.15}$$

where $G_Q^{II}(z)$ denotes the resolvent continuation to the nonphysical sheet and $z_0 = \langle\varphi| H |\varphi\rangle$ .

Thus, the problem of finding the resonance poles reduces to the following problems:

    (1) to obtain equations for $G_Q(z)$;

    (2) to continue them to nonphysical sheets, i.e. to find an equation for the resolvent $G_Q^{II}(z)$;

(3) to solve the continued equations, i.e. to find the resolvent $G_Q^{II}(z)$ itself;
(4) to solve the transcendental equation (1.14) or (1.15).

In the next section we shall discuss the general orthogonal projecting method based on the concept of penalty functions which makes it readily possible to derive the equations describing $G_Q^{II}(z)$ in the systems of an arbitrary number of particles and to simplify many formal operations substantially. After that we shall treat the general concept of the continuation of the resolvent and equations of the scattering theory. This concept is based on the construction of rigged Hilbert space (Section 4.3). We shall describe also the application of the general projecting scheme to the calculations of two-particle (Section 4.4) and three--particle resonances (Section 4.5).

We may note here that the interpretation of the resonance state as a state arising from some initial discrete state after an appropriate perturbation[3]) is introduced goes back to the classical work by Wigner and Weisskopf (see for example, [10]) in the framework of the quantum-mechanic approach, and to Titchmarsh and his predecessors in classical mathematical physics (see a good review of this material from the viewpoint of classical mathematical physics in the second volume of the well-known monograph by Titchmarsh [11]). The general formalism for describing the prepared-state decay similar to that treated in this section and the application of the formalism to examining particular examples of the decay may be found in Chapter 8 of the monograph by Goldberger and Watson [12]. Our approach in this chapter differs from that adopted by Goldberger and Watson in that it is orientated to the decay of many-particle, rather than one-particle, resonances, gives a substantially more detailed description of the procedure of analytic continuation of the resolvent, and pays much attention to the computational aspects of the problem.

## 4.2. Orthogonal Projecting Method

The orthogonal projecting method (or the method of orthogonalizing pseudo-potentials) was proposed in Refs. [13, 14] in connection with the problem of finding the solution for the Schrödinger equation in the scattering problem

$$(H - E)\psi = 0 \tag{2.1}$$

with an additional condition of orthogonality imposed on the solution for (2.1)

$$\langle \varphi \mid \psi \rangle = 0 . \tag{2.2}$$

The problem (2.1), (2.2) must be interpreted as seeking to solve the equation (2.1) in the subspace orthogonal to $\varphi$, rather than having to select the solutions

---

[3]) The role of the perturbation can be played also by a change of boundary conditions.

satisfying the condition (2.2) among all the possible solutions for (2.1), i.e. to solve the projected (onto a subspace $\mathcal{H}_Q$) Schrödinger equation

$$QHQ\psi = EQ\psi \tag{2.1a}$$

where $Q = 1 - P$ and $P = |\varphi\rangle\langle\varphi|$ is the one-dimensional projection operator $(\langle\varphi|\varphi\rangle = 1)$. For this purpose it is convenient to go over to the pseudo-Hamiltonian [13, 14]

$$\tilde{H}(\lambda) = H + \lambda P \tag{2.3}$$

and the real constant $\lambda$ should be made to approach infinity in the solution to be obtained. Indeed, it can be verified immediately that the solution of the equation

$$(\tilde{H}(\lambda) - E)\,\tilde{\psi}_E(\lambda) = 0$$

is of the form

$$|\tilde{\psi}_E(\lambda)\rangle = |\psi_E\rangle + \lambda\,\frac{G(E)\,|\varphi\rangle\,\langle\varphi\,|\,\psi_E\rangle}{1 - \lambda\,\langle\varphi|\,G(E)\,|\varphi\rangle} \tag{2.4}$$

where $\psi_E$ is the solution for (2.1); $G(E) = (E - H)^{-1}$ is the resolvent of the initial Hamiltonian. It is evident that the limit of (2.4) at $\lambda \to \infty$ is well defined:

$$|\tilde{\psi}_E\rangle = \lim_{\lambda\to\infty} |\tilde{\psi}_E(\lambda)\rangle = |\psi_E\rangle - \frac{G(E)\,|\varphi\rangle\,\langle\varphi\,|\,\psi_E\rangle}{\langle\varphi|\,G(E)\,|\varphi\rangle} \tag{2.5}$$

so that $\lim_{\lambda\to\infty}\langle\varphi|\,\tilde{\psi}_E(\lambda)\rangle = 0$ which is just required by the condition of the problem. However, the modified function $\tilde{\psi}_E$ is not solution for (2.1), but satisfies the equation:

$$(QHQ - E)\,\tilde{\psi}_E = 0 \tag{2.6}$$

where $Q = 1 - P$.

Let us write a more general form of (2.4):

$$\tilde{\psi}_E(\lambda) = \psi_E + \lambda\,G(E)\,P[P(1 - \lambda\,G(E))P]^{-1}P\psi_E \tag{2.7}$$

which is formally correct for any projector $P$. In case of the $M$-dimensional projector $P = \sum_{i=1}^{M}|\varphi_i\rangle\langle\varphi_i|$ the operator $[P(1 - \lambda\,G(E))P]^{-1}$ is defined only in the $M$-dimensional subspace $\mathcal{H}_P = P\mathcal{H}$ and equals

$$P[P(1 - \lambda\,G(E))P]^{-1}P = \sum_{ij}^{M}|\varphi_i\rangle\,(1 - \lambda\,\hat{G})_{ij}^{-1}\,\langle\varphi_j|$$

where $\hat{G}_{ij} = \langle\varphi_i|\,G(E)\,|\varphi_j\rangle$. In the limit $\lambda \to \infty$ we find from (2.7):

$$\tilde{\psi}_E = \psi_E - GP(PGP)^{-1}P\psi_E. \tag{2.7a}$$

Now we shall examine the resolvent of the pseudo-Hamiltonian $\tilde{H}$, i.e. $\tilde{G}(\lambda, z) = (z - \tilde{H}(\lambda))^{-1}$ and express it in terms of the resolvent of the initial Hamiltonian $H$. Using the convential operator identity

$$\tilde{G}(\lambda) = G + G\lambda P\,\tilde{G}(\lambda)$$

we obtain

$$\tilde{G}(\lambda) = G + GP\lambda[P(1 - \lambda G)P]^{-1}PG\,. \tag{2.8}$$

The determinant of the matrix operator $P(1 - \lambda G)P$ contained in the denominator of (2.8) is called the Weinstein-Aronszain determinant. In the limit $\lambda \to \infty$, $\tilde{G}(\lambda)$ turns out to be $G_Q$ (see (1.5)):

$$\lim_{\lambda \to \infty} \tilde{G}(\lambda) = G - GP[PGP]^{-1}PG = G_Q = Q[z - QHQ]^{-1}Q\,.$$

$$\tag{2.9}$$

Thus, the infinite-constant procedure is merely a convenient technique for inverting the operator in the subspace:

$$\lim_{\lambda \to \infty} (A + \lambda P)^{-1} = Q(QAQ)^{-1}Q\,, \tag{2.10}$$

where $P$ and $Q$ are any two mutually orthogonal projectors: $PQ = QP = 0$, $P + Q = 1$ and $A$ is an arbitrary operator (of course, if the left-hand and right-hand parts of the equation (2.10) exist). The term "orthogonal projecting method" should be understood as follows: we use a constructively prescribed projector P to the subspace $\mathscr{H}_P$ to construct the resolvent, the wave functions, etc. in the subspace $\mathscr{H}_Q$ which is an orthogonal complement to $\mathscr{H}_P$. Therefore, $G_Q$ is often called the orthogonalized (with respect to $\mathscr{H}_P$) or projected (to $\mathscr{H}_Q$) resolvent.

The basic advantage of the orthogonal projecting method is that we may make all the intermediate transformations at a *finite* value of $\lambda$ and, at the same time, remain in the full space $\mathscr{H}$. In particular, for orthogonalized resolvents $G_Q$ we can easily obtain the modified equations of scattering theory; namely, the Lippmann-Schwinger, Faddeev-Yakubovsky, etc. equations, by treating the term $\lambda P$ as a conventional operator of potential type. A more detailed mathematical basis for the $\lambda \to \infty$ transition may be found in [15, 17]. Furthermore, it may be shown [15] that the identity (2.10) is correct for any "good" operator $P$ (not only for the projector) with the range of values $\mathscr{H}_P$. In particular, if $\mathscr{H}_P$ is a union of such subspaces $\mathscr{H}_{P_i} = P_i\mathscr{H}$ that their projectors $P_i$ are not mutually orthogonal, i.e. $P_iP_j \neq 0$, and the sum $\sum_i P_i$ is not a projector, then addition of the term $\lambda \sum_i P_i$ gives rise in the limit $\lambda \to \infty$ to the orthogonalization to the whole subspace $\mathscr{H}_P$. We are faced with non-trivial problems of this kind (each $\mathscr{H}_{P_i}$ is infinite-dimensional) in describing the systems of several composite particles [14]. Thus, one of the advantages of this approach over other projection methods is that we need not construct a complete projector onto $\mathcal{H}_P$ out of non-commuting operators $P_i$.

Let us obtain the modified Lippmann-Schwinger equation for $G_Q$. If the Hamiltonian is of the form

$$H = H_0 + V$$

then the following identity (the second resolvent identity) for the resolvent $G(z) = (z - H)^{-1}$ holds:

$$G(z) = G_0(z) + G_0(z)VG(z) \tag{2.11}$$

which may be regarded as an equation for $G(z)$ if $G_0(z) = (Z - H_0)^{-1}$ is known. In the scattering theory it is called the Lippmann-Schwinger equation and its iterations give the Born series

$$G = G_0 + G_0VG_0 + G_0VG_0VG_0 + \ldots \tag{2.12}$$

Adding the term $\lambda P$ to the Hamiltonian and operating with finite values of $\lambda$, we obtain:

$$\tilde{G}(\lambda) = (z - H_0 - \lambda P - V)^{-1} = (z - H_0 - \lambda P)^{-1} +$$

$$+ (z - H_0 - \lambda P)^{-1}V(z - H_0 - \lambda P - V)^{-1}$$

i.e.

$$\tilde{G}(\lambda) = \tilde{G}_0(\lambda) + \tilde{G}_0(\lambda)V\tilde{G}(\lambda) . \tag{2.13}$$

Now we shall go over to the limit $\lambda \to \infty$ in (2.13):

$$\tilde{G} = \tilde{G}_0 + \tilde{G}_0V\tilde{G} . \tag{2.14}$$

This is the modified "orthogonalized" or "projected" Lippmann-Schwinger equation for $G_Q \equiv \tilde{G} = \lim_{\lambda \to \infty} \tilde{G}(\lambda)$ . Furthermore, we shall use the symbol $\tilde{G}$ instead of $G_Q$ for all the orthogonalized resolvents. The absence of the parameter $\lambda$ will usually mean the limit $\lambda \to \infty$. In (2.14), $\tilde{G}_0$ is the so-called free orthogonalized resolvent related to $G_0$ by (2.9):

$$\tilde{G}_0(z) = \lim_{\lambda \to \infty} (z - H_0 - \lambda P)^{-1} = G_0 - G_0P(PG_0P)^{-1}PG_0 . \tag{2.15}$$

The orthogonalized scattering wave function (2.7a) also satisfies the modified Lippmann-Schwinger equation with the kernel $\tilde{G}_0V$:

$$\tilde{\psi}_E = \tilde{\psi}_{0E} + \tilde{G}_0(E) V\tilde{\psi}_E . \tag{2.14a}$$

This equation can easily be obtained from (2.14) if we use the relation of the wave function to the resolvent on the mass shell:

$$\tilde{\psi}_E = \lim_{\varepsilon \to 0} i\varepsilon \tilde{G}(E + i\varepsilon)\, \psi_{0E}$$

where $\psi_{0E}$ is the plane wave. The inhomogeneous term in (2.14a) is the so-called "orthogonality scattering" wave function related to the plane wave by the expression analogous to (2.7a):

$$\tilde{\psi}_{0E} = \psi_{0E} - G_0(E)\, P(PG_0(E)P)^{-1}P\psi_{0E}. \tag{2.14b}$$

Below, in order to unify the notation of numerous formulae, we shall often denote by $|\tilde{p}\rangle$ the orthogonality scattering wave function $\tilde{\psi}_{0E}$.

Thus, the rule for the transition from the conventional Lippmann-Schwinger equation (2.11) to the orthogonalized equation (2.14) is as follows. We replace all the resolvents by the orthogonalized ones according to the relation (2.15). This simple general rule is also correct for the Fadeev and Faddeev-Yakubovsky equations and for other versions of the many-particle equations of scattering theory. However, the equations should initially be reduced to such a form that they contain only the potentials and resolvents of subsystems, whereupon all the resolvents are to be replaced by the orthogonalized ones.

Thus, the first problem formulated at the end of the previous section is solved simply. But how shall we find the resolvent $G_Q = \tilde{G}$? In other words, how shall we solve the resultant orthogonalized equations? Let us compare the equations (2.11) and (2.14). Their kernels differ from one another by an operator of finite rank if $P$ is finite-dimensional. Therefore, the equation for $\tilde{G}$ is no more difficult to solve than the initial equation for $G$. Moreover, the norm of the orthogonalized equation kernel $\tilde{G}_0 V$ turns out to be smaller than the norm of the kernel $G_0 V$ of the conventional Lippmann-Schwinger equation. Therefore, the orthogonalized Born series

$$\tilde{G} = \tilde{G}_0 + \tilde{G}_0 V \tilde{G}_0 + \tilde{G}_0 V \tilde{G}_0 V \tilde{G}_0 + \dots \tag{2.16}$$

converges better compared with the conventional series. The convergence of such series is studied in [8, 16] (see also the detailed discussion in [26]). If the projector $P$ contains the functions of all bound states and the functions which approximately describe the resonances, then the Born series (2.16) turns out to be convergent not only at high (as usual), but also at low energies. As the dimension of the projector $P$ increases, the norm of the kernel $\tilde{G}_0 V$ decreases and the convergence of the series (2.16) improves, so that at a sufficiently high dimension the first term $\tilde{G}_0$ already gives a good approximation. This, however, makes the transcendental equation (1.14) more difficult to solve. Therefore, to find the resonance poles, it is expedient to use only the one-dimensional projector and, in the case of poor convergence of an iterative series, to resort to summation using the Padé approximants.

The projection formalism not only makes it possible to determine the resonance energy and width, but also gives a suitable representation for the process amplitude in which the resonance term is singled out explicitly. We shall adduce one of the possible representations of the amplitudes of one-particle scattering which was proposed by Domcke [6]. Let the scattering amplitude $T(p', p)$ be expressed in terms of the complete resolvent

$$T(p', p) = \langle p' | T | p \rangle ,$$

$$T = -G_0^{-1} + G_0^{-1} G G_0^{-1} .$$

Let now the representation (1.11) for the resolvent be substituted in the latter expression considering that only the terms $\tilde{G}$ and $\tilde{G} H D H \tilde{G}$ of the transition operator contribute to the on-shell amplitude:

$$T \underset{\text{on-shell}}{\rightarrow} G_0^{-1} \tilde{G} G_0^{-1} + G_0^{-1} \tilde{G} H D H \tilde{G} G_0^{-1} ,$$

$$(\tilde{G} \equiv G_Q) . \tag{2.17}$$

We shall now introduce the "orthogonalized" scattering functions (2.7a) which satisfy the Schrödinger equation with the Hamiltonian $H_Q = QHQ$ :

$$(E - H_Q)\tilde{\psi}_p^{(\pm)} = 0 , \qquad \frac{p^2}{2m} = E .$$

These functions may be obtained from the conventional relations of scattering theory

$$\lim_{\varepsilon \to 0} i\varepsilon \tilde{G}(E + i\varepsilon) | p \rangle = |\tilde{\psi}_p^{(+)}\rangle ,$$

$$\lim_{\varepsilon \to 0} i\varepsilon \langle p | \tilde{G}(E + i\varepsilon) = \langle \tilde{\psi}_p^{(-)} | .$$

These are related to the conventional scattering functions by (2.7a) and satisfy the modified Lippmann-Schwinger equation (2.14a):

$$|\tilde{\psi}_{\hat{p}}\rangle = |\hat{p}\rangle + \tilde{G}_0 |\tilde{\psi}_p\rangle . \tag{2.18}$$

Since $G_0^{-1} |p\rangle = i\varepsilon |p\rangle$ on the energy shell, we obtain from representation (2.17)

$$T(p', p) \underset{\text{on shell}}{=} \langle p' | G_0^{-1} |\tilde{\psi}_p^{(+)}\rangle + \langle \tilde{\psi}_{p'}^{(-)} | HDH |\tilde{\psi}_p^{(+)}\rangle . \tag{2.19}$$

The first term in (2.19) is the background amplitude which may be written, considering (2.18), as the sum of two terms:

$$T_{\text{bg}}(p', p) = \langle p' | G_0^{-1} |\hat{p}\rangle + \langle \hat{p}^{(-)} | V |\tilde{\psi}_p^{(+)}\rangle =$$

$$= T_0(p', p) + T_{dir}(p', p) . \tag{2.20}$$

The first term $T_0$ is the so-called "orthogonality scattering amplitude" [4], i.e. the scattering amplitude corresponding to the "plane orthogonalized wave" $|\tilde{p}\rangle$ (see the relation (2.14b) and the comment on it).

$$|\tilde{p}^{(+)}\rangle = \lim_{\varepsilon \to 0} i\varepsilon \tilde{G}_0(E + i\varepsilon)\,|p\rangle = |p\rangle - G_0\,P(PGP)^{-1}P\,|p\rangle \, .$$

This amplitude is

$$T_0(p', p) = -\langle p'|\,(PG_0P)^{-1}\,|p\rangle \, . \tag{2.21}$$

The second term in (2.20)

$$T_{\mathrm{dir}}(p', p) = \langle \tilde{p}'^{(-)}|\,V\,|\tilde{\psi}_p^{(+)}\rangle \tag{2.22}$$

is the amplitude of the so-called direct scattering.

Finally, the second term in (2.19) is the Feshbach resonance amplitude

$$T_{\mathrm{res}}(p', p) = \langle \tilde{\psi}_{p'}^{(-)}|\,HP(P(E - H - H\tilde{G}H)P)^{-1}PH\,|\tilde{\psi}_p^{(+)}\rangle \tag{2.23}$$

which contains all resonance poles and may readily be used to derive the Breit-Wigner formula [1]. Thus, we have obtained the following general decomposition of the complete $t$-matrix:

$$T(p', p) = T_0(p', p) + T_{\mathrm{dir}}(p', p) + T_{\mathrm{res}}(p', p) \, . \tag{2.24}$$

## 4.3. Analytic Continuation of the Scattering-Theory Equations

As resonance is usually related to the poles of the S-matrix or of the Hamiltonian resolvent on nonphysical sheets, we need to know how the location of this pole can be calculated. Of course, we can calculate the matrix elements of the resolvent at real energies, or at complex energies on the physical sheet, and then continue them as analytic functions of complex variable through the cut. It is this approach that has been accepted in the overwhelming majority of textbooks on scattering theory (see, for example, [12]). However, it would be more convenient in many respects to continue the resolvent proper as an *operator* and to obtain the equations defining the continuation which would be analogous to the conventional scattering theory equations. Moreover, the residues at resonance poles of the continued resolvent are directly related to the Gamow functions of the resonance states.

Any correct mathematical description for the analytic continuation of the resolvent and of non-normalizable states, such as scattering states and the Gamow states, requires that the frames of the Hilbert space should be overcome. An appropriate mathematical basis for doing this is offered by the mathematical formalism of rigged Hilbert space (RHS) which was proposed and specified in the fourth volume [18] of the well-known series of monographs by Gelfand et al. The readers who are unlikely to go deeply into the complete theory of rigged Hilbert spaces, but wish to have only some idea of the mathematical formalism, are referred to Appendix A which briefly presents the rigged Hilbert space and some results of operator theory in RHS which are essential when describing resonances states [19, 20]. The complete quantum-mechanics formulation in RHS may be found, for example, in [21].

### 4.3.1. Continuation of the Resolvent

Any self-adjoint operator $A$ in a rigged Hilbert space $\Phi \subset \mathcal{H} \subset \Phi^*$ has a complete set of generalized eigenvectors (the Gelfand-Maurin theorem; see the Appendix). The expansion of unity of the RHS in the eigenvectors of the operator $A$ is

$$\langle \varphi \mid \psi \rangle = \sum_j \langle \varphi \mid f_j \rangle \langle f_j \mid \psi \rangle + \int_\Lambda \langle \varphi \mid R(\lambda) \rangle \langle L(\lambda) \mid \psi \rangle \, d\lambda \,,$$

$$\varphi, \psi \in \Phi \tag{3.1}$$

where $f_j \in \mathcal{H}$ are the conventional eigenvectors of the discrete spectrum; the integral is taken over the continuous spectrum $\Lambda$ of the operator $A$; $R(\lambda) = L(\lambda)$ are the right and left generalized (in the sense of RHS) eigenvectors (GEV). We assume here that $\Lambda$ coincides with the positive semiaxis. Let $R(\lambda)$ be an analytic vector-valued function in the domain $\Omega \supset \Lambda$. Then, $L(\lambda)$ is an analytic function

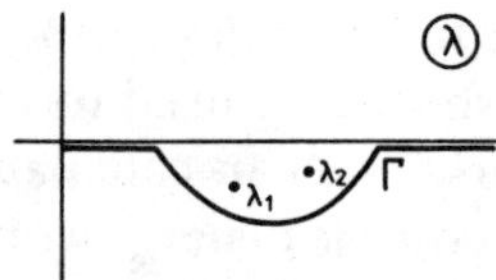


Fig. 4.1

The integration contour deformation in the expansion of unity. $\lambda_1$, $\lambda_2$ are the singularities of the integrand.

in the domain $\Omega^*$. By deforming the integration contour still within the domain $\Omega \cap \Omega^*$ as shown in Fig. 4.1, we obtain instead of (3.1):

$$\langle \varphi \mid \psi \rangle = \sum_j \langle \varphi \mid f_j \rangle \langle f_j \mid \psi \rangle +$$

$$+ \int_\Gamma \langle \varphi \mid R(\lambda) \rangle \langle L(\lambda) \mid \psi \rangle \, d\lambda + S \tag{3.2}$$

where $S$ is the contribution of the possible singularities of the integrand. In particular, if the only singularities are the first-order poles of the function $R(\lambda)$, then

$$S = -2\pi\, i \sum_l \langle \varphi \mid C_{-1}(\lambda_l) \rangle \langle L(\lambda_l) \mid \psi \rangle \qquad (3.3)$$

where $C_{-1}(\lambda_l)$ are the respective residues which are the GEV of the operator $A$. Thus, for a self-adjoint operator, we obtain an expansion in the "complete system" of the GEV corresponding to the complex eigenvalues. It should be noted that, if $R(\lambda)$ has a pole at the point $\lambda = z_R$ then $L(\lambda)$ will have a pole at the point $\lambda = z_R^*$. Therefore, they cannot have a pole simultaneously (on the real axis, $R(\lambda) = L(\lambda)$ is a continuous function). Similarly to (3.1), expansions for any analytic function $f(A)$ may be written as

$$\langle \tilde{\varphi} \mid f(A) \mid \tilde{\psi} \rangle = \sum_j f(\lambda_j) \langle \varphi \mid f_j \rangle \langle f_j \mid \psi \rangle +$$

$$+ \int_\Lambda f(\lambda) \langle \varphi \mid R(\lambda) \rangle \langle L(\lambda) \mid \psi \rangle \, d\lambda . \qquad (3.4)$$

In particular, for the resolvent $G(z) = (z - A)^{-1}$ we get (if $z$ does not belong to the spectrum) the spectrum expansion

$$\langle \varphi \mid G(z, A) \mid \psi \rangle = \sum_j \frac{\langle \varphi \mid f_j \rangle \langle f_j \mid \psi \rangle}{z - \lambda_j} +$$

$$+ \int_\Lambda \frac{\langle \varphi \mid R(\lambda) \rangle \langle L(\lambda) \mid \psi \rangle}{z - \lambda} \, d\lambda . \qquad (3.5)$$

The integration contour in (3.4) and (3.5) may also be deformed as in (3.2), so, instead of (3.5), we obtain, for example, a "spectral" representation of the resolvent including the sum over resonance states and over a part of the "complex continuum".

A great number of representations of this kind was found by Berggren [24], and by Garcia-Calderon and Peierls [25]. In this case also a peculiar condition of "completeness" [25] for the expansion of an arbitrary function in a complete set of resonance functions plus a contribution of the complex continuum may be formulated.

The representation (3.5) makes it possible to continue the matrix element of the resolvent in the states from the space $\Phi$ through the continuous spectrum to the second sheet of the Riemann energy surface. Indeed, applying the theorem for continuing the analytic function represented by the Cauchy contour integral (see Section 1.3.4), we obtain:

$$\langle\varphi|\, G^{II}(z, A)\, |\psi\rangle \;=\; \langle\varphi|\, G(z, A)\, |\psi\rangle \;-\; 2\pi\,\mathrm{i}\,\langle\varphi\,|\,R(z)\rangle\,\langle L(z)\,|\psi\rangle$$

$$(3.6)$$

where $R(z)$ and $L(z)$ are the GEV of the continuous spectrum of the operator $A$ which are continued to the analyticity domain $\Omega$. Moreover, under certain conditions imposed on the space $\Phi^4)$ (the uniform-boundedness principle is to be satisfied in this space, see e.g. [19]) there exist upper and lower limits, i.e. $\lim_{\varepsilon\to 0} G(z \pm \mathrm{i}\varepsilon)$ , for any real value of $z$ among the inner points of the continuous spectrum of $A$ and the following relation holds:

$$G(z + \mathrm{i}\varepsilon) \;=\; G(z - \mathrm{i}\varepsilon) \;-\; 2\pi\,\mathrm{i}\,|R(z)\rangle\,\langle L(z)| \;.$$

$$(3.7)$$

Here $G(z \pm \mathrm{i}\varepsilon)$ are the operators acting from $\Phi$ into $\Phi^*$. Therefore, we can also continue the resolvent $G(z, A)$ proper to the second sheet as a linear operator from $\Phi$ into $\Phi^*$ according to the formula:

$$G^{II}(z, A) \;=\; G(z, A) \;-\; 2\pi\,\mathrm{i}\,|R(z)\rangle\,\langle L(z)|$$

$$(3.7a)$$

for any $z \in \Omega$ .

It should be remembered that the resolvent $G(z, A) = (z - A)^{-1}$ as an operator in *Hilbert* space $\mathcal{H}$ is defined at all *regular* points of the complex plane $z$ (i.e. outside the spectrum) and *cannot* be continued to beyond the region of its natural definition as a bounded operator in $\mathcal{H}$. Since the spectrum of the self-adjoint operator is real, the resolvent $G(z, A)$ exist and is bounded at any $z$ outside the real axis. Therefore, from (3.7a) it is clear that all the singularities of the continued resolvent coincide with the singularities of GEV.

### 4.3.2. *Analytic Continuation of the Scattering-Theory Equations*

As shown above, the RHS formalism makes it possible to continue the resolvent of the self-adjoint Hamiltonian to the nonphysical sheet through the cut corresponding to a continuous spectrum. Naturally, the question arises: what is the equation satisfied by the continued resolvent? Such an equation may be used in practice to find the Hamiltonian resolvent on the nonphysical sheet whose pole locations and residues determine the resonance energy, width, and wave functions. Two sections are devoted to a discussion, following the original works of Pomerantsev et al. [29, 47], of the continuation of the basic equations of scattering theory to nonphysical energy sheets and of the methods for the practical solution of continued equations. In the present section we shell examine the continuation of the Lippmann-Schwinger equation which is used in the

---

4) For the meaning of the notation for the spaces used here, see Appendix $A$.

one-particle problem. In the next section we shall discuss the continuation of the Faddeev equations for three-particle resonances. Let $H = H_0 + V$ and let the corresponding resolvents of the operators $H_0$ and $H$

$$G_0(z) = (z - H_0)^{-1},$$

$$G(z) = (z - H)^{-1}$$

be bounded and defined as operators in $\mathscr{H}$ everywhere outside the spectrum. Let us further construct the RHS $\Phi^* \supset \mathscr{H} \supset \Phi$ in which $H$ and $H_0$ have complete GEV systems and the continuation of resolvents is defined according to (3.7a)

$$G_0^{II}(z) = G_0(z) - 2\pi\, i\, |R_0^{II}(z)\rangle \langle L_0(z)|, \tag{3.8}$$

$$G^{II}(z) = G(z) - 2\pi\, i\, |R^{II}(z)\rangle \langle L(z)|. \tag{3.9}$$

Here, for simplicity, we consider the spectrum to be non-degenerate (for example, we mean the radial Schrödinger equation) and assume that the potential satisfies the usual conditions ensuring the existence of wave operators, etc. Then the scattering wave functions (GEV) $R^+(z)$ and $L^+(z)$ (the superscript $(+)$ will be omitted in what follows) are related to the GEV of the operator $H_0$, i.e. to $R_0(z)$ and $L_0(z)$, as

$$|R\rangle = (1 + G_0 t)\, |R_0\rangle = (1 + GV)\, |R_0\rangle, \tag{3.10a}$$

$$\langle L| = \langle L_0|\, (1 + t G_0) = \langle L_0|\, (1 + VG). \tag{3.10b}$$

The expression (3.9) contains the GEV $|R^{II}\rangle$ continued through the cut and defined as

$$|R^{II}\rangle = (1 + G^{II}V)\, |R_0^{II}\rangle = (1 + G_0^{II}t^{II})\, |R_0^{II}\rangle. \tag{3.11}$$

Here, we have introduced the continuation of the $t$-matrix

$$t^{II} = V + VG^{II}V. \tag{3.12}$$

Now, the continuation of the Lippmann-Schwinger equation for the resolvent may be written as.

$$G^{II} = G_0^{II} + G_0^{II}VG^{II}. \tag{3.13}$$

The continuation of the resolvent $G^{II}$ defined by the relation (3.9) can be shown to actually satisfy the continued Lippmann-Schwinger equation (3.13). The relations (3.8) and (3.9) may be presented as

$$G^{II}(z) = G(z) + \Delta(z), \tag{3.14}$$

$$G_0^{II}(z) = G_0(z) + \Delta_0(z) \tag{3.15}$$

where it is evident that

$$\Delta(z) = -2\pi \, i \, |R^{II}(z)\rangle \, \langle L(z)| \tag{3.15a}$$

and the same for $\Delta_0(z)$. It should be noted that

$$\Delta(z) = (1 + G^{II}V) \, \Delta_0(z) \, (1 + VG) \,. \tag{3.16}$$

Substituting (3.8) and (3.9) in (3.13) and using (3.16), we make sure that $G^{II}$ actually satisfies (3.13). Here the continued GEV $R^{II}(z)$ also satisfies the continued Lippmann-Schwinger equation

$$R^{II}(z) = R_0^{II}(z) + G_0^{II}(z)VR^{II}(z) \,. \tag{3.17}$$

If we use another equation for $G^{II}$ , namely,

$$G^{II} = G_0^{II} + G^{II}VG_0^{II} \tag{3.18}$$

we can obtain the expression for $\Delta(z)$ which is equivalent to (3.15a)

$$\Delta(z) = -2\pi \, i \, |R(z)\rangle \, \langle L^{II}(z)| \,. \tag{3.19}$$

This shows that the GEV continued to the second sheet differs from the GEV on the first sheet only by a factor. Indeed, using (3.12) and an interesting relation for the $t$-matrix:

$$t^{II} = t + t\Delta_0 t^{II} = t + t^{II}\Delta_0 t \tag{3.20}$$

we find:

$$R^{II}(z) = (1 - 2\pi \, i \, \langle L_0| \, t^{II} \, |R_0\rangle)R(z) \,. \tag{3.21}$$

Thus, we see that the operator identity (3.13) for the resolvent continued to the nonphysical sheet holds in practice. However, the operators $G_0^{II}$ and $G^{II}$ act from $\Phi$ to $\Phi^*$, so the continuation is only possible under definite conditions imposed on the potential. Equation (3.13) is valid if $V$ transforms $\Phi^*$ into $\Phi$. This limitation is too strong, however. Evidently, the necessary condition is

$$V R(z) \in \Phi \tag{3.22}$$

which limits the region of feasible continuation of the Lippmann-Schwinger equation in $z$. The particular conditions to be imposed on the potential ensue from the explicit form of eq. (3.13) when going over from the operator equation to the integral equation.

Let us examine the continuation of the Lippmann-Schwinger equation in the case of the one-particle problem. The explicit expression for the continuation of the resolvent of the free Hamiltonian $H_0 = (-1/2\mu)\,\Delta_x$ is

$$G_0^{II}(z) = \int \frac{|k\rangle\langle k|}{z - k^2/2\mu}\,\mathrm{d}^3k \; -$$

$$- 2\pi\,\mathrm{i}\,\mu\,\sqrt{2\mu z}\int \mathrm{d}\Omega_k\,|\sqrt{2\mu z}\,\hat{k}\rangle\langle\sqrt{2\mu z}\,\hat{k}|\,, \qquad \mathrm{Im}\,\sqrt{z} < 0$$

$$(3.23)$$

where $|k\rangle \in \Phi^*$ are the (right) GEV; the corresponding eigenvalues are $k^2/2\mu$; $|2\mu z\hat{k}\rangle$ is the continuation of these vectors to the lower halfplane, $\hat{k}$ is the unit vector.

In the momentum representation the kernel the of integral operator (3.23) is

$$G_0^{II}(p, p'; z) =$$

$$= \frac{\delta(p - p')}{z - p^2/2\mu} - 2\pi\,\mathrm{i}\mu\,\sqrt{2\mu z}\,\frac{\delta(p - \sqrt{2\mu z})}{p^2}\,\frac{\delta(p' - \sqrt{2\mu z})}{p'^2}\,\delta(1 - \hat{p}\hat{p}')$$

$$(3.24)$$

Accordingly, the Lippmann-Schwinger equation on the second sheet (3.13) in the momentum representation reduces to the following integral equation:

$$G^{II}(p, p'; z) =$$

$$G_0^{II}(p, p'; z) + \frac{1}{z - p^2/2\mu}\int \mathrm{d}^3q V(p, q)G^{II}(q, p'; z) - 2\pi\,\mathrm{i}\mu\sqrt{2\mu z}$$

$$\frac{\delta(p - \sqrt{2\mu z})}{p^2}\int \mathrm{d}^3q V(\sqrt{2\mu z}\,\hat{p}, \hat{q})G^{II}(q, p'; z)\,. \qquad (3.25)$$

A momentum-representation potential, continued to the lower halfplane appeared in the kernel of the second integral term in (3.25). The iterations in (3.25) give rise to $V(\sqrt{2\mu z}\hat{p}, \sqrt{2\mu z}\,\hat{p}')$. Thus, the analyticity domain of potential defines the feasible continuation region for (3.25).

In the coordinate representation the continued resolvent $G_0^{II}$ is

$$G_0^{II}(r, r', z) =$$

$$- \frac{\mu}{2\pi}\,\frac{\exp(\mathrm{i}\,\sqrt{2\mu z}\,|r - r'|)}{|r - r'|} \; -$$

$$- 4\,\mathrm{i}\,\mu\,\sqrt{2\mu z}\,\sum_{l=0}^{\infty}\sum_{m=-l}^{l}(-1)^l j_l(\sqrt{2\mu z}r)\,j_l(\sqrt{2\mu z}r')Y_{lm}(\hat{r})\,Y_{lm}(\hat{r}')\,.$$

Actually, in order that the Lippmann-Schwinger equation be valid, it is necessary that $\langle L_0(z)|\, V\,|R(z)\rangle < \infty$ . In terms of the RHS formalism this necessitates $VR(z) \in \Phi$ . It is clear that this condition is also too restrictive. In practical applications of the continued Lippmann-Schwinger equation it is sufficient that

$$\int \sqrt{L_0(z, r)}\; V(r)R(z, r)\; \mathrm{d}^3r < \infty .$$

(3.26)

Since the asymptotic behaviour of the GEV $R$ (and $R_0$) is $R \sim e^{i\sqrt{(2\mu zr)}}$ the integral (3.26) converges at any value of $z$ if the integral $\int V(r)\, e^{-2i\sqrt{(2\mu zr)}}$ also converges. In other words, the continued integral equation may be used at any $z$ value for which the continuation of $|R^{II}(z)\rangle$ , hence of the resolvent, exists, although in this case the operator Lippmann-Schwinger equation is meaningless in terms of RHS.

In some respects it is more convenient to use the continuation of the Lippmann-Schwinger equation for the $t$-matrix

$$t^{II}(z) = V + VG_0^{II}(z)t^{II}(z)$$

(3.27)

which is of the following form in the momentum representation:

$$t^{II}(\boldsymbol{p}, \boldsymbol{p}', z) = V(p, p') + \int \mathrm{d}^3q\; V(p, q)\,\frac{1}{z - q^2/2\mu}\, t^{II}(\boldsymbol{q}, \boldsymbol{p}'; z) -$$

$$-2\pi\, i\mu\, \sqrt{2\mu z} \int \mathrm{d}\hat{\boldsymbol{q}}\; V(\boldsymbol{p}, \sqrt{2\mu z}\hat{\boldsymbol{q}})t^{II}(\sqrt{2\mu z}\hat{\boldsymbol{q}}, \boldsymbol{p}'; z) .$$

(3.28)

Expanding all quantities of (3.28) in partial waves, we obtain the equation for the partial $t$-matrix on the second sheet [28]:

$$t_l^{II}(p, p'; z) = V_l(p, p') + \int q^2\, \mathrm{d}q\; V_l(p, q)\,\frac{1}{z - q^2/2\mu}\, t_l^{II}(q, p'; z) -$$

$$-2\pi\, i\mu\, \sqrt{2\mu z}\; V(p, \sqrt{2\mu z})\; t_l^{II}(\sqrt{2\mu z}, p'; z) .$$

(3.29)

Integral equations (3.28) and (3.29) can be inferred [28] from the conventional Lippmann-Schwinger equation for the $t$-matrix without resorting to the rigorous operator formalism in the RHS. We shall use the example of (3.29) to formulate a simple rule for the continuation of the integral equations whose kernel contains sigularities [29].

The initial Lippmann-Schwinger equation for the partial $t$-matrix (the subscript $l$ will be omitted now) at the upper rim of the cut $z = E + i\varepsilon$ is of the form

$$t(p, p'; z) = V(p, p') + \int_{C_1} q^2 \, dq \, V(p, q) \frac{t(q, p'; z)}{z - q^2/2\mu} \qquad (3.30)$$

where the contour $C_1$ is as shown in Fig. 4.2.

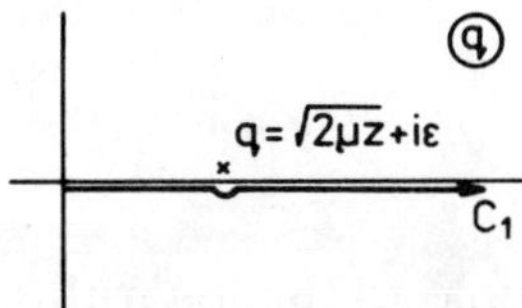


Fig. 4.2

Assuming that all integrands can be continued in $q$ at least to some region of the lower halfplane, we deform the contour of integration as shown in Fig. 4.3. According to the Cauchy theorem, the value of the integral remains unchanged in this case. Now, we shall continue (3.30) to the point $z = z_R - i z_i$, $z_i > 0$, of the lower halfplane in such a way that the pole of the integrand $q_s = \sqrt{(2\mu z)}$ would remain in the region between the real axis and the contour $C'$ (see Fig. 4.3).

The general rule holds that *in continuing the integral, the moving singularities of the integrand must not intercept* the integration contour. In other words, we must deform the contour in such a way that an intercept may be avoided. In the integral (3.30) the pole $q_s$ is the only known moving singularity. Of course, in making the deformation we may accidentally intercept the singularities of the potential $V(p, q)$ or, more likely, the singularities (depending on $z$) of unknown function $t(q, p'; z)$. To eliminate this possibility, it is convenient to use the contour $C_2$ shown in Fig. 4.4 instead of $C'$. The integral along the contour $C_2$

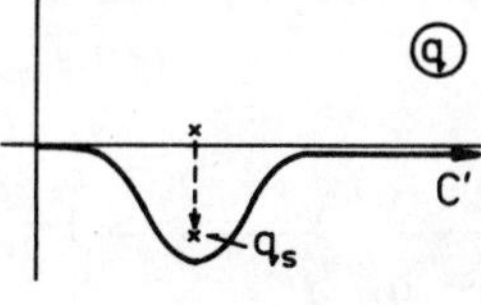


Fig. 4.3

is equal to the integral along the contour $C_1$ plus the residue at the point $q_s$. This leads us to (3.29). Such a continuation procedure is in common use and assumes that the singularities of $t(q, p'; z)$ in the $q$ plane do not intercept the real axis when $z$ moves to the lower halfplane. Of course, equation (3.29) follows directly from the theorem of the contour integral continuation (Section 1.3.4), but the general continuation rule formulated here can also be applied to other types of

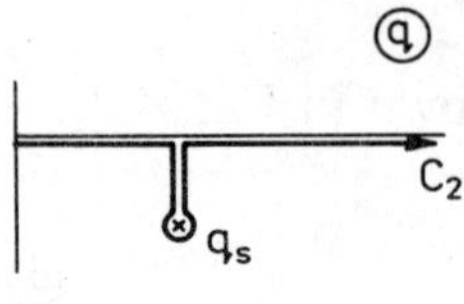


Fig. 4.4

moving singularities, for example to the logarithmic-type branch points in the Faddeev equations.

In the numerical solution the procedure of continuation is achieved even more simply. In solving the equation at the upper rim of the cut (i.e. in finding the scattering amplitude) the integral along the contour $C_1$ is presented as the sum

$$\int_{C_1} q^2 \, dq \, \frac{f(q)}{z - q^2/2\mu} = P \int_0^\infty q^2 \, dq \, \frac{f(q)}{z - q^2/2\mu} - i\pi \sqrt{2\mu z} \, \mu f\left(\sqrt{2\mu z}\right)$$

$$(3.31)$$

Furthermore, the principal value integral is taken numerically after regularization

$$P \int_0^\infty q^2 \, dq \, \frac{f(q)}{z - q^2/z\mu} = \int_0^\infty \frac{q^2 f(q) - q_s^2 f(q_s)}{z - q^2/2\mu} \, dq \,, \qquad q_s = \sqrt{2\mu z} \,.$$

$$(3.32)$$

Let $\{q_i\}$ and $\{W_i\}$ be respectively mesh points and weights of the quadrature formula. Then,

$$\int_{C_1} = \sum_{i=1}^M W_i \frac{q_i^2 f(q_i) - q_s^2 f(q_s)}{z - q_i^2/2\mu} - i \,\pi\mu q_s f(q_s) \,.$$

$$(3.33)$$

The unknown function is calculated at points $q_i$ and $q_s$, i.e. we add one more mesh point to $M$ mesh points.

With continuation in $z$ to the lower halfplane, i.e. when going over to (3.29), the integral gets regular. In this case, however, we may retain the regularization

(3.33) to improve the accuracy, because at $\mathrm{Im}\ z \ll \mathrm{Re}\ z$ the denominator of the integrand is close to zero. Then, instead of (3.31), and (3.32), we get

$$\int_{C_2} = \int_0^\infty dq\ \frac{q^2 f(q) - q_s^2(f(q_s))}{z - q^2/2\mu} + \int_0^\infty dq\ \frac{q_s^2 f(q_s)}{z - q^2/2\mu} \tag{3.34a}$$

$$- 2\pi\ i\mu q_s f(q_s) =$$

$$= \int_0^\infty dq\ \frac{q^2 f(q) - q_s^2 f(q_s)}{z - q^2/2\mu} - i\ \pi\mu q_s f(q_s)\ . \tag{3.34b}$$

Here the second integral in the right-hand side of (3.34a) is taken analytically and equals $i\pi\mu q_s f(q_s)$ because $\mathrm{Im}\ q_s < 0$. Thus, the expression for the integral term of the equation continued to the lower halfplane *coincides* with the expression on the upper rim of the cut. This evident result turns out to be very useful in continuining the more complicated equations, like Faddeev equations. The general rule for practical continuation may be formulated as follows. If, to solve the integral equation in the physical region (i.e. on the upper rim of the cut), we use the numerical-analytic procedure where the singular parts are calculated analytically, and the regularized integrals numerically, then, in making continuation through the cut to the lower halfplane we need not change the form of the linear equations to be solved, but have only to continue the *analytic* integrals correctly.

### 4.3.3. *Continuation of the Faddeev Equations*

Let us consider the problem of analytic continuation of the Faddeev equations using the example of the simplest equations for three identical spinless particles with separable interactions. All the characteristic singularities of the kernel are preserved in this case.

In the case of the separable potential

$$V = \lambda\ |g\rangle\ \langle g| \tag{3.35}$$

the two-particle $t$-matrix is calculated in the explicit form

$$t(z) = |g\rangle\ \tau(z)\ \langle g| \tag{3.36}$$

where

$$\tau(z) = \lambda^{-1} - \int d^3p\ \frac{|g(p)|^2}{z - p^2/2\mu}\ . \tag{3.37}$$

In the examined case there exists a single symmetrized scattering amplitude $X$ (including the scattering with rearrangement) which satisfies the equation (see [30])

$$X = 2Z + 2Z\tau X \qquad (3.38)$$

where $Z = \langle g_\alpha | G_0 | g_\beta \rangle$, $\alpha \neq \beta$ is the matrix element of the free three-particle resolvent in terms of the potential from-factors. (The indices $\alpha$, $\beta$ distinguish the two-body subsystems).

The explicit form of the Faddeev equation for $X$ as an integral equation in the momentum representation [30] is

$$X(\boldsymbol{q}, \boldsymbol{q}'; z) = 2Z(\boldsymbol{q}, \boldsymbol{q}'; z) +$$

$$+ 2\int d^2q'' \, Z(\boldsymbol{q}, \boldsymbol{q}''; z) \, \tau\left(z - \frac{q''^2}{2\mu}\right) X(\boldsymbol{q}'', \boldsymbol{q}'; z) \qquad (3.39)$$

where $\mu = \frac{1}{2}m$ and $M = \frac{2}{3}m$ are the reduced masses and $m$ is the mass of the particles. After series-expanding in the Legendre polynomials $P_L(\cos \vartheta)$ (where $\vartheta$ is the angle between $\boldsymbol{q}$ and $\boldsymbol{q}'$), we obtain the one-dimensional equations for partial amplitudes:

$$X_L(q, q'; z) = 2Z_L(q, q'; z) +$$

$$+ 8\pi \int_0^\infty dq'' q''^2 Z_L(q, q''; z) \, \tau\left(z - \frac{3}{4}\frac{q''^2}{m}\right) X_L(q'', q'; z) \,.$$

$$(3.40)$$

Let us examine the singularities of the kernel as functions of $q''$. The two-particle propagator $\tau(z)$ has a pole at the energy of the two-particle bound state $z = \varepsilon_b$. Therefore, at the real values of $z > \varepsilon_b$ the kernel of (3.40) contains a pole in $\tau$ at

$$q''_s = \sqrt{\tfrac{3}{4}m(z - \varepsilon_b)} \,.$$

The function $Z_L(q, q''; z)$ gets singular at $z > 0$. The complete function

$$Z(\boldsymbol{q}, \boldsymbol{q}'; z) = \frac{g^*(\boldsymbol{q}' + \tfrac{1}{2}\boldsymbol{q}) \, g(-\tfrac{1}{2}\boldsymbol{q}' - \boldsymbol{q})}{z - \dfrac{q^2}{2m} - \dfrac{q'^2}{2m} - \dfrac{(\boldsymbol{q} + \boldsymbol{q}')^2}{2m}}$$

has poles at definite values of $q$, $q'$ and $\cos \vartheta$. After integration over $\cos \vartheta$ with the Legendre polynomial, the poles turn into logarithmic singularities. These logarithmic branch points appear in $Z_L(q, q''; z)$ under the condition

$$mz - (q \pm \tfrac{1}{2}q'')^2 - \tfrac{3}{4}q''^2 = 0$$

i.e. at the points $q'' = \pm q_{1,2}$, where

$$q_1 = \frac{q}{2} + \sqrt{mz - \tfrac{3}{4}q^2}, \tag{3.41a}$$

$$q_2 = -\frac{q}{2} + \sqrt{mz - \tfrac{3}{4}q^2}. \tag{3.41b}$$

Fig. 4.5 shows the locations of the logarithmic branch points $q''(q)$ and of the propagator pole $\tau$. The function $Z_L$ is complex between two arcs of ellipses in

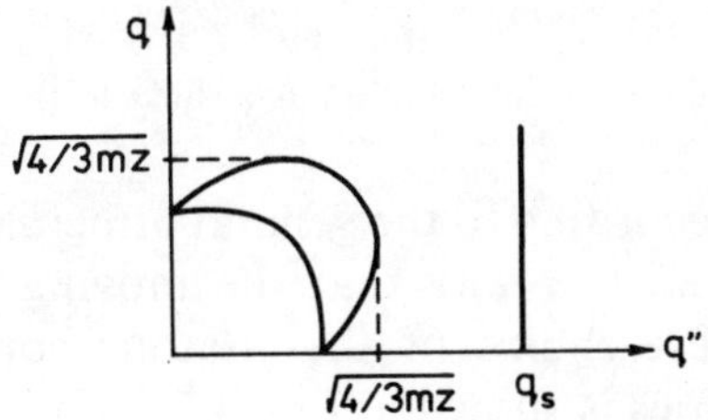


Fig. 4.5

The positions of the singularities of the Faddeev equation kernel $(Z_L\tau)$ in the $qq''$-plane.

Fig. 4.5 and real in the rest of the $qq''$ plane. The singular part of $Z_L$ is proportional to the expression

$$Z_L^S(q, q''; z) \sim \ln \frac{(q'' - q_1(q))\,(q'' + q_2(q))}{(q'' + q_1(q))\,(q'' - q_2(q))}. \tag{3.42}$$

Two of four branch points $\pm q_{1,2}$ (i.e. $q_1, -q_1, q_2, -q_2$) lie on the integration contour (positive semiaxis), and to avoid them we add $i\varepsilon$ to the real energy $E$. Let us now examine the location of singularities at $z = E + i\varepsilon$, $E > 0$ and at a fixed value of $q < \sqrt{\tfrac{4}{3}mE}$ in the complex plane. Let $q_2 > 0$, then the branch points are located as shown in Fig. 4.6. It is convenient to connect in pairs the branch points by the cuts as shown in Fig. 4.6. Then the integration contour will not intercept the cuts.

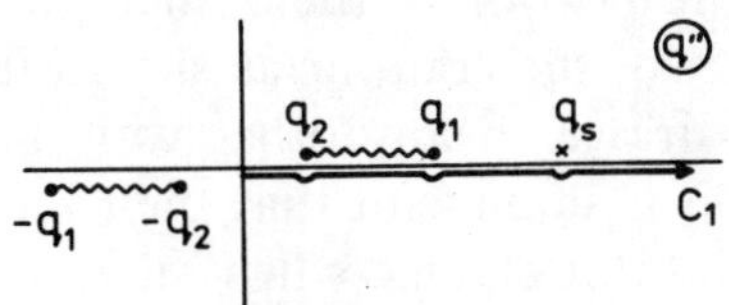


Fig. 4.6

The positions of the Faddeev equation kernel singularities in the plane of integration variable $q''$ at $z = E + i\varepsilon$ and a fixed value of $q < \sqrt{(4mE/3)}$ for $q_2(q) > 0$.

We shall now examine what will happen to the singularities when the equation (3.40) is continued in $z$ to the lower halfplane. In the case shown in Fig. 4.6. the singularities $q_2$, $q_1$ and $q_s$ intercept the real axis and are shifted to the lower $q''$

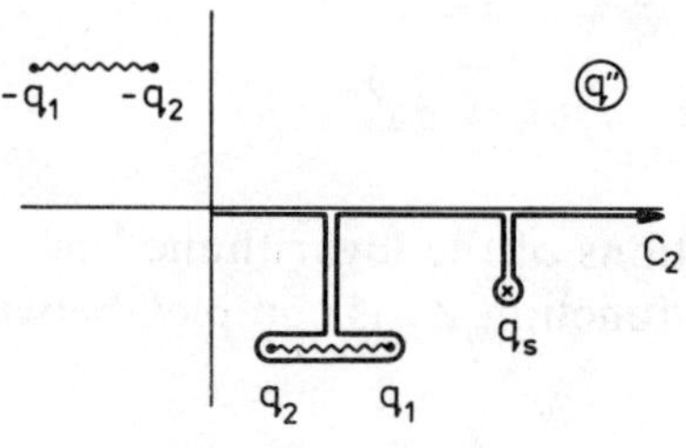

Fig. 4.7

The integration contour deformation as $z$ shifts to the lower halfplane.

halfplane (see (3.41)). According to the general principle, we have to deform the integration contour in such a way that the moving singularities would not intercept it. The simplest choice of the contour corresponding to the new location of the singularities is shown in Fig. 4.7. Thus, to the integral along the real axis we must add the residue at point $q_s$ and the integral of the integrand discontinuity along the logarithmic cut:

$$X^{II}(q, q'; z) = 2Z(q, q'; z) +$$

$$+ 8\pi \int\limits_0^\infty d\, q'' q''^2 Z(q, q''; z)\tau(z - \frac{3}{4}\frac{q''^2}{m})X^{II}(q'', q'; z) +$$

$$+ 2\pi i \left[ \mathrm{res}\, \tau \right]_{q_s} 8\pi q_s^2 Z(q, q_s; z)X^{II}(q_s, q'; z) +$$

$$+ 8\pi \int\limits_\gamma dq''^2 \, \mathrm{disc}\, \left[ Z(q, q''; z) \right] \times$$

$$\times \tau(z - \frac{3}{4}\frac{q''^2}{m})X''(q'', q'; z) . \tag{3.43}$$

In the case of a different location of the branch points $(q_2 < 0)$ we have to carry out the deformation of the contour as shown in Fig. 4.8. Of course, the logarithmic cuts can be drawn in any other way, in particular, through the integration contour, but it is important that their location with respect to the integration contour should not change when making the continuation.

Apart from the examined singularities, the Faddeev equation kernel may also contain the poles in the function $\tau$ which correspond to the two-particle resonances. These poles correspond to the cuts running parallel to the real axis in the fourth quadrant on the second logarithmic sheet of the $z$-plane. If, when making the continuation in $z$, such a pole crosses the real axis in the $q''$ plane, then its residue has to be added to the equation. This corresponds to the continuation in $z$ through the two-particle resonance cut.

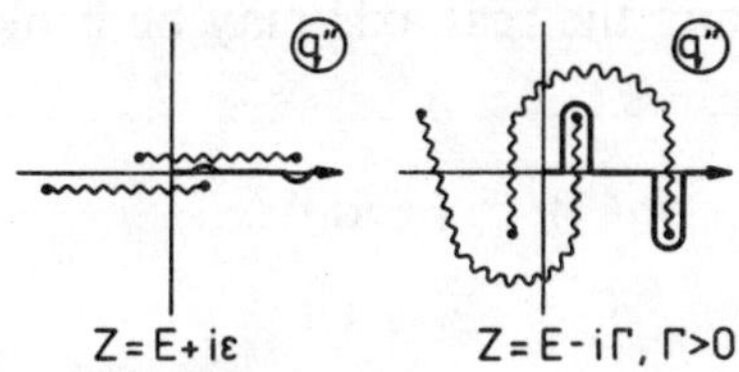


Fig. 4.8

Disposition of logarithmic singularities and the integration contour deformation at $q_2(q) > 0$.

Two methods are used to solve the Faddeev equations numerically in the physical domain $z = E + i\varepsilon$ , namely the integration contour deformation proposed by Hetherington and Schick [31] (see also [32], [33]) and the combined numerical-analytic integration. These methods may also be used to solve the

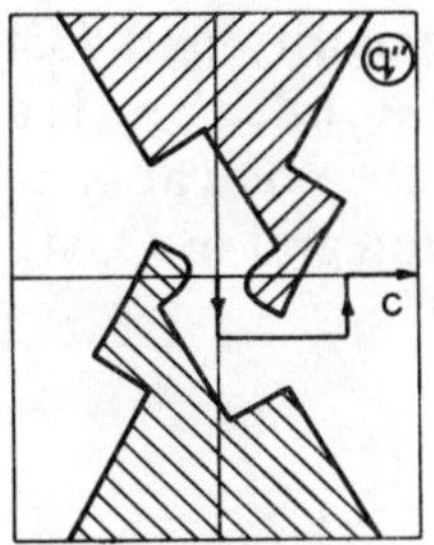


Fig. 4.9

An example of the deformed integration contour used in [34] to continue the Faddeev equations through logarithmic cut. The shaded areas are logarithmic singularities in $Z_L$.

Faddeev equations on the nonphysical sheet. In Refs. [34], [35] the method of contour deformation was used to calculate the trajectories of resonance poles for the system of three identical particles with separable potentials. The choice of the integration contour when solving the continued equations is a rather complicated problem because the location of singularities in the $q''$ plane depends on $q$ and $z$. The main difficulty in this case is due to the singularities of the sought amplitude proper. The example of the contour used in Ref. [34] is shown in Fig. 4.9.

The continuation procedure described above and based on the contour deformation shown in Figs. 4.7 and 4.8 is more suitable for use in the numerical-analytic method for solving the Faddeev equations [36, 37]. We shall begin by describing the solution procedure for the real energies $z = E + i\varepsilon$. First of all, allowance should be made for the fact that the fixed pole singularities corresponding to two-particle bound state ($q_s$ in Fig. 4.5) are always separated from the region of the logarithmic singularities $(q_s > \sqrt{(\tfrac{4}{3}mz)}$ and $q_1 \leq \sqrt{(\tfrac{4}{3}mz)})$. This also holds true in the general case of different particles. Therefore, the integral along the real axis may be broken into two parts:

$$\int_0^\infty dq'' = \int_0^{\sqrt{4/3mz}} dq'' + \int_{\sqrt{4/3mz}}^\infty dq'' .  \tag{3.44}$$

The second integral contains only the fixed poles and is regularized exactly in the same way as in the case of the one-particle Lippmann-Schwinger equation. As a result, several (as many as the number of bound states in the subsystem) points $q_s$ get added to the mesh points $\{q_i''\}$. In the first integral we separate the singular part $Z^S$ from $Z_L$:

$$Z_L(q, q''; z) = Z_L^R(q, q''; z) +$$

$$+ Z^S(q, q''; z)R_L(q, q''; z) .  \tag{3.45}$$

Here $Z^S$ is the logarithmic singularity part (3.42); $Z_L^R$ and $R_L$ are the regular functions of $q''$. The amplitude $X_L(q'', q'; z)$ is also a regular function [36] and the propagator $\tau$ contains only the poles at $q > \sqrt{(\tfrac{4}{3}mz)}$. In accordance with the separation (3.45), the first integral in (3.44) may be presented as the sum

$$\int_0^{\sqrt{4/3mz}} = I_R + I_S$$

$$I_S = \int_0^{\sqrt{4/3mz}} Z^S(q, q''; z)R_L(q, q''; z)\tau(q'', z)X_L(q'', q'; z)q''^2 \, dq'' .$$

The regular part $I_R$ is calculated using the quadrature formula with nodes $\{q_{ij}''\}$ and the factor at $Z^S$ in the singular part $I_S$ is approximated by an interpolation polynomial running through the points $q_i''$. Then, the problem reduces to calculating the singular integrals

$$I_n^S(q, z) = \int_0^{\sqrt{4/3mz}} Z^S(q, q''; z)q''^n \, dq''  \tag{3.46}$$

which are taken analytically (see formula (3.42) for $Z^S$). As a result, we arrive at a set of linear equations whose coefficients are analytic functions of $z$. When making the continuation in $z$ through the logarithmic cut we need only to correctly continue the integrals (3.46) by choosing the necessary branch of the logarithm. Here the integral over the discontinuity along the logarithmic cut is added automatically if the approximation of the regular part of the integrand by polynomials remains correct at the cut. The problem of the analytic properties of the solution $X_L(q, q'; z)$ proper as a function of $q$ and $q'$ is very complicated. Brayshaw [38] has shown that at "real" $z = E + i\varepsilon$ the half-off--shell amplitude $X_L(q, q_s; z)$ is an analytic function in sector $q = \varrho\, e^{i\varphi}$ for

$$\varphi < \arctan \sqrt{-3\varepsilon_b/(z - \varepsilon_b)}\,.$$

When using the contour depicted in Figs. 4.7 and 4.8 it is of importance that the singularities of solution $X_L$ in the $q$-plane should not be close to the logarithmic cut and should not intercept it when $z$ shifts to the lower halfplane.


## 4.4 One-Particle Resonances

### 4.4.1. Equation for the Resonance Energy and Width

Let us now describe the application of general projection formalism to the calculation of one-particle resonances. As shown in Section 4.1, the resonance energy and width $z_R = E_R - i\Gamma/2$ are found by solving the transcendental equation (1.15):

$$z_R = \langle\varphi|\, H\, |\varphi\rangle + \langle\varphi|\, H\tilde{G}^{II}(z_R)H\, |\varphi\rangle \tag{4.1}$$

where $\tilde{G}^{II}$ is the continuation of the orthogonalized resolvent to the nonphysical sheet. We have seen how $\tilde{G}^{II}$ should be found (Section 4.2) and continued to the nonphysical sheet (Section 4.3). We shall now discuss the method for solving the transcendental equation (4.1). We shall henceforth omit the superscript II and everywhere imply the continuation of the resolvent to the nonphysical sheet.

The most natural way of solving (4.1) is to use the iterative method, i.e.

$$z^{(k+1)} = z_0 + \langle\varphi|\, H\tilde{G}(z^{(k)})H\, |\varphi\rangle \tag{4.2}$$

where $z_0 = \langle\varphi|\, H\, |\varphi\rangle$ is the mean value of the Hamiltonian in the initial state $\varphi$ (assumed to be normalized to unity). Let us find out the conditions which permit a convergent iterative solution for (4.1). The equation (4.1) may be presented as

$$z = z_0 + F(z) \tag{4.3}$$

where

$$F(z) = \langle\varphi|\, H\tilde{G}H\, |\varphi\rangle = z - z_0 - \frac{1}{\langle\varphi|\, G\, |\varphi\rangle}\,. \tag{4.4}$$

Here we have used the relation (1.9c) which in the case of a one-dimensional projector, is of the form

$$\langle\varphi|\, z - H - H\tilde{G}H\, |\varphi\rangle = \frac{1}{\langle\varphi|\, G\, |\varphi\rangle}. \tag{4.5}$$

In order that (4.3) permit an iterative solution, we must satisfy the condition

$$\left|\frac{dF}{dz}\right| \leq p < 1 \tag{4.6}$$

for all $z$ in some region containing the fixed point $z_R$, i.e. the solution sought; here, all the iterations $z_0$, $z_1 = z_0 + F(z_0)$, ... must remain in this region. It is clear that $z_0$ must be sufficiently close to the exact value of $z_R$. Since $z_0$ is real (we consider $H$ to be a self-adjoint operator), the iteration process for the poles located at a great distance from the real axis may not converge because of the poor initial approximation.

Let us find out now what condition is necessary for the convergence of the iterations of (4.3). It is evident that the condition (4.6) must be satisfied at the fixed point $z_R$ proper, i.e. the point must be attractive. At this point the resolvent $G(z)$ continued as an operator from $\Phi$ to $\Phi^*$ has a pole on the nonphysical sheet:

$$G(z) \underset{z\to z_R}{\sim} \frac{|\varphi_R\rangle\langle\bar{\varphi}_R|}{z - z_R}. \tag{4.7}$$

Here $\varphi_R$ and $\bar{\varphi}_R$ are the Gamow functions of the resonance state. In terms of the RHS, $\varphi_R$ is the right-hand GEV of the operator H coresponding to the eigenvalue $z_R$, and $\bar{\varphi}_R$ is the left-hand GEV corresponding to the eigenvalue $z_R^*$, i.e. $\bar{\varphi}_R(z_R) = \varphi_R(z_R^*)$. In the coordinate representation the asymptotic behaviour of these functions is

$$\varphi_R(r) \underset{r\to\infty}{\sim} \frac{e^{ik_R r}}{r}, \qquad \bar{\varphi}_R(r) \sim \frac{e^{-ik^*_R r}r}{r}, \tag{4.8}$$

$$G(r, r'; z) \underset{z\to z_R}{\sim} \frac{e^{ik_R(r-r')}}{rr'},$$

where

$$k_R = \sqrt{2\mu z_R}, \qquad k_R = k_1 - ik_2, \quad k_1 > 0, \quad k_2 > 0. \tag{4.9}$$

Thus, the necessary condition for the iterations to converge is

$$\left|\frac{dF}{dz}\right|_{z=z_R} = \left|1 - \frac{1}{\langle\varphi\,|\,\varphi_R\rangle\langle\bar{\varphi}_R\,|\,\varphi\rangle}\right| \leq p < 1. \tag{4.10}$$

It is evident that the closer the overlap value $\langle \varphi \mid \varphi_R \rangle$ (and $\langle \bar{\varphi}_R \mid \varphi \rangle$) is to unity, the smaller is the convergence parameter $p$, hence the faster is the convergence of the iterations and the larger is the convergence region. The Gamow functions entering (4.7) are assumed to be normalized by the condition

$$\langle \bar{\varphi}_R \mid \varphi_R \rangle = 1 .$$

Their normalization is discussed in Sections 1.1. and 2.2. Therefore, the proximity of the quantity $\langle \varphi \mid \varphi_R \rangle$ to unity defines the proximity of the function $\varphi$ to the exact Gamow function $\varphi_R$ describing the required resonance state. The closer $\varphi$ is to $\varphi_R$, the better is the convergence of the iterations.

On the other hand, it is clear that $\varphi$ can always be so selected that $\langle \varphi \mid \varphi_R \rangle$ would be small. In this case the fixed point $z_R$ gets repulsive and the iterations do not converge. Now, if $\varphi$ is taken to be orthogonal to $\varphi_R$ in this sense, i.e. $\langle \varphi \mid \varphi_R \rangle = 0$ , then the matrix element $\langle \varphi \mid G \mid \varphi \rangle$ will not contain the pole at $z_R$ and, therefore, $z_R$ will not, generally, be a solution for (4.1). Thus, for successful application of the described projection formalism, the function $\varphi$ must be selected to be "close" to the function of the resonance sought. If the resonance is sufficiently narrow, i.e. if $E_R \gg \Gamma$ then a good approximation will be a function obtained by diagonalizing the Hamiltonian on the basis of quadratically integrable functions. In this case the initial approximation $z_0$ will also be a good approximation of $z_R$. Essentially, it is this circumstance that justifies the well-known procedure of finding the excited (i.e. resonance) states in atoms, molecules, and nuclei by diagonalizing the matrix of the Hamiltonian on a suitable $L_2$-basis (for more details see Chapter 7 and Section 4.6).

Let us again rewrite the expression (4.1):

$$z = z_0 + \langle \varphi \mid H \, \tilde{G}(z) \, H \mid \varphi \rangle = z_0 + \Delta(z) - i \, \Gamma(z)/2 =$$

$$= E_R - i \, \Gamma(z)/2 . \tag{4.11}$$

Here $E_R(z) = z_0 + \Delta(z)$ is the "energy", and $\Gamma(z)$ the "width" of the resonance; $\Delta(z)$ is the real energy shift. Let us now write down schematically the spectral representation for the resolvent

$$\tilde{G}(z) = \sum_{\alpha} \oint \int_{\Lambda_\alpha} \frac{\mid \tilde{\psi}_\alpha(\varepsilon_\alpha) \rangle \, \langle \tilde{\psi}_\alpha(\varepsilon_\alpha) \mid}{z - \varepsilon_\alpha} \, d\varepsilon_\alpha +$$

$$+ \sum_i \frac{\mid \tilde{\psi}_i \rangle \, \langle \tilde{\psi}_i \mid}{z - \varepsilon_i} \tag{4.12}$$

where $\displaystyle\sum_{\alpha}\!\!\!\!\!\int$ means the sum and the integral (in case of the many-particle problem) over all the branches of the continuous spectrum; $\sum_i$ includes the full discrete spectrum. Then the first iteration (4.11) yields ($z_0$ is real):

$$\Gamma^{(1)}(z_0) = -2\,\mathrm{Im}\,(\langle\varphi|\,H\,\tilde{G}(z_0 + i\varepsilon)\,H\,|\varphi\rangle =$$

$$= \sum_{\alpha}\!\!\!\!\!\int \Gamma_\alpha^{(1)}(z_0) \tag{4.13}$$

where

$$\Gamma_\alpha^{(1)}(E) = 2\pi\,|\langle\varphi|\,H\,|\tilde{\psi}_\alpha(E)\rangle|^2\,. \tag{4.14}$$

Here $\tilde{\psi}_\alpha(E)$ is the orthogonalized continuum wave function introduced in Section 4.2; the sum (integral) in (4.13) is taken over all open channels.

The expression (4.14) gives the partial decay width, i.e. the probability for the decay of the prepared state $\varphi$ to channel $\alpha$ with energy $E$ and is a smooth function of energy. The full probability of decay is the sum (integral) of all partial widths.

The shift $\Delta(E)$ is related to the width of $\Gamma(E)$ at real energies $E$ through the dispersion-like relation, obtained in [39]:

$$\Delta^{(1)}(E) = \sum_{\alpha}\!\!\!\!\!\int \Delta_\alpha^{(1)}(E) + \Delta_{\mathrm{disc}}\,, \tag{4.15a}$$

$$\Delta_\alpha^{(1)}(E) = \frac{1}{2\pi}\,P\int_{\Lambda_\alpha} d\varepsilon\,\frac{\Gamma_\alpha^{(1)}(\varepsilon)}{E - \varepsilon}\,. \tag{4.15b}$$

The integral in (4.15b) is taken over the spectrum in channel $\alpha$ and the sum (4.15a) contains all the channels, including the closed ones; $\Delta_{\mathrm{disc}}$ denotes the contribution of the discrete spectrum from the last sum in the spectral representation (4.12). If the function $\varphi$ is close to the resonance function, and the energy $z_0$ to the resonance energy $z_R$ (this is possible only in case of narrow resonances), then the quantity $\Gamma_\alpha^{(1)}(z_0)$ has the meaning of partial resonance width, and $\Gamma^{(1)}(z_0)$ of total width. In this case,

$$\Gamma \ll E_R$$

$$\Delta(z_R) \sim \Delta(E_R)\left(1 + O\left(\frac{\Gamma}{E_R}\right)\right)$$

$$\Gamma(z_R) \sim \Gamma(E_R)\left(1 + O\left(\frac{\Gamma}{E_R}\right)\right)$$

and the total resonance width is

$$\Gamma = \Gamma(E_R) = \sum_{\substack{\text{over open} \\ \text{channels}}} \Gamma_\alpha(E_R) \qquad (4.16)$$

where the partial width in channel $\alpha$ is

$$\Gamma_\alpha = \Gamma_\alpha(E_R) = 2\pi \, |\langle \varphi| \, H \, |\tilde{\psi}_\alpha(E_R)\rangle|^2 \qquad (4.17)$$

which corresponds to Fermi's "golden rule" [40].

### 4.4.2. Iterative Inference of Resonance Parameters and Decay Amplitude

All the formulae given hitherto are general. Consider, now the one-particle problem with the Hamiltonian

$$H = H_0 + V \qquad (4.18)$$

To define $\tilde{G}$, we use the projected Lippmann–Schwinger equation (2.14):

$$\tilde{G} = \tilde{G}_0 + \tilde{G}_0 V \tilde{G}$$

generating the iterative series (2.16). In Section 4.2 it was noted that this series could be made rapidly convergent if sufficiently many functions are included in the projector. Here we wish to limit ourselves to the case of a one-dimensional projector. In this case the resonance location is inferred from the solution of the transcendental equation (4.1).

$$z_R = z_0 + \langle \varphi| \, H \, \tilde{G}(z_R) \, H \, |\varphi\rangle \, . \qquad (4.19)$$

The eigenfunction of the bound state of the Hamiltonian $H_0 + \beta V$ at the value of $\beta > 1$ for which the resonance would be a bound state is taken here to be the function $\varphi$. If the Hamiltonian has no other bound states, the series (2.16) will be convergent. Furthermore, the quantity $\langle \varphi| \varphi_R \rangle$ is not very different from unity if $\beta$ is not too large compared with unity. Therefore, equation (4.19) can be solved by iterations:

$$z^{(k)} = z_0 + \langle \varphi| \, H \, \tilde{G}(z^{k-1}) \, H \, |\varphi\rangle \, . \qquad (4.20)$$

Substitution of the series (2.16) in (4.20) yields

$$z^{(k)} = z_0 + \sum_{n=0}^{N} \langle \varphi| \, H \, \tilde{G}_0(z^{(k-1)}) \, [V \tilde{G}_0(z^{(k-1)})]^n \, H \, |\varphi\rangle \, . \qquad (4.21)$$

The number $N$ of terms in the series is determined by the convergence rate of the series (2.16). Various schemes of the double iteration process may be used; at $N = k - 1$, in particular, we obtain the diagonal iterations which reduce the computation involved.

Beginning from $k = 2$, the energies in the right-hand sides of (4.20) and (4.21) get complex, so it is necessary to add an appropriate residue to the integrals because $\tilde{G}_0(z)$ must be taken on the second sheet. However, in the case of narrow resonances $\Gamma \ll E_R$, $E_R$ may be used instead of $z_R$ in the right-hand side of (4.20)

$$z_R = E_R - i\Gamma/2 = E_0 + \langle \varphi | \, H \, \tilde{G}(E_R + i\varepsilon) \, H \, | \varphi \rangle \, .$$

This leads to the following iterations for the resonance energy:

$$E_R^{(k)} = E_0 + \mathrm{Re} \, \langle \varphi | \, H \, \tilde{G}(E_R^{(k-1)}) \, H \, | \varphi \rangle \tag{4.22}$$

and to Fermi's "golden rule" for the width:

$$\Gamma = 2\pi \int |\langle \varphi | \, H \, | \tilde{\psi}_p \rangle|^2 \, \delta \left( E_R - \frac{p^2}{2m} \right) \mathrm{d}^3 p \tag{4.23}$$

where $\tilde{\psi}_p$ is the orthogonalized scattering function corresponding to the incoming wave with momentum $\boldsymbol{p}$ and real energy $E_R = p^2/2m$. If the potential is spherically symmetric, the resonance state has a certain angular momentum $l$. Therefore, $\varphi$ must correspond also to this value of $l$. Then, the integral over angles in (4.23) is removed and the following simple relation remains:

$$\Gamma_l = 2\pi \, |\langle \varphi_l | \, H \, | \tilde{\psi}_l(E_R) \rangle|^2 \, . \tag{4.24}$$

It is expedient here to make the following important remarks. As follows from the formulae obtained the application of the projecting formalism, in particular, of the exact scattering functions $\tilde{\psi}_p$ in the subspace $\mathcal{H}_Q$, is absolutely necessary when calculating the prepared-state decay rate. In the opposite case, the final--state function is not orthogonal to the initial-state function and, as a rule, a contribution of the so-called "false" transitions appears (for example, at $H = 1$ in (4.24)). Therefore, the replacement of $\tilde{\psi}_p$ by a plane wave $|\boldsymbol{p}\rangle$ in terms of the simple perturbation theory will inevitably include a contribution of the false transitions. The correct first approximation for $\Gamma_l$ arises from the replacement of $\tilde{\psi}_l(E_R)$ by $\tilde{\psi}_l^0(E_R)$, i.e., by a "plane" wave in the subspace $\mathcal{H}_Q$, which is known in the literature [41] as the orthogonality scattering function (see (2.14b)).

Inded, the perturbation series for the scattering wave function in $\mathcal{H}_Q$ corresponding to the series (2.16) is of the form:

$$\tilde{\psi} = \tilde{\psi}_0 + \tilde{G}_0 V \tilde{\psi}_0 + \tilde{G}_0 V \tilde{G}_0 V \tilde{\psi}_0 + \ldots \tag{4.24a}$$

(where for simplicity we have omitted the subscript $l$).
This series can be readily obtained either by iterations of the corresponding Lippmann-Schwinger equation for $\tilde{\psi}$ in $\mathcal{H}_Q$, i.e. $\tilde{\psi} = \tilde{\psi}_0 + \tilde{G}_0 V \tilde{\psi}$, or by using (2.16).

In many cases the above mentioned Born series in $\mathscr{H}_Q$ is the simplest and most natural way of finding $\tilde{\psi}(E_R)$ and, thereby, the decay probability. The corresponding perturbation series for the decay amplitude $A = \langle \varphi | H | \tilde{\psi}(E_R) \rangle$ is

$$A = \langle \varphi | H | \tilde{\psi}_0 \rangle + \langle \varphi | H \tilde{G}_0 V | \tilde{\psi}_0 \rangle + \dots$$

$$= A^{(0)} + A^{(1)} + A^{(2)} + \dots \tag{4.24b}$$

Since on imposing the perturbation which results in the prepared-state decay (i.e. on imposing the coupling between the subspaces $\mathscr{H}_P$ and $\mathscr{H}_Q$ through the Hamiltonian $H$), the inner part of the initial state changes only slightly (at $\Gamma/E_R \ll 1$), then the rescattering of the decay particles in the peripheral region of the interaction potential corresponds physically to the perturbation series (4.24b). This effect manifests itself in a most interesting way in the problem of many-particle resonance decay where characteristic resonances may exist in some pairs. Then, the series of the type of (4.24b) will correspond to successive rescatterings of the decay particles in the final state (see below).

We shall now derive the expression (4.23) for the resonance width using the unitarity relation for the $t$-matrix instead of the spectral representation of the complete resolvent (4.12). Such a method is also applicable to the many-particle case and is advantageous in that only the states on the mass shell are used (as is the case in the unitarity relations [12]). According to (4.11) the width is

$$\Gamma = \Gamma(z_R) = -2 \, \mathrm{Im} \, \langle \varphi | H \, \tilde{G}(z_R) \, H | \varphi \rangle \tag{4.25}$$

In the case of narrow resonances, $\Gamma(z_R) \approx \Gamma(E_R)$, i.e.

$$\Gamma = -2 \, \mathrm{Im} \, \langle \varphi | H \, \tilde{G}(E_R + i\varepsilon) \, H | \varphi \rangle$$

$$= i \, \langle \varphi | H \, \Delta \tilde{G}(E_R) \, H | \varphi \rangle \tag{4.26}$$

where

$$\Delta \tilde{G}(E) = \tilde{G}(E + i\varepsilon) - \tilde{G}(E - i\varepsilon) \equiv \tilde{G}^{(+)} - \tilde{G}^{(-)}$$

is the resolvent discontinuity through the cut. The resolvent discontinuity may be expressed in terms of the $t$-matrix discontinuity which is in turn determined by the unitarity relation. Before presenting the relevant formulae, it should be noted that we need a discontinuity of the *orthogonalized* resolvent, i.e.

$$\lim_{\lambda \to \infty} \Delta \, \tilde{G}(\lambda) \, .$$

However, we may readily go over to the limit $\lambda \to \infty$ in the final relation for the width. This is one of the advantages of the projecting method described here, compared with the conventional Feshbach approach. For the sake of simplicity we shall write $G(z)$ instead of $\tilde{G}(z, \lambda) = (z - H - \lambda P)^{-1}$.

The relation between the $t$-matrix and the resolvent $t = V + VGV$ may be formally inverted:

$$G = -V^{-1} + V^{-1}tV^{-1} .\qquad (4.27)$$

Here,

$$V^{-1} = G_0(E + i\varepsilon) + t^{-1}(E + i\varepsilon) = G_0(E - i\varepsilon) + t^{-1}(E - i\varepsilon) \qquad (4.28)$$

is the Hermitian operator (if the potential is Hermitian). Although the operators $V^{-1}$ and $t^{-1}$ are not defined in all cases, the *formal* operations with them can be carried out (see [30]). Because of the hermiticity of $V^{-1}$ we have

$$\Delta G = V^{-1} \Delta t\, V^{-1} \qquad (4.29)$$

where

$$\Delta t = t(E + i\varepsilon) - t(E - i\varepsilon) \equiv t^{(+)} - t^{(-)} .$$

From (4.28) we obtain

$$t^{-1}(E + i\varepsilon) - t^{-1}(E - i\varepsilon) =$$
$$= -G_0(E + i\varepsilon) + G_0(E - i\varepsilon) = -\Delta G_0(E) . \qquad (4.30)$$

Multiplying (4.30) by $t(E + i\varepsilon)$ on the right, and by $t(E - i\varepsilon)$ on the left, we find:

$$\Delta t = t^{(+)} \Delta G_0 t^{(-)} . \qquad (4.31)$$

The discontinuity of the free resolvent through the cut may be found using the first resolvent identity

$$\Delta G_0 = G_0(E + i\varepsilon) - G_0(E - i\varepsilon) =$$
$$= -2\, i\varepsilon G_0(E + i\varepsilon)\, G_0(E - i\varepsilon) =$$
$$= -2\, i\varepsilon \left[ (E - H_0)^2 + \varepsilon^2 \right]^{-1} .$$

From here, with the help of the relation

$$\pi\, \delta(x) = \lim_{\varepsilon \to 0} \varepsilon (x^2 + \varepsilon^2)^{-1}$$

we obtain in the limit $\varepsilon \to 0$

$$\Delta G_0(E) = -2\pi\, i\delta(E - H_0) . \qquad (4.32)$$

Thus,

$$\Delta t(E) = -2\pi\, it^{(+)}\delta(E - H_0)t^{(-)} . \qquad (4.33)$$

This is the unitarity relation for the $t$-matrix[5]. Substituting it into (4.29) and using (4.28), we obtain

$$\Delta G = -2\pi i (1 + G_0^{(+)} t^{(+)}) \, \delta(E - H_0) \, (1 - t^{(-)} G_0^{(-)}) \, .$$

Let us use now the spectral representation for $H_0$

$$\delta(E - H_0) = \int d^3 p \, \delta\left(E - \frac{p^2}{2m}\right) |p\rangle \langle p| \qquad (4.34)$$

and the relation

$$|\psi_p\rangle = (1 + G_0^{(+)} t^{(+)}) \, |p\rangle$$

where $|p\rangle$ is the "plane" wave in $\mathcal{H}_Q$. Then we obtain finally:

$$\Delta G = -2\pi i \int d^3 p \, \delta\left(E - \frac{p^2}{2m}\right) |\psi_p\rangle \langle \psi_p| \qquad (4.35)$$

and, thus,

$$\Gamma = 2\pi \int |\langle \varphi | H | \tilde{\psi}_p \rangle|^2 \delta\left(E - \frac{p^2}{2m}\right) d^3 p \, . \qquad (4.36)$$

In the final expression (4.36) we have gone over to the limit $\lambda \to \infty$ thereby giving rise to the orthogonalized scattering function $\tilde{\psi}_p$. Of course, to obtain (4.35), we could immediately use the expression for $\Delta G$ similar to (4.32), namely, $\Delta G = -2\pi i \delta(E - H)$ and the spectral representation of the operator $H$. In the many-particle case, however, the spectral representation of the total Hamiltonian is difficult to write. The basic idea of the derivation presented here is that we use the unitarity relation for the transition operator (4.33) and the spectral representation (4.34) for the asymptotic Hamiltonian (the asymptotic Hamiltonians in the many-particle case).

In the case where the complex resonance energy $z_R$ has already been found, we can calculate the resonance wave function (i.e. the Gamow wave function) using the homogeneous (non-projected) Lippmann–Schwinger equation continued to the second sheet, i.e.

$$\varphi_R = G_0^{II}(z_R) V \varphi_R \, . \qquad (4.37)$$

If the examined resonance corresponds to the maximum eigenvalue of the kernel, i.e. bound states and other near resonances are absent in the system, it is convenient to solve (4.37) by iterations at the fixed energy $z_R$ found by the method described above:

$$\varphi_R^{(k)} = G_0^{II}(z_R) V \varphi_R^{(k-1)} \, .$$

---

[5] We must emphasize that all relations here are taken for finite $\lambda$. By $H_0$ we understand $H_0(\lambda) = H_0 + \lambda P$ .

### 4.4.3. Examples

Let us examine two examples illustrating the effectiveness of the projecting formalism described above.

(1) As the first example we shall examine an exactly solvable model of resonance, namely a particle of mass $m$ with angular momentum $l = 0$ in a rectangular potential well of depth $V_w$ and width $R_w$ confined by a rectangular barrier of height $V_B$ and width $R_B - R_w$ (see Fig. 4.10). This example is often used (see, for example, [42], [43]) for illustrative purposes.

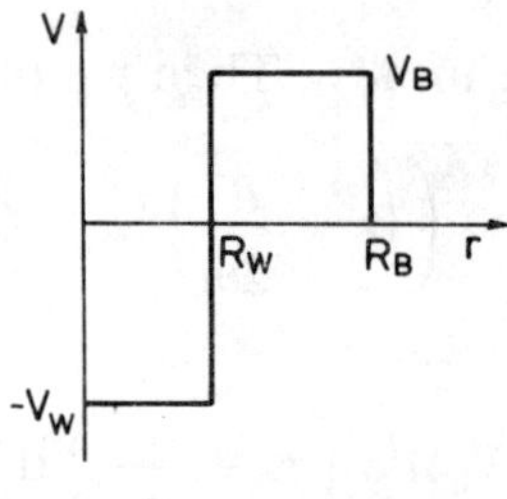


Fig. 4.10

The parameters of the potential must be so selected that the well would not contain levels other than the required quasistationary state, whereupon the one-dimensional projector $P = |\varphi\rangle \langle\varphi|$ may be used. The eigenfunction of the Hamiltonian with infinitely wide barrier, i.e. at $R_B = \infty$ , which corresponds to a completely confined particle will be taken to be $\varphi$. It is clear that in the case of a sufficiently wide barrier the function $\varphi$ in the inner region $r < R_w$ is very close to the resonance function. Table 4.1 presents the results of calculating the

Table 4.1

The convergence of resonance energies and widths for a particle confined by a rectangular potential barrier with the following parameters: $R_W = 3.5$ , $V_W = 0.5$ , $V_B = 4.5$ (see Fig. 4.10).

| $R_B$ | $z_R$ exact | Direct summation | Number of energy iterations | Number of terms of perturbative series | Padé approximation |
|---|---|---|---|---|---|
| 3.75 | 0.783 6–1.00 i | 0.789 1–1.01 i | 12 | 8 | 0.783–1.00 i |
| 4.00 | 1.250 –0.605 i | 1.254 –0.608 i | 6 | 16 | 1.254–0.607 i |
| 4.25 | 1.419 –0.380 i | 1.415 –0.375 i | 3 | 53 | 1.417–0.377 i |
| 4.50 | 1.500 –0.116 i | 1.501 –0.201 i | 2 | 72 | 1.500–0.118 i |
| 5.50 | 1.592 –0.046 6 i | divergent | 2 | – | 1.592–0.046 9 i |
| 5.75 | 1.598 –0.030 9 i | divergent | 2 | – | 1.597–0.031 5 i |

resonance energy and width at various values of the barrier width [7]. The transcendental equation for $z_R$ was solved by iterations using the convergent series (2.16). The table shows the convergence of the iterations (4.20) and the number N of the series terms necessary for the prescribed precision to be obtained. The resonance width $\Gamma$ decreases as the barier width increases because $\Gamma$ is proportional to the barrier penetrability. In this case the initial function $\varphi$ becomes an ever improving approximation for the resonance function, so the iterations in $z$ converge more rapidly. On the other hand, the series for $\tilde{G}$ converges less and less well with increasing $R_B$ until it becomes divergent. This divergence is due to the appearance of a bound state in the mirror Hamiltonian $H_0 - V$. Therefore, a negative eigenvalue the magnitude of which exceeds unity appears in the kernel $\tilde{G}_0 V$. It is this eigenvalue that causes the divergence of the series (2.16). To speed up the convergence of the series, it is convenient to use the Padé approximant technique (see Chapter 1) which also yields good results in case of a divergent series (see the last column of Table 4.1).

(2) As the second example we shall examine the nuclear resonances in the systems consisting of an $\alpha$-particle and a neutron which interact through the Woods–Saxon potential with spin-orbit coupling [44]:

$$V(r) = V_1(r) + \frac{\Delta}{r} \frac{\mathrm{d}}{\mathrm{d}r} V_2(r) \tag{4.38a}$$

where

$$V_i(r) = \frac{C_i}{1 + \exp\left((r - R_i)/a\right)} . \tag{4.38b}$$

The parameters of the potential are presented in Table 4.2. To simplify the calculations, the potential (4.38) was approximated by a sum of Gaussians and all the calculations were carried out with this approximate potential. The results

Table 4.2

Parameters of the Woods-Saxon potential for the system $\alpha$ + n.

|       | $C$(MeV) | $a$(fm) | $R$(fm) | $\Delta$(fm$^2$) |           |
|-------|----------|---------|---------|------------------|-----------|
| $V_1$ | $-43$    | 0.7     | 2       | 0.5 for          | $j = 3/2$ |
| $V_2$ | 40       | 0.35    | 1.5     | $-1$ for         | $j = 1/2$ |

for the exact potential (4.38) differ somewhat from those presented in Table 4.3. The bound state function in the potential $\beta V (\beta > 1)$ calculated by the variational method was used as the initial function $\varphi$. The results of the calculations for the states $l = 1$, $j = 1/2$ and $j = 3/2$ (the low-lying levels $J^\pi = 1/2^-$

and $3/2^-$ of $^5$He nucleus correspond to these resonances) and the convergence of the iterations (4.20) are presented in Table 4.3. The Padé approximants were used in all cases to sum up the series (4.18). From the Table it may be seen that

Table 4.3

Convergence of the iterations in energy for the resonance parameters (i.e. Re $z_R$ and Im $z_R$) for the $j = 3/2(p_{3/2})$ and $j = 1/2(p_{1/2})$ states in the $\alpha + n$ system.

| $k$ | $j = 3/2$, $\beta = 1.9$ | | $j = 1/2$, $\beta = 1.3$ | |
|---|---|---|---|---|
| | Re $z_R$ | Im $z_R$ | Re $z_R$ | Im $z_R$ |
| 0 | 2.3377 | 0.0 | 3.842 9 | 0.0 |
| 1 | 1.2766 | −0.661 2 | 2.490 5 | −1.651 4 |
| 2 | 0.998 9 | −0.615 2 | 1.957 3 | −2.021 0 |
| 3 | 0.934 2 | −0.539 0 | 1.743 2 | −2.215 4 |
| 4 | 0.937 9 | −0.499 0 | 1.652 0 | −2.363 8 |
| 5 | 0.951 2 | −0.489 6 | 1.627 6 | −2.486 3 |
| 6 | 0.957 4 | −0.491 2 | 1.650 0 | −2.584 0 |
| PA summation | 0.957 6 | −0.494 4 | 1.812 4 | −2.521 0 |

even in the case of a very broad resonance $l = 1$, $j = 1/2$, $z_R = 1.812\ 4 - i\ 2.521\ 0$ (MeV) this method permits reliable calculations of location and width. The last line of Table 4.3 shows the results of extrapolating the sequence $\{z^{(k)}\}$ by the Padé approximants of type II. Thus, the double application of the Padé approximant technique has substantially reduced the calculations. As in the previous example, the requirement of simultaneous good convergence of the Born series (2.16) and of the iterations (4.20) does not prove to be realizable. Thus, as $\beta$ increases the convergence of the series improves, but $\varphi(\beta)$ approximates the resonance function $\varphi_R$ less well and the convergence of the iterations in z deteriorates. Therefore, we must select a compromise initial function $\varphi$ in order to produce convergence of the iterations (4.20) and convergence (though not very rapid) of the rescattering series (2.16).

A method similar to that discussed above has been recently proposed by Domcke and collaborators [46] in the context of electron molecule scattering.

## 4.5. Resonances in Three-Particle System

It is clear that, since the particles emerging as a result of the three (and more)-particle system decay will be strongly rescattered by each other under certain conditions, the Faddeev theory for many-body scattering has to underlie any mathematical treatment of the process. The relevant many-particle equations must be written, however, in a certain subspace $\mathcal{H}_Q$ orthogonal to the prepared-state function. The projecting method described above is very convenient for just this type of problem (it should be noted that the traditional projecting formalism may be faced here with appreciable difficulties). It is also evident that to study the decay of the genuine three-body resonance which can decay in both two- and three-particle channels is a difficult task, irrespective of the examination method use.

Let us examine a three-particle system with two-body potentials $V_\alpha$, where $\alpha = 1, 2, 3$ are different two-cluster decompositions $\alpha + (\beta\gamma)$. All the general formulae of Section 4.4.1 are valid in the case of an arbitrary system, but the projected resolvent $\tilde{G}$ in the three-particle case must be defined in terms of the Faddeev equations. The complete resolvent is expressed as the sum of three Faddeev components

$$\tilde{G} = \tilde{G}_0 + \tilde{G}^{(1)} + \tilde{G}^{(2)} + \tilde{G}^{(3)} \tag{5.1}$$

which satisfy the set of "projected" Faddeev equations [8]:

$$\tilde{G}^{(\alpha)} = \tilde{G}_\alpha - \tilde{G}_0 + \tilde{G}_\alpha V_\alpha (\tilde{G}^{(\beta)} + \tilde{G}^{(\gamma)}) \tag{5.2}$$

$$(\alpha\,\beta\,\gamma) = 123,\, 231,\, 312.$$

All the orthogonalized Green functions, i.e. $\tilde{G}_\alpha$ and $\tilde{G}_0$, are related to the conventional Green functions through (2.15) in the following way:

$$\tilde{G}_i = G_i - G_i\, P(PG_i\, P)^{-1} PG_i\,, \qquad i = 0, 1, 2, 3$$

and $G_\alpha$ are the channel resolvents, i.e. the "two-particle" resolvents in three--particle space:

$$G_\alpha(z) = (z - H_0 - V_\alpha)^{-1}\,.$$

The iterative series for $\tilde{G}$ which may be inferred take a more amenable form if we introcuce the orthogonalized $t$-matrices satisfying the relations

$$\tilde{G}_\alpha V_\alpha = \tilde{G}_0 \tilde{t}_\alpha\,,$$

$$\tilde{G}_\alpha - \tilde{G}_0 = \tilde{G}_0 t_\alpha \tilde{G}_0\,. \tag{5.3}$$

Iterating the system (5.2) and substituting the components $\tilde{G}^{(\alpha)}$ in the sum (5.1), we obtain the iterative series for a complete orthogonalized resolvent:

$$\tilde{G} = \tilde{G}_0 + \sum_{\alpha=1}^{3} \tilde{G}_0 \tilde{t}_\alpha \tilde{G}_0 + \sum_{\alpha \neq \beta} \tilde{G}_0 \tilde{t}_\alpha \tilde{G}_0 \tilde{t}_\beta \tilde{G}_0 + \dots \tag{5.4}$$

(see the relevant explanations in Chapter 1). The conditions for the convergence of the series (5.4) are formulated in [8]. As in the case of one-particle scattering, the series can be made convergent by including sufficiently many functions in the projector. Here we again restrict ourselves to the one-dimensional projector. If the series (5.4) diverges, we may use the Padé-summation technique or solve the system (5.2) by direct matrix inversion. Substituting (5.4) in (4.1), we obtain an equation for finding the three-body resonance energy, thereby generalizing the two-particle case (4.20) directly:

$$z_R = z_0 + \langle \Phi| H \, \tilde{G}_0 \, H \, |\Phi\rangle + \sum_{\alpha=1}^{3} \langle \Phi| H\tilde{G}_0 \, \tilde{t}_\alpha \, \tilde{G}_0 H \, | \Phi\rangle$$

$$+ \sum_{\alpha \neq \beta} \langle \Phi| HG_0 \, t_\alpha \, F_0 \, t_\beta \, G_0 H \, |\Phi\rangle + \ ...$$

Now, an expression for the width of a narrow resonance $(\Gamma \ll E_R)$ will be obtained using again the unitarity relations [30]. (In contrast to the two-particle case, the use of a three-particle relation of unitarity here is of decisive importance). According to (4.26),

$$\Gamma \approx \Gamma(E_R) = \mathrm{i} \, \langle \Phi| \, H \, \Delta\tilde{G}(E_R)H \, |\Phi\rangle \, . \tag{5.5}$$

Let us define the transition operators from channel $\alpha$ to channel $\beta$:

$$U_{\beta\alpha}^{(+)}(z) = \overline{V}_\beta + \overline{V}_\beta \, G(z) \, \overline{V}_\alpha \, , \tag{5.6a}$$

$$U_{\beta\alpha}^{(-)}(z) = \overline{V}_\alpha + \overline{V}_\beta \, G(z) \, \overline{V}_\alpha \tag{5.6b}$$

where $\overline{V}_\alpha = V - V_\alpha$, $V = V_1 + V_2 + V_3$; $\alpha, \beta = 0, 1, 2, 3$. Here $\alpha = 0$ denotes the three-particle channel, so that $\overline{V}_0 \equiv V$. Let the complete resolvent $G$ be formally expressed in terms of the diagonal operator $U_{\alpha\alpha} = U_{\alpha\alpha}^{(+)} = U_{\alpha\alpha}^{(-)}$

$$G = -\overline{V}_\alpha^{-1} + \overline{V}_\alpha^{-1}U_{\alpha\alpha}\overline{V}_\alpha^{-1} \, . \tag{5.7}$$

The operator $\overline{V}_\alpha^{-1}$ is Hermitian and related to $U_{\alpha\alpha}^{-1}$ as

$$\overline{V}_\alpha^{-1} = U_{\alpha\alpha}^{-1}(E + \mathrm{i}\varepsilon) + G_\alpha(E + \mathrm{i}\varepsilon) = U_{\alpha\alpha}^{-1}(E - \mathrm{i}\varepsilon) + G_\alpha(E - \mathrm{i}\varepsilon)$$

$$\tag{5.8}$$

The relation (5.8) follows from the Lippmann–Schwinger equation for $U_{\alpha\alpha}$

$$U_{\alpha\alpha} = \overline{V}_\alpha + \overline{V}_\alpha G_\alpha U_{\alpha\alpha} \, .$$

All the energy-dependent operators are taken on the upper rim of the cut at $E + \mathrm{i}\varepsilon$ and the operators labelled (*) are taken at $E - \mathrm{i}\varepsilon$ . The following expression for the complete resolvent discontinuity follows from (5.7) and from the hermiticity of $\overline{V}_\alpha$:

$$\Delta G = \overline{V}_\alpha^{-1} \, \Delta U_{\alpha\alpha} \, \overline{V}_\alpha^{-1} \, . \tag{5.9}$$

Here, as in the two-particle case we shall again treat the three-particle projector $\lambda|\Phi\rangle \langle \Phi|$ as included in the free three-body Hamiltonian and go over to the limit

$\lambda \to \infty$ only in the final expression. Use will be made now of the three-particle unitarity relation for the transition operators which is derived in the same way as the relation (4.33) (see [30]):

$$\Delta U_{\alpha\alpha} = -2\pi i \left\{ \sum_{\gamma=1}^{3} \sum_{n} U_{\alpha\gamma} \Delta_{\gamma n} U_{\gamma\alpha}^{(*)} + U_{0\alpha} \Delta_0 U_{0\alpha}^{(*)} \right\}. \tag{5.10}$$

Here

$$\Delta_{\gamma n} = \int d\mathbf{q}_\gamma \, |\varphi_{\gamma n, \, \mathbf{q}_\gamma}\rangle \, \delta\left( E - \frac{q_\gamma^2}{2\mu_\gamma} - \varepsilon_{\gamma n} \right) \langle \varphi_{\gamma n, \, \mathbf{q}_\gamma}| \tag{5.11}$$

is the contribution to the discontinuity $\Delta G_\gamma$ from the bound state of the subsystem; $\varphi_{\gamma n}$ is the channel wave function in channel $\gamma n$ (see the formulae appearing after (5.21));

$$\Delta_0 = -\frac{1}{2\pi i} \Delta G_0 = \delta(E - H_0)$$

$$= \int |\varphi_{\mathbf{q}_\gamma \mathbf{p}_\gamma}\rangle \, \delta\left( E - \frac{q_\gamma^2}{2\mu_\gamma} - \frac{p_\gamma^2}{2m_\gamma} \right) \langle \varphi_{\mathbf{q}_\gamma \mathbf{p}_\gamma}| \tag{5.12}$$

is the discontinuity of the free three-particle resolvent $G_0$. The operators are taken on the mass shell, so that

$$\bar{V}_\alpha^{-1} U_{\alpha\gamma} |\varphi_{\gamma n, \, \mathbf{q}_\gamma}\rangle = \bar{V}_\alpha^{-1} U_{\alpha\gamma}^{(+)} |\varphi_{\gamma n, \, \mathbf{q}_\gamma}\rangle =$$

$$= (1 + G\bar{V}_\gamma) |\varphi_{\gamma n, \, \mathbf{q}_\gamma}\rangle = |\psi_{\gamma n, \, \mathbf{q}_\gamma}\rangle \, ,$$

$$\langle \varphi_{\gamma n, \, \mathbf{q}_\gamma}| \, U_{\gamma\alpha}^{(*)} \bar{V}_\alpha^{-1} = \langle \psi_{\gamma n, \, \mathbf{q}_\gamma}| \, ,$$

$$\bar{V}_\alpha^{-1} U_{\alpha 0} |\varphi_{\mathbf{p}_\gamma \mathbf{q}_\gamma}\rangle = |\psi_{\mathbf{p}_\gamma \mathbf{q}_\gamma}\rangle \, ,$$

$$\langle \varphi_{\mathbf{p}_\gamma \mathbf{q}_\gamma}| \, U_{0\alpha}^{(*)} \bar{V}_\alpha^{-1} = \langle \psi_{\mathbf{p}_\gamma \mathbf{q}_\gamma}| \tag{5.13}$$

where $|\psi_{\gamma n, \, \mathbf{q}_\gamma}\rangle$ is the scattering wave function in the two-particle channel $\gamma n$, i.e. the function corresponding to the incoming wave $|\varphi_{\gamma n, \, \mathbf{q}_\gamma}\rangle$; $|\psi_{\mathbf{p}_\gamma \mathbf{q}_\gamma}\rangle$ is the scattering wave function of the three-particle channel. Substituting (5.10) in (5.9) and considering (5.13), we obtain[6]

$$\Delta G = -2\pi i \left\{ \sum_{\gamma=1}^{3} \sum_{n} \int d\mathbf{q}_\gamma \, |\psi_{\gamma n, \, \mathbf{q}_\gamma}\rangle \delta\left( E - \frac{q_\gamma^2}{2\mu_\gamma} - \varepsilon_{\gamma n} \right) \langle \psi_{\gamma n, \, \mathbf{q}_\gamma}| + \right.$$

$$\left. + \int\int d\mathbf{p}_\gamma d\mathbf{q}_\gamma \, |\psi_{\mathbf{p}_\gamma \mathbf{q}_\gamma}\rangle \delta\left( E - \frac{p_\gamma^2}{2m_\gamma} - \frac{q_\gamma^2}{2\mu_\gamma} \right) \langle \psi_{\mathbf{p}_\gamma \mathbf{q}_\gamma}| \right\} \tag{5.14}$$

---

[6]) We remind the reader that, for the present, this relation is written in complete space, i.e. at a finite value of the orthogonalizing constant $\lambda$ .

Thus, in the limit $\lambda \to \infty$ , we find the expression for the total width of a narrow resonance to be of the form of the sum of partial widths:

$$\Gamma = \sum_{\gamma=1}^{3} \sum_{n} \Gamma_{\gamma n} + \Gamma_0 \tag{5.15}$$

where

$$\Gamma_{\gamma n} = 2\pi \int d\boldsymbol{q}_\gamma \, \delta \left( E - \frac{q_\gamma^2}{2\mu_\gamma} - \varepsilon_{\gamma n} \right) |\langle \Phi \,|\, H \,|\, \tilde{\psi}_{\gamma n, \, \boldsymbol{q}_\gamma} \rangle|^2 \tag{5.16}$$

is the partial width of the decay into the two-particle channel $\gamma n$, i.e. the probability of the decay into the bound state of the pair $(\alpha\beta)$ with energy $\varepsilon_{\gamma n}$ and a third particle $\gamma$ .

$$\Gamma_0 = 2\pi \int\!\!\int d\boldsymbol{p}_\gamma \, d\boldsymbol{q}_\gamma \, \delta \left( E - \frac{q_\gamma^2}{2\mu_\gamma} - \frac{p_\gamma^2}{2m_\gamma} \right) |\langle \Phi \,|\, H \,|\, \tilde{\psi}_{\boldsymbol{p}_\gamma \boldsymbol{q}_\gamma} \rangle|^2 \tag{5.17}$$

is the probability of the genuine three-particle decay. Because of the presence of $\delta$-functions, the two-particle channels contribute to the total width at $E > \varepsilon_{\gamma n}$, while the three-particle channel makes such a contribution only at $E > 0$ . In the case of a three-particle channel it is possible to introduce the partial widths

$$\Gamma_0^\gamma(\boldsymbol{p}_\gamma, \, \boldsymbol{q}_\gamma) = 2\pi \int d\Omega_{\boldsymbol{p}_\gamma} \, d\Omega_{\boldsymbol{q}_\gamma} \, |\langle \Phi \,|\, H \,|\, \psi_{\boldsymbol{p}_\gamma \boldsymbol{q}_\gamma} \rangle|^2 \tag{5.18}$$

corresponding to the probability of the decay to a three-particle state $\boldsymbol{p}_\gamma \boldsymbol{q}_\gamma$, where the relative momentum of the first two particles in channel $\gamma$ falls within the *unit* interval around the values of $p_\gamma$, while the third particle appears in the unit interval of momenta around the value of $q_\gamma$, the momentum of the third particle in the common centre-of-mass system. Then the corresponding composite probability is

$$\Gamma_0(E) = \int p^2 \, dp q^2 \, dq \, \Gamma_0^\gamma(p, q) \, \delta \left( E - \frac{q^2}{2\mu_\gamma} - \frac{p^2}{2m_\gamma} \right) \tag{5.19}$$

All widths $\Gamma_{\gamma n}(E)$ and $\Gamma_0^\gamma(p, q)$ are continuous functions of energy and define the probability of the decay of the prepared state $\varphi$ to the respective channels. In the given case, the dispersion relation for the energy shift (4.14) is

$$\Delta(E) = \frac{1}{2\pi} \sum_{\gamma, \, n} P \int\limits_{\varepsilon \in \Lambda_{\gamma n}} \frac{\Gamma_{\gamma n}(\varepsilon)}{E - \varepsilon} \, d\varepsilon +$$

$$+ \frac{1}{2\pi} \int p^2 \, dp \left\{ P \int q^2 \, dq \, \frac{\Gamma_0^\gamma(p, q)}{E - \dfrac{p^2}{2m_\gamma} - \dfrac{q^2}{2\mu_\gamma}} \right\} + \Delta_{\text{disc}} \, . \tag{5.20}$$

The scattering wave functions $\tilde{\psi}$ satisfy the modified (orthogonalized) Faddeev equations. For example, the components $\tilde{\psi}^{(i)}$ of the projected three-particle wave function

$$\tilde{\psi}_{\gamma n} = \sum_{i=1}^{3} \tilde{\psi}_{\gamma n}^{(i)}$$

are determined by the set

$$\tilde{\psi}_{\gamma n, \, \boldsymbol{q}_\gamma}^{(i)} = \delta_{i\gamma} \tilde{\varphi}_{\gamma n, \, \boldsymbol{q}_\gamma} + \tilde{G}_0 \tilde{t}_i (\tilde{\psi}_{\gamma n, \, \boldsymbol{q}_\gamma}^{(j)} + \tilde{\psi}_{\gamma n, \, \boldsymbol{q}_\gamma}^{(k)}) , \qquad i \neq j \neq k \qquad (5.21)$$

where

$$|\tilde{\varphi}_{\gamma n, \, \boldsymbol{q}_\gamma}\rangle \equiv |\varphi_{\gamma n, \, \boldsymbol{q}_\gamma}\rangle - G_\gamma |\Phi\rangle \langle \Phi | G_\gamma | \Phi\rangle^{-1} \langle \Phi | \varphi_{\gamma n, \, \boldsymbol{q}_\gamma}\rangle .$$

If the iterations of the system (5.21) converge, we use the perturbation theory series to find the width $\Gamma_{\gamma n}$:

$$\Gamma_{\gamma n} = 2\pi \int |A_{\gamma n}^{(0)} + A_{\gamma n}^{(1)} + A_{\gamma n}^{(2)} + ...|^2 \delta \left( E - \frac{q_\gamma^2}{2\mu_\gamma} - \varepsilon_{\gamma n} \right) d\boldsymbol{q}_\gamma$$

$$(5.22)$$

where

$$A_{\gamma n}^{(k)} = \langle \Phi | H | \tilde{\psi}_{\gamma n, \, \boldsymbol{q}_\gamma}^{(k)}\rangle \qquad (5.23)$$

is the decay amplitude corresponding to the rescattering of the $k$-th order in the final state

$$\tilde{\psi}_{\gamma n, \, \boldsymbol{q}_\gamma}^{(k)} = \sum_{\alpha_i \neq \alpha_{i-1}} \prod_{i=1}^{k} [\tilde{G}_0 \tilde{t}_{\alpha_i}] \tilde{\varphi}_{\gamma n, \, \boldsymbol{q}_\gamma} . \qquad (5.24)$$

In (5.24) the sum is taken over all such products containing $k$ factors $\tilde{G}_0 \tilde{t}_\alpha$ that any two neighbouring $t$-matrices pertain to different subsystems. For example,

$$\psi_{1n, \, \boldsymbol{q}_1}^{(2)} = (\tilde{G}_0 \tilde{t}_2 \tilde{G}_0 \tilde{t}_3 + \tilde{G}_0 \tilde{t}_3 \tilde{G}_0 \tilde{t}_2 +$$

$$+ \, G_0 \tilde{t}_1 \tilde{G}_0 \tilde{t}_2 + \tilde{G}_0 \tilde{t}_1 \tilde{G}_0 \tilde{t}_3) \, \tilde{\varphi}_{1n, \, \boldsymbol{q}_1} . \qquad (5.24\text{a})$$

From the expressions obtained it is evident that the higher the values of the pair (off-mass-shell) scattering amplitude i.e. the higher the number of singularities (bound states and resonances) near the three-particle threshold which exist in the two-particle subsystems forming the decaying three-particle system, the more significant is the influence of the terms corresponding to high-multiplicity rescatterings. At the same time, when using a "good" function of the initial state $\Phi$ approximating the true Gamow function inside a sufficiently large hypersphere, as well as at $\Gamma/E_R \ll 1$ , the contribution of the simple rescattering terms will be substantial, whereas the contribution of higher scatterings will be small, except in those situations resembling the Efimov effect [45]. In this sense, the convergence of the rescattering series is apparently better in the case of many--particle resonances than in two-particle case.

For broader resonances or for obtaining a higher accuracy, we have to use the complete equation

$$z_R = z_0 + \langle \Phi| \, H \, \tilde{G}(z_R) \, H \, |\Phi\rangle \tag{5.25}$$

where $z_0 = \langle \Phi| \, H \, |\Phi\rangle$ is the mean value of the three-particle Hamiltonian. In solving (5.25), it is necessary to continue the matrix elements of the three-particle resolvent $\tilde{G}$ to nonphysical sheets.

Publications giving accurate calculations of three-particle resonances (outside the scope of the first order of conventional perturbation theory) are very scanty. The authors know only a few such works and quite rare studies concerning three-body resonance rescatterings in the final state. But even the available works have usually been based on simple model interactions [34, 35]. Nevertheless, as these model examples already contain many essential features of the true pattern, the results inferred from them seen to be of importance also in many realistic situations.

As an example we shall examine here a simple model of three identical particles with the separable $s$-wave Yamaguchi potential

$$V(p, p') = -\lambda \, g(p) \, g(p') \,,$$

$$g(p) = (p^2 + \beta^2)^{-1} \,, \qquad \beta = 1.304 \text{ fm}$$

for the total angular momentum of the three-particle system $L = 1$ or $2$. Furthermore, we shall limit ourselves to the region below the three-particle threshold $z < 0$.

As the two-particle binding energy $z_2$ decreases (i.e. as the two-particle threshold approaches the three-particle one), we get abrupt enhancement of the exchange interaction (resulting from the transition of the third particle from the bound state with the first particle to the bound state with the second particle) [48, 49] due to the identity of the particles and to the "swelling" of the two--particle subsystems. This gives rise to long-range interaction which in turn results at $L = 0$ in the appearance of an infinitely large number of near-threshold three-particle levels (at $z_2 < 0$), the so-called Efimow effect [45]. In the case of $L > 0$ this effect weakens, but at $z_2 \to 0$ the closeness of the three-particle threshold shows itself as the characteristic shape (loop) of the resonance trajectory. This model was studied in [35] where the trajectory of the resonance pole with varying interaction constant $\lambda$ was calculated. At $\lambda > \lambda_1$ a bound state appears in the system. As $\lambda$ decreases this state turns into a resonance and the corresponding pole goes to the second sheet through the root cut arising from the two-particle bound state. As the two-particle interaction strength $\lambda$ decreases, the pole $z_R$ describes a trajectory in the comples plane $z$ on the second sheet coupled to the physical sheet by the root cut and, simultaneously, the root branch point $z_2$ moves along the negative real axis towards the origin. At the threshold value $\lambda \to \lambda_0$ the two-particle bound state "sits down on the

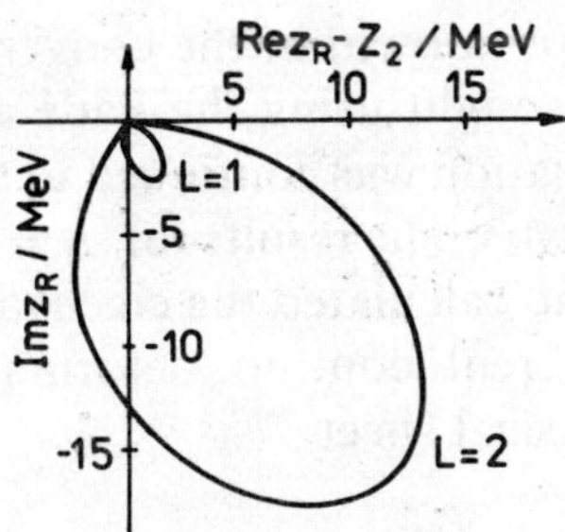


Fig. 4.11

The trajectories of the three-particle resonance pole $z_R$ in the $z_R - z_2$ plane with varying the constant $\lambda(\lambda > \lambda_0)$ for the total orbital momentum $L = 1$ and 2.

threshold" $(z_2(\lambda_0) = 0)$ and the resonance pole returns to the origin on the real axis. As the constant $\lambda$ decreases further, the two-particle cut goes to the nonphysical (second) "logarithmic" sheet coupled to the physical sheet by the logarithmic cut, while the pole $z_R$ has to remain on the second ("root") sheet.

Table 4.4

Convergence of simple iterations $z^{(k+1)} = z_0 + f\left(z^{(k)}\right)$ for the resonances in a three-particle system with the total orbital momentum $L = 1$ below the three-particle threshold for the two-particle interaction constant $\lambda/\lambda_0 = 2.13(\lambda_0 = 2.82$ fm$^{-3}$ is the threshold value of the two-body force constant).

| $k$ | Re $z^{(k)}$ (MeV) | Im $z^{(k)}$ (MeV) |
|---|---|---|
| 0 | 6.366 | 0.0 |
| 1 | $-10.506$ | $-8.069$ |
| 2 | $-14.659$ | $-4.859$ |
| 3 | $-14.497$ | $-3.013$ |
| 4 | $-14.113$ | $-2.594$ |
| 5 | $-13.749$ | $-2.546$ |
| 6 | $-13.711$ | $-2.596$ |
| 12 | $-13.735$ | $-2.637$ |

This region, $\lambda < \lambda_0$, has not yet been investigated and the pole trajectory at $\lambda < \lambda_0$ is not known. Since, as the constant $\lambda$ varies, the branch point moves also, it is more convenient to depict the trajectory in the $z - z_2$ plane. In this plane the pole $z_R$ describes a closed trajectory, as shown in Fig. 4.11.

In [47] the pole trajectories $z_R$ were calculated at $\lambda_1 > \lambda > \lambda_0$ for $L = 1$ and 2 (Fig. 4.11) using the projecting method. For points close to the real axis equation (5.25) was solved by iterations. An example of the convergence is shown in Table 4.4. For more distant points the simple iterations (5.25) do not converge because $z_0$ does not fall within the convergence region. In such cases the solution for (5.25) was sought using the Padé approximant technique (see Table 4.5). The Faddeev equation was continued to the nonphysical sheet by the method described in Section 4.3. The results for $L = 1$ coincided with the results obtained by Glöckle [35] who calculated the eigenvalues of the Faddeev equation kernel and used a different contour deformation method to make the continuation to the nonphysical sheet.

Table 4.5

Convergence of the iterations of the same system as in Table 4.4 when seeking for the solution through the Padé approximants $\gamma/\gamma_0 = 1.42$.

| $k$ | Re $z^{(k)}$ (MeV) | Im $z^{(k)}$ (MeV) |
|---|---|---|
| 0 | 35 | 0.0 |
| 1 | $-1.000^*$) | $-0.010^*$) |
| 2 | $-1.000^*$) | $-3.000^*$) |
| 3 | $-2.084$ | $-0.748$ |
| 4 | $-3.182$ | $-2.197$ |
| 5 | $-2.927$ | $-1.172$ |
| 6 | $-2.958$ | $-1.011$ |
| 7 | $-2.940$ | $-0.974$ |

$^*$) These values are fixed "by hands" to avoid the divergence of the iteration terms at the first iterative steps at positive energies.

The projecting method described in this chapter has proved to be very convenient and effective in describing resonances. It is based on the transcendental equation (5.25) for finding the resonance parameters. This equation is exact and can be applied to any many-particle system. Moreover, it permits the resonance width expansion to be obtained in any order of rescattering multiplicity in the final state. In this case the total width may be presented as a sum of partial widths determining the decay probability in each of the possible many--particle channels. The projecting approach can readily be used to develop the perturbation theory with respect to the number of rescatterings. Of course, as the number of interacting particles increases the problem usually gets complicated. This is associated first of all with the complex character of the $N$-particle equations determining the resolvent $\widetilde{G}$. The structure of the Riemannian energy surface, on which the resolvent continuation is specified also gets more complicated with increasing $N$. Nevertheless, it is highly probable that, given a "good" initial-state function $\Phi$ (see the comment after formula (5.24a)), the convergence of the multiple-scattering series (5.24) for many-particle systems gets *improved*.

The projecting formalism described above makes it possible not only to study the decay of a two- or many-particle prepared state but also to have a consistent and practically realizable description of the whole many-particle scattering process including many-particle resonances [9].

There also exist other approximate approaches to describing resonance states which are relevant to the projecting methods. We shall examine some of them in the next section.

Now we shall briefly discuss the physical effects which are observed in experiments with three-particle resonance state decays.

First of all, much experimental evidence has been accumulated as a result of studying photonuclear reactions and nuclear electrodisintegration reactions (see the reviews and compilations [50, 51]) especially on light nuclei $^6$Li, $^7$Li, $^9$Be, $^{12}$C, etc., for which the energy and angular correlations of the charged products of the reactions have been studied in detail [52, 53]. At high (on a nuclear scale) energies the so-called stellar (i.e. many- body) decays of nuclei become perceptible and, because of a high density of states in the given excitation-energy range ($E^* \sim 20$–$30$ MeV) and the large width of the relevant states, it is the resonances rather than the so-called direct decays that are predominant. A large and relatively well-localized group of highly-excited states with excitation energies $E^* \sim 20$–$25$ MeV which are excited intensively by tens-of-MeV $\gamma$-quantum absorption (and also in other nuclear reactions) is called giant (dipole) resonance and, in principle, exhibits dipolar behaviour with respect to the ground state. In other words, this group of states is characterized by the selection rules as to the electric dipole transitions. An enormous number of calculations made in terms of the shell model using the diagonalization procedure (works [54–56] are model examples) have shown that even on a purely discrete shell basis one can succeed in describing the basic features of the regularities observed experimentally, such as:

(1) locations of the basic maxima of the excitation function for reactions $(\gamma, n)$, $(\gamma, p)$, $(\gamma, \alpha)$, etc.;

(2) general character of the spectra of outgoing particles;

(3) relative populations of states of the daughter nuclei, etc.

On the other hand, the angular and energy correlations of outgoing particles can hardly be described adequately in terms of such an approach. The formalism explained in the last sections is most probably more suitable for the purpose.

Furthermore, a broad class of many-body resonance states are excited in the low-energy nuclear reactions [58, 59] of the type

$$p + {}^{11}B \rightarrow 3\alpha ; \tag{5.26}$$

$$\alpha + {}^6Li \rightarrow 2\alpha + n + p \tag{5.27}$$

or in the inelastic scattering of high-energy hadrons by light nuclei, for example

$$n \; + \; {}^{9}Be \quad \rightarrow \quad n' \; + \; {}^{9}Be^* \searrow_{2\alpha \, + \, n} . \qquad (5.28)$$

In particular, the last reaction is one of the basic sources of neutron multiplication in thermonuclear reactor blankets and in the outer shells of conventional nuclear reactor. In such many-particle processes the angular and energy correlations of the product particles depend strongly on the rescattering effects in the final state. In particular, many resonance maxima [57] corresponding to the highly excited states of the $^{12}C$ nucleus with $E^* > 15$ MeV manifest themselves in the excitation function of the reaction (5.26) at the ingoing-proton energies $E_p \sim 1\text{–}3$ MeV. Detailed study of the two-particle energy correlation using coincidence measurements has shown that the well-known low-lying $\alpha$–$\alpha$ resonances $(0^+$ and $2^+)$ manifest themselves in the correlation function; however their energy, and more especially width, differ appreciably from those observed in the case of free $\alpha$–$\alpha$ scattering. As shown in [60] (see also some details in [30]) these distortions arise from the identity of the three $\alpha$ particles and from multiple rescatterings in the final state.

In the high-energy region, a lot of reactions are accompanied by formation of many-meson final states whose decays are determined to a substantial degree by the effects of meson rescattering in the final state. Such effects may be consistently calculated using, apart from the tradition N/D method, a formalism similar to that explained above (with a possible generalization to make allowance for relativism, etc.).

Finally, in atomic and molecular physics the autoionization states are essentially three-particle resonances whose decays are governed totally by the effect of electron rescattering (see Chapter 7).

Thus, from the above it is seen that many-particle resonance states are widespread in various regions of present-day physics and their further study both theoretical and experimental yields appreciable information about the dynamics of the processes and about the character of elementary particle interactions.


## 4.6. Treatment of Resonance States in the Shell-Model Description of Nuclear Reactions

In this section we shall briefly review the methods used to make allowance for resonance states when describing the numerous types of nuclear reactions in terms of the shell model with continuous spectrum.

More specifically, we mean an allowance for the one-particle resonances contained in the Hamiltonian of the self-consistent field of the nucleus (i.e. contained in the shell-model Hamiltonian of the nucleus) when the shell-model basis functions are used to calculate the reaction amplitudes. Examples of such resonances are numerous in nuclear physics, in particular in the systems

$$
\begin{aligned}
{}^{4}\text{He} &+ \text{p(n)} \qquad (\text{p}_{1/2} \text{ and } \text{p}_{3/2} \text{ orbitals}), \\
{}^{12}\text{C} &+ \text{p(n)} \qquad (\text{p}_{1/2} \text{ orbital}), \\
{}^{16}\text{O} &+ \text{p(n)} \qquad (\text{d}_{3/2} \text{ orbital}).
\end{aligned}
$$

The orbitals given in brackets are resonances. Replacing these resonance orbitals in the calculations of nuclear reactions by bound orbitals leads in many cases to appreciable inaccuracies because in this case the actual decay channels of the system with emission of one or several particles turn out to be closed. On the other hand, the one-particle potential of the self-consistent field of a nucleus $U_{sp}(r)$ (in nuclear physics this potential is usually approximated by the Woods- -Saxon potential $V_0/(1 + \exp((r - R)/a))$ is finite at $r \to \infty$ (i.e. $U_{sp}(r) = 0$, $r \to \infty$) and the discrete orbitals of bound states $\{\varphi_n(r)\}$ fail to form a complete basis. To be completed, the basis $\{\varphi_n(r)\}$ has to include the contribution from the continuous spectrum corresponding to $U_{sp}(r)$, i.e. the expansion of the wave function $\psi(r)$ describing the scattering has to include, apart from $\{\varphi_n(r)\}$, the integral over the continuous spectrum:

$$
\psi(r) = \sum_{n=1}^{N} C_n \varphi_n(r) + \int_{0}^{\infty} dk c(k) \varphi_k^{(+)}(r)
$$

$$
= P\psi + Q\psi^{7)} \tag{6.1}
$$

with the appropriate condition of completeness

$$
\sum_{n} \varphi_n^{*}(r)\varphi_n(r') + \int dk \varphi_k^{(\pm)*}(r)\varphi_k^{(\pm)}(r') = \delta(r - r') . \tag{6.2}
$$

However, in most real nuclear problems direct application of the expansion (6.1) results in substantial numerical difficulties because the integrals over $k$ arising in describing nuclear reactions are singular and the allowance for one-particle resonances leads to the appearance of poles near the integration contour (see below). In this case the density of the mesh points in $k$ (or $\varepsilon$) must be sufficiently high so that the dimension of the resultant algebraic set of equations is also high.

First, we shall examine the general ideas using as example the one-particle problem, whereupon we shall show how the formalism can be generalized to the many-particle problem with channel coupling. It should be noted that there are numerous publications devoted to allowing for one-particle resonance states in

---

describing nuclear reactions (see the book [61] and the references therein to the works published by the late sixties). For later publications see the recent review [62] and the references therein.

### 4.6.1. One-Particle Problem

Let us examine the scattering of a particle by a central field $V(r)$ such that the particle can be captured by it in a one-particle resonance state (the so-called shape resonances).

The partial-wave representation of the Schrödinger equation for wave functions of a particle in the discrete and continuous spectra may be written as

$$(\varepsilon_{nl} - h_l)\, \tilde{\varphi}_{nl}(r) = 0 , \qquad n = 0, 1, ..., M , \tag{6.3a}$$

$$(\varepsilon - h_l)\, \tilde{\varphi}_{\varepsilon l}(r) = 0 , \qquad \varepsilon > 0 \tag{6.3b}$$

where (6.3a) is related to the discrete branch of the spectrum of $h_l$, and (6.3b) to the continuous branch. Here the notation $\tilde{\varphi}_{nl}$ ($n$ is the radial quantum number, $l$ the orbital momentum) and $\tilde{\varphi}_{\varepsilon l}$ is self-evident (for simplicity, we neglect here the spin variables) and

$$h_l = \frac{\hbar^2}{2m_r}\left(-\frac{\mathrm{d}}{\mathrm{d}r^2} + \frac{l(l+1)}{r^2}\right) + V(r) \tag{6.4}$$

$m_r = (1 - 1/A)m_N$ is the reduced mass of the nucleon + nucleus system.

In the general case the potential $V(r)$ may also depend on the quantum numbers of state, i.e. $V(r) \rightarrow V_{nl}(r)$

Our aim is to include the narrow one-particle resonances in a subset of $M$ discrete states (6.3a). Thus, the new discrete basis (we shall call the corresponding configurations the $b$ configurations) will include the states which have a large amplitude inside the nucleus. Therefore, the interaction between the respective one-particle configurations (in a many-particle system) must be treated as accurately as possible. On the other hand, the wave functions of the resultant modified continuum orthogonal to the discrete basis extended owing to resonances will have relatively small amplitudes inside the nucleus. Therefore, in the many-particle problem the interaction between the configurations (called $c$ configurations henceforth) including the one-particle state of the modified continuum is relatively weak and may be allowed for in terms of perturbation theory.

It should be noted that a similar idea underlies the method for orthogonal rearrangement of the Born series (see Section 4.2). Indeed, the series (2.16), (4.24a) and (4.24b) are essentially the perturbation series in powers of interaction between the states of the orthogonalized continuum $\tilde{\psi}_0(p)$, i.e. the series in powers of the operator $\tilde{G}_0 V$. From this it is clear that the higher the dimension of the projector $P$ consisting of $L_2$-functions localized in the inner region, the smaller will be the amplitude of the modified scattering functions in the inner region and, thus, the faster will be the convergence of the Born series rearranged in such a way. Simultaneously, the dimension of the finite-dimensional, i.e. algebraic, part of the problem will, of course, increase.

We may also proceed analogously in the shell-model theory of nuclear reactions [61–63]. This means that we may reasonably allow exactly for the interaction of the many-particle configurations containing only discrete orbitals (the normalized resonance orbitals inclusive, see below) or including a transition of one particle from the configuration "b" to "c" and completely neglect the interaction matrix elements including two $c$-configurations [63], i.e. we may allow exactly for the matrix elements of the type of $b$–$b$ and $b$–$c$ and treat those of the type of $c$–$c$ using the perturbation method.

Also, we may make use of the fact that the states of the new (i.e. orthogonalized) continuum are smooth functions of energy (because the one-particle resonances have been excluded from them) and solve also this part of the problem exactly by reducing it to the coupled-channel method [61, 62, 64].

The following two methods are mainly used to approximate one-particle resonances by quadratically-integrable functions.

(i) The method for diagonalizing the one-particle (i.e. shell-model Hamiltonian) in an oscillator basis [63, 65] $\{\varphi_{nl}(r)\}$ (of common use in nuclear physics). Here we have, generally, to include in the basis a slightly larger number of oscillator functions compared with the number of bound states in a given partial wave. Then, $M + N$ ($N$ is the number of the examined one-particle resonances) lowest discrete states obtained by diagonalizing $h_l$ can approximately describe the bound and resonance orbitals of the one-particle Hamiltonian (6.4), i.e. $M + N$ one-particle orbitals

$$\Phi_{il}(r) = \sum_{n=1}^{N+M} c_n^i \varphi_{nl}(r), \qquad i = 1, 2, ..., M + N \tag{6.5}$$

form a modified discrete basis. It is of importance here that the approximate functions $\Phi_{il}(r)$ prove automatically to be orthonormalized. For a sufficient, but not very high dimension of the oscillator basis $N + M$, the quality of the resonance wave function approximation in the inner region will be very good (for details see Chapter 7).

(ii) The cut-off method proposed by Shakin and Wang [66]. In this approach, the approximate quadratically-integrable functions describing one-particle resonances are obtained by cutting-off the exact scattering functions $\varphi_{pl}$ taken at resonance energies $\varepsilon_{pl}$ (for narrow resonances these are well-defined quantities). Then, the cut-off functions are normalized to unity and orthogonalized to the exact functions of bound-state orbitals. Thus, instead of the system of functions

$\overline{\varphi}_{nl}(r)$ and $\overline{\varphi}_{\varepsilon l}(r)$ determined by solving (6.3), we construct a new set of discrete orbitals:

$$\Phi_{nl}^{c.o.}(r) = \overline{\varphi}_{nl}(r) , \qquad n = 1, ..., M , \tag{6.6a}$$

$$\Phi_{pl}^{c.o.}(r) = N_{pl}\,\vartheta(R - r)\varphi_{pl}(r) + \sum_{n=1}^{M} N_{nl}\,\overline{\varphi}_{nl}(r) , \tag{6.6b}$$

$$p = M + 1 , ..., M + N$$

where $\vartheta(R - r)$ is the step function which cuts off scattering functions $\varphi_{pl}(r)$ at $r = R$; $N_{nl}$ and $N_{pl}$ are the constants providing for the orthonormality of the basis $\{\Phi_{il}^{c.o.}(r)\}$ $(i = 1, ..., N + M)$. After the extended discrete basis $\{\Phi_{il}\}$ is constructed, the orthogonalized continuum function $\widetilde{\varphi}_{\varepsilon l}$ can be found by solving the following Schrödinger equation projected onto the subspace $\mathcal{H}_Q^{sp}$:

$$\left(\varepsilon - Q^{sp}h_l\,Q^{sp}\right) \widetilde{\varphi}_{\varepsilon l} = 0 \tag{6.7}$$

where the Hamiltonian $h_l$ is given by (6.4) and the projector $Q^{sp} = 1 - P^{sp}$ realizes the projection onto the states orthogonal to those included in the projector $P^{sp}$

$$P^{sp} = \sum_{i=1}^{M+N} \Phi_{il}^{c.o.}\rangle \langle \Phi_{il}^{c.o.}| , \tag{6.8}$$

generated by the set of orthonormalized orbitals (6.6).

It is quite evident that the new one-particle basis $\{\Phi_{il}^{c.o.}, \widetilde{\varphi}_{\varepsilon\lambda}\}$ can be so normalized as to satisfy the standard conditions of orthonormality and completeness

$$\int dr\; \Phi_{il}^{c.o.}(r)\Phi_{i'l}^{c.o.}(r) = \delta_{ii'} , \tag{6.8a}$$

$$\int dr\; \Phi_{il}^{c.o.}(r)\widetilde{\varphi}_{\varepsilon l}(r) = 0 , \tag{6.8b}$$

$$\int dr\; \widetilde{\varphi}_{\varepsilon l}^{*}(r)\, \varphi_{\varepsilon'l}(r) = \delta(\varepsilon - \varepsilon') , \tag{6.8c}$$

$$\sum_{i=0}^{M+N} \Phi_{il}^{c.o.}(r)\, \Phi_{il}^{c.o.}(r') + \int_{0}^{\infty} d\varepsilon\; \widetilde{\varphi}_{\varepsilon l}(r)\widetilde{\varphi}_{\varepsilon l}^{*}(r') = \delta(r - r') . \tag{6.8d}$$

It can easily be understood that the new one-particle spectrum will be generated by the operator $\overline{h}_l$:

$$\overline{h}_l = \sum_{i=1}^{M+N} j|\Phi_{il}^{c.o.}\rangle\, \varepsilon_{il}\, \langle\Phi_{il}^{c.o.}| + Q^{sp}h_l Q^{sp} . \tag{6.9}$$

### 4.6.2. Many-Particle Problem

Here we introduce the very important concept of a quasi-bound state embedded in a continuum (QBSEC) [2, 3]. Such a state is defined as one in which all the $A$ particles of the system (of nucleons) occupy the bound and quasi-bound (i.e. $\Phi_{il}^{c.o}(r)$) one-particle orbitals. Thus, the initial many-particle Hamiltonian generated by QBSEC is defined as

$$\bar{H}_0 = \sum_{k=1}^{A} \bar{h}(r_k) \tag{6.10}$$

where $\bar{h}(r_k)$ is the one-particle Hamiltonian corresponding to (6.9) in the partial wave representation. Now, the total $A$-particle Hamiltonian $H$ can be defined as

$$H = \bar{H}_0 + V_{\text{res}} \tag{6.11}$$

where $V_{\text{res}} = \sum_{i>j}^{A} V_{ij}$ is the so-called residual interaction among all $A$ particles. It is this interaction that leads to mixing of various many-particle configurations and, hence, to the QBSEC decay. The QBSEC are constructed to be antisymmetrized products of $N + M$ one-particle orbitals given by (6.6), i.e.

$$\Phi_i(1, 2, \ldots A) = \frac{1}{A!} \, \mathscr{A}\!\left(\Phi_{i_1}^{c.o.}(r_1)\, \Phi_{i_2}^{c.o.}(r_2)\, \ldots\, \Phi_{i_A}^{c.o.}(r_A)\right)^{8)} \tag{6.12}$$

where $\mathscr{A}$ is the antisymmetrization operator and the set $\{i = i_2, i_2, \ldots i_A\}$ gives the distribution of particles in the accessible one-particle states.

Let us now construct the projection operator on QBSEC

$$P = \sum_i |\Phi_i(1, \ldots, A)\rangle \, \langle \Phi_i(1, \ldots, A)| \tag{6.13}$$

and the projector $Q = 1 - P$ complementary to the former (so that $QP = PQ = 0$). In the subspace $\mathscr{H}_Q$ we are to allow for such states of $A$ particles that only one of them will be found in the c-state (i.e. in the modified continuum). This restriction makes it possible to substantially simplify the solution of the subsequent equations. The version of the model permitting the allowance for the configurations where two particles may be simultaneously present in the continuum is treated in the next chapter. However, in contrast to the approach described here the calculations in terms of that model have not been made yet. In accordance with the introduced projectors $P$ and $Q$, we have the parts of the total Hamiltonian $H$, i.e. $H_{PP}$, $H_{PQ}$, $H_{QP}$ and $H_{QQ}$, governing the transitions between respective configurations.

---

[8)] The exact treatment of the motion of the centre-of-mass of the entire system of $A$ particles is unessential here.

Then, for the complete many-particle Green function we obtain a decomposition given by the formula (1.11) where the Hamiltonian is determined by the expression (6.11). It is not difficult to show (see, for example, [2, 3]) that the decomposition of the complete wave function of the system corresponding to the decomposition of the resolvent (1.11) is

$$\psi = \tilde{\psi}_E + (P + G_Q H_{QP})\, D(E)\, H_{PQ} \tilde{\psi}_E \tag{6.14}$$

where the function $\tilde{\psi}_E$ is (2.6) with an appropriate boundary condition. In the case of the problems of elastic and inelastic scattering of a particle by the system of $A - 1$ particles we have an incoming wave in the entrance channel and outgoing waves in all channels. In this case the operator $D(E)$ is given by (1.12) with $G_Q$ taken at the energy $E + i\varepsilon$, i.e. $G_Q^{(+)} = G_Q(E + i\varepsilon)$.

The decomposition (6.14) is used in most of the studies based on the projecting formalism. The difference is only in the method for calculating $\tilde{\psi}_E$ and the operator $G_Q H_{QP} D(E) H_{PQ}$. In terms of the approach described in Sections (4.2) and 4.4, $\tilde{\psi}$ and $\tilde{G}_Q(E)$ are found by iterations on the basis of the Born series (2.16) and (4.24a) in the subspace $\mathcal{H}_Q$. However, if we limit ourselves to the many-particle configurations $Q$ in which only one of the $A$ particles is in the continuum, then it is convenient to use [64] the coupled-channel method which gives a set of differential (or integro-differential) equations [61, 64] of the type of

$$\sum_{c'} \left( E_c \delta_{cc'} - Q^{s.p.} H_{cc'} Q^{s.p.} \right) \psi_{c'} = 0 \tag{6.15}$$

where $c$ is the index of a channel; $c \equiv (i_1, i_2, \dots i_{A-1})$ and

$$H_{cc'}(\mathbf{r}, \mathbf{r}') =$$

$$= \left\langle \mathcal{A}\left(\Phi_{i_1}, \dots \Phi_{i_{A-1}} \delta(\mathbf{r}_A - \mathbf{r})\right) \middle| H \middle| \mathcal{A}\left(\Phi_{i'_1}, \dots \Phi_{i'_{A-1}} \delta(\mathbf{r}_A - \mathbf{r}')\right) \right\rangle$$

is the matrix element of the channel coupling which depends on the variable of only the particle which resides in the continuum; the integration is made over all other $A - 1$ one-particle orbitals.

Since the projector $P$ is of a finite rank, the decomposition (6.14) may be written as

$$\psi = \tilde{\psi} + \sum_{ij} (\Phi_i + \omega_i^{(+)})\, (E - M)_{ij}^{-1} \left\langle \Phi_j \middle| H_{PQ} \middle| \tilde{\psi} \right\rangle \tag{6.14a}$$

where the wave functions $\omega_i^{(+)}$ are defined to be solutions for the equation

$$\omega_i^{(+)} = G_Q(E + i\varepsilon) H_{QP}\, \Phi_i \tag{6.16a}$$

or

$$(E - QHQ)\omega_i^{(+)} = QHP\Phi_i \tag{6.16b}$$

and describe the decay of the states $\Phi_i$ belonging to the $P$-space into the continuum, i.e. into the states of the $Q$ space produced by the operator $H_{QP}$.

The operator $M$ in (6.14a) is determined by a finite-dimensional matrix with elements

$$M_{ij} = \langle \Phi_i | (H + H G_Q^{(+)} H) | \Phi_i \rangle =$$

$$= \langle \Phi_i | H | \Phi_j \rangle + \langle \Phi_i | H | \omega_j^{(+)} \rangle . \tag{6.17}$$

Here we have used (6.16a). It should be remembered that $H_{ij} = \langle \Phi_i | H | \Phi_j \rangle$ is the matrix of the Hamiltonian in the shell-model basis. In the standard many--particle shell-model calculations the spectrum of the excited states (of a nucle-us) is obtainable by diagonalizing the symmetric matrix $H_{ij}$.

Let us examine how the *many-particle* resonance states appear in the appro-ach described above. The matrix $M$ is complex and symmetric and may be diagonalized using the orthogonal matrix $\hat{O}$ [64]

$$\sum_{ij} \hat{O}_{mi} M_{ij} \hat{O}_{jn} = (\tilde{E}_n - i\tilde{\Gamma}_n/2)\delta_{mn} \tag{6.18}$$

where $\tilde{E}_n - i\tilde{\Gamma}_n/2$ are the complex diagonal elements obtained by diagonalizing the matrix $M$. And, since the matrix $M$ depends (parametrically) on the total energy $E$, also

$$\tilde{E}_n = \tilde{E}_n(E) , \tag{6.19a}$$

$$\tilde{\Gamma}_n = \tilde{\Gamma}_n(E) . \tag{6.19b}$$

When making the $\hat{O}$ transformation (6.18), the wave function (6.14a) may be written as

$$\psi = \tilde{\psi} + \sum_n \bar{\Omega}_n^{(+)} \frac{1}{E - \tilde{E}_n + i\Gamma/2} \langle \bar{\Phi}_n^* | H | \tilde{\psi} \rangle . \tag{6.20}$$

The transformed wave functions are

$$\bar{\Omega}_n^{(+)} = \bar{\Phi}_n + \bar{\omega}_n^{(+)} \tag{6.21a}$$

where

$$\bar{\Phi}_n = \sum_i \hat{O}_{ni} \Phi_i , \qquad \bar{\omega}_n^{(+)} = \sum_i \hat{O}_{ni} \omega_i^{(+)} . \tag{6.21b}$$

Finally, the "diagonalized" form for the complete Green function (1.11) is

$$G^{(+)} = G_Q + \sum_n |\bar{\Omega}_n^{(+)}\rangle \frac{1}{E_n - \tilde{E}_n + i\tilde{\Gamma}_n/2} \langle \bar{\Omega}_n^{(+)}| . \tag{6.22}$$

As regards the form, this representation resembles the expansion of the Green function in the Gamow states corresponding to resonance poles, as discussed in Chapter 2. It should be noted that the "orthogonalized" part of the Green function in (6.22), i.e. $G_Q$, has no near poles (see the discussion in Section 4.2). However, this analogy is only superficial because in our case the energy $E$ is taken for convenience to be everywhere real and governs $\tilde{E}_n$ and $\tilde{\Gamma}_n$ (see (6.19)). Therefore, we have to seek the energy $E_n$ on many-particle resonances by solving the transcendental equations

$$E_n = \tilde{E}_n(E_n) \,. \tag{6.23}$$

Then, the resonance width is defined as

$$\Gamma_n = \tilde{\Gamma}_n(E_n) \,.$$

There exists also a close analogy of the expansions (6.20) and (6.22) with the Kapur–Peierls procedure discussed in Chapter 2. However, the projectors $P$ and $Q$ are treated here in Hilbert space, whereas the corresponding Kapur–Peierls projection procedure is realized in *configuration* space. The states $\bar{\Omega}_n^+$ also depend parametrically on the total (real) energy $E$. Their asymptotic behaviour corresponds to outgoing waves in all the channels (see (6.21) and (6.16b)), i.e. in this respect they are also analogous with the many-particle Gamow states. However, since the energy is assumed to be real, some difficulties with the normalization of such states arise (see also [67]).

It is clear that these analogies with the Gamow states are correct only for the states $\bar{\Omega}_n^{(+)}$ with narrow widths i.e. at $\Gamma_n/E_n \ll 1$. In this case also the amplitude of the function $\omega_n^{(+)}$ in (6.21a) which defines an outgoing wave will be relatively small in the inner region and, therefore, $\bar{\Omega}_n^{(+)}$ will be in this case very similar to the real function $\bar{\Phi}_n$, i.e. to the function found by diagonalizing the matrix $M$. And finally, we shall present the expression for the multi-channel $S$ matrix [61] describing the process of inelastic scattering of a nucleon by a nucleus, which corresponds to the transition of the system from the initial channel $c_0$ to the exit channel $c$.

For this purpose, the representation (6.20) can be rewritten as

$$\psi_c^{(c_0)} = \tilde{\psi}_c^{(c_0)} + \sum_n \bar{\Omega}_c^{(n)} \frac{1}{E - \tilde{E}_n + i\Gamma/2} \langle \bar{\Phi}_n^*| \, H \, |\tilde{\psi}_c^{(c_0)}\rangle \tag{6.24}$$

where the superscript $c_0$ means that the incoming wave is located only in the channel $c_0$; $\tilde{\psi}_c^{(c_0)}$ is found by solving the coupled channel problem (6.15) with the necessary boundary conditions. In the asymptotic region $r \to \infty$ the solution $\tilde{\psi}_c^{(c_0)}$ behaves in the standard way [3]:

$$\tilde{\psi}_c^{(c_0)}(r) \to \frac{i}{2}\left( I_l(r) - \frac{S_{cc_0}^{(0)}}{\sqrt{k_c k_{c_0}}} \, O_l(r) \right) \tag{6.25}$$

where $I_l$ and $O_l$ are the incoming and outgoing waves, respectively (if allowance is made for the Coulomb interaction, these will be Coulomb wave functions); $k_c$ and $k_{c_0}$ are the wave vectors in the entrance and exit channels, respectively; $S_{cc_0}^{(0)}$ is the so-called background scattering matrix (this term has arisen from the fact that the corresponding phase shifts do not exhibit resonance behaviour). To find the asymptotic behaviour of $\psi$ from (6.24), we have to determine the behaviour of $\bar{\Omega}_c^{(n)}$ for $r \to \infty$ . It may be shown [61, 64] that at $r \to \infty$ we get

$$\bar{\Omega}_c^{(n)} \to \bar{\omega}_c^{(n)} \to \exp\left(i\delta_c^0\right) \sqrt{\frac{m_r}{\hbar^2 k_c}} \, \gamma_c^{(n)} O_l(r) \tag{6.26}$$

where $O_l(r)$ is the outgoing wave and the background phase shift $\delta_c^0$ is defined in the standard way: $S_{cc}^{(0)} = \exp\left(2i\delta_c^0\right)$ (i.e. when *both* incoming and outgoing waves in (6.25) are in the channel $c$). The factors $\gamma_c^{(n)}$, which are called the amplitudes of partial widths, may be found by solving the set of equations for coupled channels (6.15). As a result, we obtain [61, 62] the following expression for the many-particle $S$ matrix:

$$S_{cc_0} = S_{cc_0}^{(0)} - i \exp\left[i(\delta_c^0 + \delta_{c_0}^0)\right] \sum_n \frac{\gamma_c^{(n)} \gamma_{c_0}^{(n)}}{E - \tilde{E}_n + \frac{1}{2} i \tilde{\Gamma}_n} \tag{6.27}$$

where the basic parameters $\gamma_c$, $\tilde{E}_n$ and $\tilde{\Gamma}_n$ are energy-dependent; $\gamma_c^{(n)}$ turn out to be complex quantities owing to the channel coupling and to the QBSEC mixing in the diagonalization of the matrix $M$.

Now, if we use (6.27) and define the decay rate for the resonance state $\bar{\Omega}_c^{(n)}$ in the channel $c$ in terms of its squared amplitude in the asymptotic region multiplied by the outgoing particle velocity, we can easily find the expression for the decay partial width:

$$\Gamma_c^{(n)} = |\gamma_c^{(n)}|^2 . \tag{6.28}$$

It can be shown [64] that, because of the condition $\sum_i |\hat{O}_{ni}|^2 \geqslant 1$ for the complex matrix $\hat{O}$, the total resonance width

$$\tilde{\Gamma}_n \leqslant \sum_c |\gamma_c^{(n)}|^2 \tag{6.29}$$

although the departure of $\sum_c |\gamma_c^{(n)}|^2$ from $\tilde{\Gamma}_n$ *is not as a rule substantial in real calculations* [64].

During the last decade numerous calculations of the elastic and inelastic scattering of nucleons by light nuclei (see, for example [62]) were carried out within the approximate framework of the formalism described above. On the whole, this technique is very promising.

## 4.7. Summary

The projecting formalism for describing the resonance processes and states discussed in this chapter is a very powerful, practically effective, and universal means of resolving many types of problems. The basic physical idea underlying this approach is that, as a zero-order approximation, the resonance wave function $\psi$ (at real or complex energies) is replaced by the well-known quadratically integrable wave function $\Phi$ showing a large overlap with the exact wave function in the inner region, which is possible, of course, only in the case of sufficiently narrow resonances ($\Gamma/E \ll 1$). In other words, we assume that $\psi_{res} = \Phi + \chi$, so that the correction $\chi$ can be considered in a sense to be small and orthogonal to $\Phi$. Furthermore, this correction describing the decay of the initial state may be calculated either using the perturbation method (as carried out in Sections 4.2–4.4) or "exactly", but in terms of a certain model, for example by the coupled channel method. However, since the construction of the initial wave function is a finite-dimension algebraic problem (and such a problem can be solved very reliably and rapidly with modern computers even in the case of high dimensions) and because any increase in the $P$-subspace dimension and an extension of the region of the space where $\psi \simeq \Phi$ inevitably entail a substantial suppression of the amplitude of the correction wave function $\chi$ in the inner region (and hence the respective acceleration of iteration convergence in the subspace), it is probably preferable to use the perturbation approach with iterations of the dynamic (Schrödinger, Faddeev-Yakubovsky, etc.) equations in the $Q$-subspace, but with a sufficiently high dimension of the $P$-subspace, so that the number of iterations would be minimal. This method will be especially effective for describing the many-particle resonance processes where iterations are involved in the calculations of multidimensional singular integrals.

*References*

1. FESHBACH H., Ann. Phys. *5* (1958) 357; *19* (1962) 287.
2. FONDA L., NEWTON R. G., Ann. Phys. *10* (1960) 490.
3. NEWTON R. G., Scattering Theory of Waves and Particles, 2$^{nd}$ ed., Springer, Berlin Heildeberg 1982.
4. WANG W. L., SHAKIN C. M., Phys. Lett. *32B* (1970) 421.
5. SHEERBAUM R. R., SHAKIN C. M., THALER R. M., Ann. Phys. *76* (1973) 333; LEV A., BERES W. P., Phys. Rev. C *14* (1976) 354.
6. DOMCKE W., Phys. Rev. A *28* (1983) 2 777.
7. KUKULIN V. I. at al., J. Phys. G.: Nucl. Phys. *8* (1982) 1671.
8. KUKULIN V. I., POMERANTSEV V. N., Ann. Phys. *111* (1978) 330.
9. KUKULIN V. I., POMERANTSEV V. N., Teor. Mat. Fiz. *73* (1987) 443 (in Russian).
10. WIGNER E. P., WEISSKOPF V. F., Zs. Phys. *63* (1930) 62.
11. TITCHMARSH E. C., Eigenfunction Expansion Associated with Second-Order Differential Equations, Part, II, Clarendon Press, Oxford 1958.
12. GOLDBERG M. L., WATSON K. M., Collision Theory, John Wiley, N. Y. 1964.

13. KRASNOPOL'SKY V. M., KUKULIN V. I., Yad. Fiz. *20* (1974) 883. [Sov. J. Nucl. Phys. *20* (1975) 470].
14. KUKULIN V. I. et al., J. Phys. G.: Nucl. Phys. *4* (1978) 1409.
15. ALEKSANDROV A.P., DMITRIEV B. U., In Computational Methods and Programming, (in Russian), Vol. 24, Moscow University, Moscow 1975, p. 23.
16. KUKULIN V. I., POMERANTSEV V. N., Theor. Mat. Fiz. *27* (1976) 373;
    POMERANTSEV V. N., Theor. Mat. Fiz. *29* (1976) 94.
17. KLAUS M., SIMON B., Ann. Phys. (N. Y.) *130* (1980) 251.
18. GEL'FAND I. M. et al., Generalized functions, Vol. 4, Academic Press, N. Y. 1964.
19. PARRAVACINI G., GORRINI V., SUDARSHAN E. C. G., J. Math. Phys. *21* (1980) 2 208.
20. GORRINI V., PARRAVACINI G., Unstable Quantum States and Rigged Hilbert Spaces. In: Group Methods in Physics, Lecture Notes in Physics, Vol. 94, Springer, N. Y. 1979, p. 219.
21. BÖHM A., Boulder Lec. Theory Phys. *9A* (1966) 255;
    The Rigged Hilbert Spaces and Quantum Mechanics, Lecture Notes in Physics, Vol. 78., Springer, N. Y. 1978.
22. SCHAEFER H. H., Topological Vector Spaces, Macmillan Co., N. Y. 1966.
23. MAURIN K., General Eigenfunction Expansion and Unitary Representation of Topological Groups, PWN, Warsaw 1968.
24. BERGGREN T., Nucl. Phys. *109A* (1968), 265.
25. GARCIA-CALDERON G., PEIERLS R., Nucl. Phys. *A265* (1976) 443.
26. KUKULIN V. I., NEUDATCHIN V. G., OBUHKOVSKY I. I., SMIRNOV Yu. F., Clusters as Subsystems in Nuclei, Vieweg, Braunschweig, 1982.
27. SCHWARTZ L., Méthodes mathématiques pour les sciences physiques, Hermann, Paris 1961.
28. ORLOV Yu. V., ZhETF Letters *33* (1981) 380 [JETP Lett. *33* (1981) 363].
29. KLIMOV V. I., POMERANTSEV V. N., Preprint FEI-1342, Obninsk 1982.
30. SCHMID E. W., ZIEGELMANN H., The Quantum Mechanical Three-Body Problem, Pergamon Press, Braunschweig 1974.
31. HETHERINGTON J. H., SHICK L. H., Phys. Rev. *B137* (1965) 2 265.
32. EBENHÖH W., Nucl. Phys. *A191* (1972) 97.
33. AVISHAI Y., Phys. Rev. D. *3* (1971) 3 232.
34. MÖLLER K., Preprints of Zentralinstitut für Kernforschung Rossendorf, Dresden, ZfK-327, 1977; ZfK-351, 1978.
35. GLÖCKLE W., Phys. Rev. C, *18* (1978) 564.
36. SOHRE F., ZIEGELMANN H., Phys. Lett. *34B* (1971) 579.
37. DOLESCHALL P., Nucl. Phys. *A201* (1972) 264.
38. BRAYSHOW D. D., Phys. Rev. *176* (1968) 1 855.
39. GARSIDE L., MACDONALD W. M., Phys. Rev. *54* (1965) 393.
40. REED M. C., SIMON B., Methods of Modern Mathematical Physics, Vol. IV: Analysis of Operators, Academic Press, N. Y. 1978.
41. SHAKIN C. M., WEISS M. S., Phys. Rev. *C11* (1975) 756.
42. TOBOCMAN W., Phys. Rev. *C17* (1978) 2 205.
43. GIRAUD B. G., MIHAILOVIČ M. V., LOVAS R. G., NAGARAJAN M. A., MITIČ D. V., Preprint Daresburg Lab. DL/Nuc/P 1 341, 1981 to be published in Ann. Phys.
44. KUKULIN V. I. et al., Yad. Fiz. *37* (1983) 862. [Sov. J. Nucl. Phys. *37* (1983) 514].
45. EFIMOV M. V., Phys. Lett. *33B* (1970) 563; Nucl. Phys. *A210* (1973) 157.
46. BERMAN M., MÜNDEL C., DOMCKE W., Phys. Rev. *31A* (1985) 641.
    DOMCKE W., BERMAN M., MÜNDEL C., MEYER H.-D., Phys. Rev. *33A* (1986) 222.
47. EMELYANOV V. G., KLIMOV V. I., POMERANTSEV V. N., Preprint FEI-1566, Obninsk 1984.
48. AARON R. D., AMADO R. D., Yam Y. Y., Phys. Rev. *136* (1964) B650.
49. KUKULIN V. I., Czech J. Phys. *B21* (1971) 923.
50. ISHKHANOV B. S., KAPITONOV I. M., NEUDATCHIN V. G., ERAMZHYAN R. A., EChAYa *12* (1981) 905. [Sov. J. Part. Nucl. 12 (1981) 362].

51. ERAMZHAN R. A. et al., Phys. Rep. *136* (1986) 229.
52. DENISOV V. P., Yad. Fiz. , *27* (1978) 882 [Sov. J. Nucl. Phys. *27* (1978) 469].
53. ISHKHANOV B. S., MOKEEV V. I., NOVIKOV Yu. A., PISKAREV I. M., Yad. Fiz. *32* (1980) 11. [Sov. J. Nucl. Phys. *32* (1980) 5].
54. MAJLING L., KUKULIN V. I., SMIRNOV F. Yu., Czech. J. Phys. B *18* (1968) 1 563; Phys. Lett. *27B* (1967) 487.
55. IAKOBSEN M. J., Phys. Rev. *123* (1961) 229.
56. ČUJEC B., Nucl. Phys. *37* (1962) 396; BARKER F. C., Nucl. Phys. *18* (1968) 1563.
57. AJZENBERG-SELOVE F., Nucl. Phys. *A248* (1975) 62, 63, 69.
58. TREADO M., Nucl. Phys. *A198* (1972) 21.
59. SEGEL R. E., HANNA S. S., ALLAS R. G.. Phys. Rev. *139B* (1965) 818.
60. SUFFERT M. A., FELDMAN R. G., HANNA S. S., Part. and Nuclei *4*(1972) 175.
61. MAHAUX C., WEIDENMÜLLER H. A., Shell-Model Approach to Nuclear Reactions, North-Holland, Amsterdam 1969.
62. ROTTER I., EChaYa (in Russian) *15* (1984) 762. [Sov. J. Part. Nucl. *15* (1984) 341].
63. MIKLINGHOFF M., Fizika *9* (1977) 57.
64. BARZ H. W., ROTTER I., HÖHN J., Nucl. Phys. *A275* (1977) 111.
65. BIRKHOLTZ J., Nucl. Phys. *A189* (1972) 385, BIRKHOLTZ J., HEIL V., Nucl. Phys. *A225* (1974) 429.
66. WANG W. L., SHAKIN C. M., Phys. Rev. *C5* (1972) 1898.
67. MAHAUX C., SARIUS A. M., Nucl. Phys. *A117* (1921) 103.

# Chapter 5

# Theory of Resonance States and Processes Based on Analytical Continuation in the Coupling Constant

This chapter deals with a new approach to the theory of resonance states and processes, namely, the method of analytic continuation in the coupling constant (ACCC) proposed by the authors earlier [1–4]. In the previous chapters it was noted repeatedly that the most accurate method for finding the resonances and their characteristics was based on analytic continuation. Since the resonant features of the scattering amplitude occur on *non-physical* energy sheets and the whole theoretical apparatus used (dynamical equations, unitarity relations, completeness, orthogonality, etc) is formulated as a rule on the *physical* sheet, it is only natural to define all the resonance parameters of amplitudes via analytic continuation in energy (or in $k$) from the physical to nonphysical sheets. The analytic continuation in energy may be replaced by analytic continuation in the coupling constant of the interaction, which is more convenient and physically more lucid in many cases.

It is intuitively clear that the resonance may be treated as such an eigenstate of the Hamiltonian of a system which arises from a purely bound state when the intensity of the attractive part of the interaction decreases. In the language of analytical properties, this means that a resonance $S$-matrix pole on the nonphysical sheet is defined as the analytic continuation of a bound-state pole on the physical sheet in the coupling constant of the attractive part of the total Hamiltonian. In the case of potential scattering, such a definition of the resonance is mathematically equivalent to the conventional definition through poles [5]. In the particular case of the three-body problem, it may also be shown that such a definition is equivalent to the conventional definition through poles for the three-particle $S$-matrix which emerges from the solution of the Faddeev equations. In the general case this equivalence is probable but it has not been proved rigorously yet. A characteristic feature of the present approach is that the above-mentioned obvious physical idea is realized in the theory of resonance states and the processes involving the resonance states. This theory makes it possible to describe in a unique way a broad class of phenomena in quantum-mechanical systems by studying their evolution in the coupling constant. The constructive character of this theory may be counted among its advantages. In contrast to some of the approaches treated in the previous chapters, this theory entails a simple calculation method which makes it possible to calculate the characteristics of the resonance states and processes (the resonance energies, widths, wave functions, the cross sections of the reactions involving resonances, etc.), proceeding only from the bound-state calculations. Such simplicity arises for two reasons: (i) careful allowance for the analytical properties of the examined parameters as functions of the coupling constant and energy; (ii) application of the Padé approximant technique which gives an effective calculation algorithm involving analytic properties in the process of analytic continuation.

## 5.1. Two-Particle Resonances in Real and Complex Potentials

In this section we shall show how the examined approach is used to construct the theory of the resonance states in a system of two spinless particles. The interaction potential (real or complex) between them is assumed to be finite.

### 5.1.1 Analytic Properties of the Energy Eigenvalues as Functions of the Coupling Constant

Allowance for the analytic behaviour of energy eigenvalue near threshold as a function of the coupling constant is of decisive importance in making a correct and stable analytic continuation to the resonance region. We shall consider the Schrödinger equation for the $l$-th partial wave (for brevity, the index $l$ will be omitted):

$$\left[ -\frac{d^2}{dr^2} + \frac{l(l+1)}{r^2} + \frac{2m}{\hbar^2} \lambda V(r) \right] \psi_\lambda(k, r) = k^2(\lambda) \, \psi_\lambda(k, r) \qquad (1.1)$$

where $\lambda$ is the coupling constant; the interaction operator $V$ is assumed to be negatively definite (a more general case see below). It is known [7] that in the vicinity of a threshold (say, at $k^2 \sim 0$) the conventional series of perturbation theory is meanigless (because of the proximity of the continuum) and should be so modified as to allow for the continuum. One of the ways of making such a modification is to treat, instead of the eigenenergy $E$, the evolution of the zero of the Jost function corresponding to a near-threshold bound state. In this case the pole trajectory of the $S$-matrix defined as

$$S_l(k) = \frac{f_l(\lambda, -k)}{f_l(\lambda, k)}$$

is obtainable from the equation

$$f_l(\lambda, k(\lambda)) = 0 \,. \qquad (1.2)$$

In case of finite-range potentials, i.e. potentials vanishing beyond a definite radius, the function $S_l(k)$ is a meromorphic function throughtout the $k$-plane [5] and has a finite number of poles located on only the positive imaginary semiaxis in the upper half-plane and an infinite number of poles in the lower half-plane.

The $S$-matrix poles on the positive imaginary semiaxis $k$ correspond to bound states, while the poles on the negative imaginary semiaxis correspond to anti-bound states (whose wave function increases exponentially at $r \to \infty$). The poles in the third and fourth quadrants are grouped in pairs symmetrical with respect to the imaginary axis. The poles in the fourth quadrant near the real axis are usually regarded as associated with resonance states.

In Ref. [8] it is shown for a potential of finite radius that all poles lying on the imaginary axis $(k_n = i\varkappa_n)$ may be classified to form two groups:

(i) $$\varkappa_1 > \varkappa_2 > \varkappa_3 > ... > \varkappa_N > \bar{\varkappa}$$

and

(ii) $$\varkappa'_2 < \varkappa'_3 < ... < \varkappa'_N < \bar{\varkappa}$$

where $\varkappa_N$ can be both positive and negative.
Further,

$$-\varkappa_1 < \varkappa'_2 < -\varkappa_2 < ... < -\varkappa_{N-1} < \varkappa'_N < \bar{\varkappa} .$$

If $V(r) \to +0$ with increasing $r$, then another pole appears in the lower $k$-halfplane at $\varkappa'_1 < -\varkappa_1$ and the total number of poles is even. If $V(r) \to -0$ with increasing $r$, the additional pole at $\varkappa'_1$ is absent and the total number of poles is odd. As the coupling constant varies, the poles $\varkappa_n$ and $\varkappa'_n$ move in the following way [8]: as $\lambda$ decreases, poles at $\varkappa_n > \bar{\varkappa}$ move downwards along the imaginary axis towards the poles $\varkappa_n < \bar{\varkappa}$ and merge in pairs at the point $\bar{\varkappa}$ at a certain value $\lambda = \lambda_0$, whereupon they move apart symmetrically with respect to the imaginary axis. The odd pole (if any) moves downwards along the imaginary axis and remains always above $\bar{\varkappa}$. The position of $\bar{\varkappa}$ depends on the type of potential. In the case of a finite potential we get $\bar{\varkappa} = 0$ , at $l \neq 0$, whereas at $l = 0$ the position of $\bar{\varkappa}$ depends, generally, on the eigenvalue number $n$, so the determination of $\bar{\varkappa}$ (hence of $\lambda_0$) is a difficult task (see Section 5.1.3).

If the representation of the Jost function $f_l(\lambda, k)$ (at a fixed value of $\lambda$) [5]

$$f_l(\lambda, k) = (a_0 + a_2 k^2 + a_4 k^4 + ...) + i k^{2l+1}(b_0 + b_2 k^2 + ...)$$

(which is correct at $k \to 0$) is applied and the analytical properties of $f_l(\lambda, k)$ with respect to $\lambda$ [5] are used, then we obtain after tedious, but straightforward, calculation [7]:

$$k_l(\lambda) = \sum_{j=1}^{l} A_j x^{2j-1} + \sum_{j=2l}^{\infty} B_j x^j \tag{1.3}$$

where $x = \pm\sqrt{(\lambda - \lambda_0)}$ , $\lambda_o$ is the threshold value of the coupling constant, i.e. $k_l(\lambda_0) = i\bar{\varkappa}$ . We shall present below a simpler way of deriving the expansion

(1.3) on the basis of the Hellmann-Feynman theorem [9]. This theorem may be written as

$$\frac{\partial E_l}{\partial \lambda} = \frac{\hbar^2 k_l}{m} \frac{\partial k_l}{\partial \lambda} = \frac{\langle \psi_\lambda | \, V \, | \psi_\lambda \rangle}{\langle \psi_\lambda | \, \psi_\lambda \rangle}. \tag{1.4}$$

For simplicity, the potential $V$ is assumed to vanish at a certain radius exceeding $R^1$). Then (omiting for simplicity the index $l$),

$$\frac{\hbar^2 k}{m} \frac{\partial k}{\partial \lambda} = \frac{\displaystyle\int_0^R |\psi_\lambda(k, r)|^2 V(r)\, dr}{\displaystyle\int_0^R |\psi_\lambda(k, r)|^2 \, dr + \int_R^\infty |\psi_\lambda(k, r)|^2 \, dr}. \tag{1.5}$$

We shall use the following well-known analytic properties of the wave function $\psi_\lambda(k, r)$:

$$\psi_l(k, r) = \begin{cases} \varphi_1(k, r) = \hat{c}_1 (kr)^{l+1} \displaystyle\sum_{i=0}^\infty a_i (kr)^{2i}, & r < R \\[2ex] \varphi_2(k, r) = \hat{c}_2 k^{l+2} K_l(ikr), & r > R \end{cases}$$

where $K_l(ikr)$ is the McDonald function [10]. (It should be stressed that $\psi_\lambda(k, r)$ is here the bound-state eigenfunctions at any $\lambda$).

Substituting this expression in (1.5), we obtain:

$$\frac{\hbar^2 k}{m} \frac{\partial k}{\partial \lambda} = \frac{\displaystyle\int_0^R |\varphi_1(k, r)|^2 V(r)\, dr}{\displaystyle\int_0^R |\varphi_1(k, r)|^2 \, dr + \int_R^\infty |\varphi_2(k, r)|^2 \, dr}. \tag{1.6}$$

At $l = 0$, from (1.6) it follows that

$$\frac{\hbar^2 k_0}{m} \frac{\partial k_0}{\partial \lambda} = k_0 \frac{\displaystyle\sum_{j=0}^\infty d_j \, k_0^{2j}}{\displaystyle\sum_{j=0}^\infty f_j \, k_0^{j}}$$

---

$^1$) This assumption is not necessary for our derivation, but leads to simplification of the derivation proper.

whence, retaining only the terms with $j = 0$ on the right hand side at small $k_0$ values we get

$$k_0(\lambda) \sim (\lambda - \overline{\lambda}_0), \quad k_0(\overline{\lambda}_0) = 0$$

so the collision point $\overline{\varkappa}_0$ of the poles is on the negative imaginary semiaxis: $\overline{\varkappa}_0 < 0$. At $l > 0$, we get

$$\frac{\hbar^2 k_l}{m} \frac{\partial k_l}{\partial \lambda} = \frac{\sum\limits_{j=0}^{\infty} g_j \, k_l^{2j}}{\sum\limits_{j=0}^{l-1} h_j \, k_l^{2j} + \sum\limits_{j=2k}^{\infty} q_j k_l^{j-1}} \tag{1.6a}$$

(where the coefficients $d_j$, $f_j$, $g_j$, $h_j$, and $q_j$ are some radial integrals) and, again retaining the leading term, we find:

$$k_l(\lambda) \simeq \pm \text{const} \, . \, \sqrt{\lambda - \lambda_0} \quad \text{at} \quad \lambda \to \lambda_0 \, . \tag{1.7}$$

This will be the leading term of the expansion of $k_l(\lambda)$ near the threshold. The complete $k_l(\lambda)$ series arises if we make one iteration in (1.6a), i.e. if we substitute (1.7) in the right-hand side of (1.6a). This gives immediately (1.3). In case of a potential barrier and $l = 0$ the analytic behaviour of the function $k_l(\lambda)$ may be expressed more accurately [7]:

$$k_0(\lambda) \underset{\lambda \to \lambda_0}{\sim} i\overline{\varkappa}_0 \pm i \sqrt{\lambda - \lambda_0} \, ; \qquad \lambda_0 < \overline{\lambda}_0 \tag{1.8}$$

where we have already $k_0(\lambda_0) = i\overline{\varkappa}_0$ and $\overline{\varkappa}_0 < 0$, while the position of the branch point $\lambda_0$ depends on the potential barrier penetrability. When the barrier is absent, we get $\overline{\varkappa}_0 \sim -1/R$, where $R$ is the radius at which the potential is truncated (a more detailed discussion of the branch-point position see below). It should be noted that the series (1.3) does not contain some powers of $k$ (the number of the missing terms increases with $l$). The use of the Hellmann-Feynman theorem is also convenient because it permits a simple generalization to the case of "distorted" waves, i.e. the analytic continuation is made in the partial, rather then total, coupling constant. Consider, for example an arbitrary two--particle Hamiltonian $H = H_0 + V$, where $H_0$ is the kinetic energy operator. Let the potential $V$ be separated into a sum of two sign-definite operators: $V = V_1 + V_2$ where $V_1$ is the negatively definite operator; $V_2$ is the positively definite operator. We can now study the evolution of the Hamiltonian eigenstates

$$H(\lambda) = (H_0 + V_1) + \lambda V_2 \equiv H_1 + \lambda V_2$$

with respect to the partial coupling constant $\lambda$ of the repulsive interaction $V_2$. For this purpose, $V$ in (1.4) is replaced by $V_2$. As $\lambda$ rises from $\lambda = 0$, the bound state level $E(\lambda)$ moves towards the threshold and may (but need not) become a resonance or a virtual state at a certain $\lambda_0$. If the quantity $k_l(\lambda)$ does not exhibit other singularities in the vicinity of $\lambda_0$ (or if the operator $V_2$ exhibits "good" analytic properties), all the analytic properties of the function $k_l(\lambda)$ inferred above will also be preserved in this case.

We shall discuss now the behaviour of the near-threshold state energy in the spaces of different dimensions. In the three-dimensional case for a short-range interaction potential, the behaviour of energy eigenvalue $E_l$ for the $l$-th partial wave may readily be obtained as a function of the coupling constant $\lambda$ on the basis of the series (1.3) [7, 38]:

$$E_l(\lambda) = (c_1(\lambda - \lambda_0) + c_2(\lambda - \lambda_0)^2 + \ldots + c_l(\lambda - \lambda_0)^l) +$$
$$+ d_0(\lambda - \lambda_0)^{l+1/2} + c_{l+1}(\lambda - \lambda_0)^{l+1} + d_2(\lambda - \lambda_0)^{l+3/2} + \ldots$$

$$(1.9a)$$

This means that, up to the $l$-th order inclusively, the series (1.9a) for perturbed energy eigenvalue (we consider the nonperturbed value of energy $E_l(\lambda_0) = 0$ to be located at the threshold) coincides with the standard series of perturbation theory. However, we deal further with the terms with semi-integer powers of $(\lambda - \lambda_0)$ which, as the threshold is traversed (i.e. $\lambda \to \lambda_0$), get imaginary and result eventually in a complex perturbed energy value whose imaginary part gives the width of the quasistationary state:

$$\Gamma_l(\lambda) = 2 \operatorname{Im} E_l(\lambda) \simeq \text{const} \,.\, (\lambda - \lambda_0)^{l+1/2}, \qquad \lambda < \lambda_0. \qquad (1.9b)$$

Or, considering that $k \sim (\lambda - \lambda_0)^{1/2}$, we obtain the (1.9b) dependence of the width of the near-threshold resonance on its energy and orbital momentum:

$$\Gamma_l(k) \simeq \text{const} \,.\, k^{2l+1}. \qquad (1.9c)$$

In the case of the $S$-wave at $\lambda \to \lambda_0$, we get

$$E_0(\lambda) \simeq A_0(\lambda - \lambda_0)^2 \qquad (1.9d)$$

i.e. the term touches the threshold and, after that, again goes to the negative semiaxis of energy, but on the *nonphysical* sheet.

It is important to emphasize that the character of the threshold features of the function $E(\lambda)$ depends on the confining potential barrier form and on the space dimension. The dependences presented above relate to the short-range spherically-symmetric potential effective in three-dimension space and to the centrifugal confining barrier. In the case of the Coulomb confining barrier whose penetrability at low energies is known [5] to be exponentially low, it may readily be shown that (see e.g. Section 5.2.2)

$$\Gamma_l(E) \underset{E \to 0}{\simeq} \text{const} \,.\, \exp\left(-\frac{a}{\sqrt{E}}\right)$$

at all values of $l$ . In this case an essential singularity appears in the function $E_l(\lambda)$ at the threshold due to the fact that the Coulombic Jost function contains a cut in the $k$-plane extending from zero to $-i\infty$, i.e. occupying the entire imaginary semiaxis.

In the one-dimensional problem with a short-range potential [35] we get

$$E(\lambda) \sim \lambda^2 \qquad (1.9e)$$

and the total series is of the form:

$$E(\lambda) = \sum_{n=2}^{\infty} b_n \lambda^n$$

In other words, as $\lambda$ decreases, the discrete level fails to turn into a resonance level, while the bound state exists for arbitrarily weak potential.

In the two-dimensional problem, it is possible to show (as it was "by fingers" in Landau and Lifshitz's monograph; for a rigorous approach see Klaus and Simon [35]) that

$$E(\lambda) \sim \exp\left(-\frac{c}{\lambda}\right) \qquad (1.9f)$$

i.e. the binding energy of a near-threshold state is exponentially small compared with the intensity of the interaction potential $\lambda V(r)$. The analytical properties of energy as a function of the coupling constant was mathematically studied in detail for various space dimensions in [34, 35]. The forms of the threshold laws (1.9c, e, f) and the similar forms found in other cases [35] are of great importance when studying the properties of the conductivity electrons in one-dimensional periodic systems, in laminary media and numerous other problems of the theory of condensed media [42].

### 5.1.2. Analytic Continuation in Coupling Constant and the Determination of the Parameters of the Bound, Resonance, and Anitibound States

From the previous section it follows that in the complex $\lambda$ – plane the function $k_l(\lambda)$ has a root branch point on the positive real axis $\lambda$. In the case of $l \neq 0$ the branch point corresponds to the value of the function $i\overline{\varkappa}_l = k_l(\lambda_0) = 0$ , i.e. it occurs at the threshold. At $l = 0$ the branch point $\lambda_0$ does not correspond to the threshold of the elastic scattering, so $i\overline{\varkappa}_0 = k_0(\lambda_0) \neq 0$ . Starting from the point $\lambda_0$, a cut appears in the $\lambda$ plane. Going over to the uniformizing variable $x$ :

$$x = \sqrt{\lambda - \lambda_0}$$

makes it possible to single out two analytic functions $k_l(x)$ and $k_l(-x)$ which are two branches of the initial function $k_l(\lambda)$. As $\lambda$ changes, the branch $k_l(x)$

describes the motion of the $S$-matrix pole corresponding to a bound state from the positive imaginary semiaxis $k$, through the point $k = i\bar{\varkappa}$ to the third quadrant of the $k$-plane, while the branch $k_l(-x)$ describes the motion of an antibound state from the negative imaginary semiaxis $k$ to the fourth quadrant (see below).

Using the analyticity of the functions $k_l(x)$ and $k_l(-x)$, we can determine them in the region $\lambda < \lambda_0$ (the resonance region) via analytic continuation from the region $\lambda > \lambda_0$ (bound-state region). The analytic continuation may be made using the series (1.3) (a detailed discussion of the analytical continuation technique see in Section 1.1.3): however, the calculations of its coefficients are faced with serious difficulties. Besides, the direct use of the expansion (1.3) is probably justified only near the threshold $k = 0$ because the series (1.3) diverges at greater distances from the point $k = 0$ and has to be rearranged, thereby invoking the analytic continuation of the function $k_l(x)$ to higher values of $x$. The analytic continuation can be carried out effectively using the Padé approximant (PA) technique (see Section 1.1.2).

The PA may used to advantage because a PA sequence converges throughout the analyticity domain of the function to be continued. However, if PA are constructed via the coefficients of the series (1.3), i.e. the type I PA we are again faced with difficulties in calculating the coefficients of the series (a very good example of such difficulties including, among other things, a numerical instability see in ref. [48]).

As the calculations of the high-order coefficients in the series (1.3) are rather difficult, we shall use the type II or III PA, i.e. the approximants constructed on the basis of the numerical values of the function $k_l(\lambda)$. Thus, we construct the PA:

$$k_l(x) \simeq k_l^{[N,M]}(x) = \frac{P_N(x)}{Q_M(x)} = i \frac{c_0 + c_1 x + c_2 x^2 + \dots + c_N x^N}{d_0 + d_1 x + d_2 x^2 + \dots + d_M x^M}.$$

$$(1.10)$$

In conformity with the convergence theorems (see Section 1.1.2) for $N$, $M \to \infty$, the sequence of the PA $k_l^{[N,M]}(x)$ converges to an accurate value of the function $k_l(x)$ at all $x$ belonging to the analyticity domain of the function $k_l(x)$. The coefficients of the polynomials $P_N$ and $Q_M$ are real because they are calculated in the bound-state region. In case of $\lambda > \lambda_0$, $x$ is real and $\mathrm{Re}\, k_l = 0$, which corresponds to a bound-state pole. At $\lambda < \lambda_0$, $x$ is purely imaginary and the function $k_l(x) = k_1 - ik_2$ is complex. Accordingly, the resonance energy and width can be inferred from the relation (1.10):

$$E_R - i\Gamma/2 = \frac{\hbar^2(k_1 - ik_2)^2}{2m}.$$

Thus, in the case of varying $\lambda$ the PA(1.10) prescribes the trajectory of a given $S$-matrix pole in the complex $k$-plane to be expressed by an analytic rational--fractional function. The coefficients of the PA (1.10) can be found as follows. Let the resonance of interest to us correspond to $\lambda = 1$ in (1.1). We shall take advantage of the fact that, at $\lambda > \lambda_0$, $k_l^2 < 0$ is merely the bound-state energy $E(\lambda)$ which may be inferred from the equation (1.1) and this can readily be solved in this case. Let a sequence of the coupling-constant values $\{\lambda_i\}$ be defined such that all the $\lambda_i (i = 1, 2 ..., p)$ values exceed $\lambda_0$. Further, by solving equation (1.1) for each value of $\lambda_i$ with the bound- state boundary conditions, we find the respective eigenvalues $\{E_i\}$ (all $E_i < 0$). After that we go over to the reference points $\{k_i = \sqrt{(2m/\hbar^2)}\, E_i\}$ and $\{x_i = \sqrt{(\lambda_i - \lambda_0)}\}$ (for the method of finding $\lambda_0$ see Section 5.1.3) and, having substitued them in the PA(1.10), find the coefficients $\{c_j\}$ and $\{d_j\}$ (here, $N + M + 1 \leq p$). Assuming now that $\lambda = 1$, i.e. $x = \sqrt{1 - \lambda_0}$ in (1.10), we obtain the resonance parameters from the bound-state parameters $\{E_i\}$ and $\{\lambda_i\}$. It should again be emphasized that this is only possible if the analytic properties of the function $E(\lambda)$ are allowed for. In other words, we are to go over from the sets $\{E_i\}$ and $\{\lambda_i\}$ to the sets $\{k_i\}$ and $\{x_i\}$, i.e. to the function $k(x) = \sqrt{(2m/\hbar^2)}\, E$. By going over to $k = \sqrt{(2m/\hbar^2)}\, E$ we allow for the fact that there exist physical and nonphysical sheets, while by going over to the uniformizing variable $x$ we define the way for the trajectory to pass to the complex $k$-plane. Fig. 5.1 compares the resonance trajectories calculated

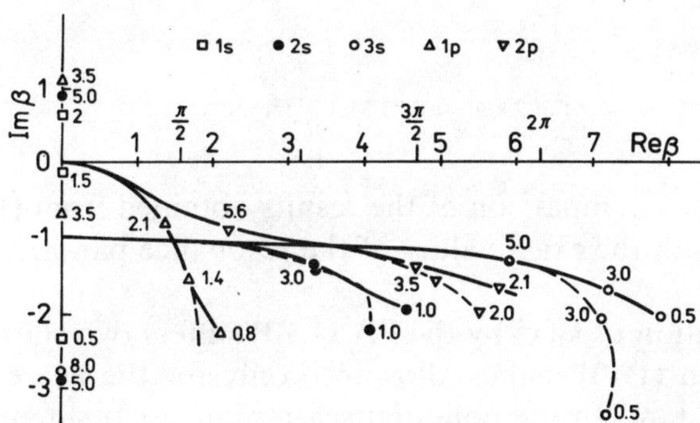


Fig. 5.1

Trajectories of the $S$-matrix poles $k_l(\lambda)$ $(l = 0,1)$ in the $\beta$ – plane $(\beta = kR_0)$ for the 1s, 2s, 3s, 1p and 2p states for rectangular well. The dotted line represents the exact trajectories [8], the solid line the trajectories obtained using the [5,5] PA. The numerals at the curves are the values of the parameter $\beta_0 = (2mV_0\lambda/\hbar^2)^{1/2} R_0$.

in the above-mentioned way with the exact trajectories obtained by solving (1.1) *directly* in the lower $k$ half-plane for the square-well potential. Raising the approximant order enlarges the region where the PA (1.10) describes the exact trajectory (however, only subject to a sufficient accuracy of the initial $\{E_i\}$ data).

Table 5.1

Positions of antibound ($a$) and resonance ($r$) $S$-matrix poles for the rectangular well $V_0\,\Theta(r - R_0)$ ($l = 0$ and 1); $n$ is the number of the level.

| $l$ | $n$ | PA order | [1, 1] | [2,2] | [3, 3] | [4, 4] | [5, 5] | [6, 6] | Exact value |
|---|---|---|---|---|---|---|---|---|---|
| | $2a$ | Im $(k\,R_0)$ | $-2.744$ | $-2.847$ | $-2.873$ | $-2.633$ | $-2.907$ | $-2.903$ | $-2.855$ |
| 0 | $3r$ | Re $(k\,R_0)$ | 5.963 | 5.913 | 5.911 | 5.910 | 5.903 | 5.901 | 5.894 |
| | | Im $(k\,R_0)$ | $-1.241$ | $-1.276$ | $-1.328$ | $-1.331$ | $-1.333$ | $-1.331$ | $-1.321$ |
| | $4r$ | Re $(k\,R_0)$ | 9.708 | 9.735 | 9.615 | 9.722 | 9.715 | 9.698 | 9.653 |
| | | Im $(k\,R_0)$ | $-1.229$ | $-1.312$ | $-1.383$ | $-1.550$ | $-1.547$ | $-1.618$ | $-1.597$ |
| | $1a$ | Im $(k\,R_0)$ | $-1.048$ | $-1.934$ | $-0.788$ | $-2.289$ | $-0.652$ | $-0.658$ | $-0.644$ |
| | | $\lambda_0$ | 0.822 8 | 0.809 5 | 0.807 3 | 0.806 6 | 0.806 3 | 0.806 4 | 0.805 7 |
| 1 | $2r$ | Re $(k\,R_0)$ | 4.764 | 5.138 | 5.127 | 4.985 | 4.955 | 4.895 | 4.791 |
| | | Im $(k\,R_0)$ | $-1.503$ | $-1.280$ | $-1.282$ | $-1.526$ | $-1.544$ | $-1.482$ | $-1.399$ |
| | | $\lambda_0$ | 3.237 2 | 3.226 4 | 3.224 4 | 3.223 7 | 3.223 3 | 3.223 3 | 3.222 7 |
| | $3r$ | Re $(k\,R_0)$ | 8.870 | 8.832 | 8.798 | 8.814 | 8.806 | 8.802 | 8.452 |
| | | Im $(k\,R_0)$ | $-1.749$ | $-1.195$ | $-1.252$ | $-1.432$ | $-1.427$ | $-1.424$ | $-1.769$ |
| | | $\lambda_0$ | 7.267 1 | 7.255 1 | 7.253 0 | 7.252 2 | 7.251 8 | 7.251 5 | 7.251 1 |

$k = (2mE/\hbar^2)^{1/2}$ ;

$$V_0 = \begin{cases} 25 \text{ MeV}, \; l = 0 \\ 12.25 \text{ MeV}, \; l \neq 0; \; R_0 = 2 \text{ fm}; \; 2m/\hbar^2 = 0.25 \text{ Mev}^{-1}fm^{-2} \, . \end{cases}$$

Table 5.1 presents a comparison of the results obtained from (1.10) at different values of $N$, $M$ with the exact values of the resonance parameters for the same potential.

The representation of $k_l(x)$ by the PA (1.10) suffers two shortcomings. First, the representation (1.10) makes allowance only for the nearest singularity in the coupling constant. In the potential scattering case a singularity arises also at the point $\lambda = 0$ [8], so that at small $\lambda$ $(\lambda/\lambda_0 \ll 1)$ the representation (1.10) becomes inadequate. However, such small values of $\lambda$ are not usually of physical interest because the corresponding poles are located very far from the physical region [8]. Secondly, superfluous terms appear at small $x$, as compared with (1.3) (the PA [$N$, $M$] contains all the integer powers of $x$ up to $N + M + 1$). This may be easily remedied by going over to the PA of the following non-standard form:

$$k_l^{[N,M]}(x) = ix\,\frac{P_L(x) + ix^{2l+1}Q_{N-L}(x)}{R_M(\lambda)} \tag{1.11}$$

where $x = \pm \sqrt{(\lambda - \lambda_0)}$ as above. The approximant structure chosen in (1.11) contains only the necessary powers of $x$ at $\lambda \to \lambda_0$ (cf. (1.3)) and should prove to be more usable than the form (1.10). However, the practical calculations have shown that, despite a certain inadequacy, the form (1.10) provides for a fairly high stability and for a good convergence to the exact result, so that the form (1.11) need not, generally, be used.

It was noted above that at $\lambda < \lambda_0$ the bound-state branch $k_l(x)$ ($x$ is purely imaginary) appeared in the 3rd quadrant of the $k$-plane. Therefore, by substituting $x = (1 - \lambda_0)$ in (1.10), we find, generally, the position of the antiresonance $k'_R$ in the 3rd quadrant. However, the antiresonance and resonance positions are related to each other as

$$k'_R = -k^*_R$$

i.e.

$$\operatorname{Im} k_R = \operatorname{Im} k'_R \qquad \text{at} \qquad \operatorname{Re} k_R = -\operatorname{Re} k'_R .$$

One of the advantages of the representation (1.10) is that the Padé approximant at the pure imaginary values of $x$ exhibits the necessary symmetry, i.e.

$$k_l^{[N,M]}(-x) = \left(-k_l^{[N,M]}(x)\right)^* .$$

Thus, the resonance and antiresonance branches differing in signs of $x$ appear automatically to be symmetric, so it is sufficient to treat just one of them. In going backwards through the threshold, i.e. at $\lambda > \lambda_0$, the symmetry is violated, so $k_l^{[N,M]}(x)$ and $k_l^{[N,M]}(-x)$ begin describing the asymmetrical branches of the bound and virtual (antibound) states. In other words, by substituting $-x$ for $x$ in the Padé approximant (1.10) from which we inferred the resonance parameters, and going further to $\lambda > \lambda_0$, we obtain the virtual-state branch. Table 5.1 compares the virtual-state positions so obtained with the respective exact values in the case of a rectangular-well potential.

Thus, we have arrived at a surprising result, namely, given only the bound-state evolution in the coupling constant (i.e. the function $E_B(\lambda)$ at $\lambda > \lambda_0$) even on a descrete set of points and knowing the nearest $E_B(\lambda)$ singularities, we may still readily use (1.10) to find *all other branches* of polar trajectories; i.e. the resonance, antiresonance, and virtual-state branches.

We have examined the analytic continuation of bound-state solutions to the resonance and virtual state region. The alternative of proceeding from the scattering region is also usable. This approach was tackled, for example, in [11] where the direct analytic continuation of scattering amplitude to the region of discrete spectrum was made.

For this purpose, the partial $t$-matrix $t_l(z)$ (for definition see (2.1.18)), may be expressed in terms of the perturbation series in the coupling constant (this can be done, for example, by iterating the Lippmann-Schwinger equation (1.5.13)):

$$t_l(\lambda, z) = \lambda t_l^{(1)}(z) + \lambda^2 t_l^{(2)}(z) + \ldots \tag{1.12}$$

This series is called also the Neumann or Born series. Now, let the series (1.12) be transformed into the corresponding Padé approximant in $\lambda$:

$$t_l^{[N,M]}(\lambda, z) = \frac{P_N(\lambda, z)}{Q_M(\lambda, z)} \tag{1.13}$$

where the coefficients of the polynomials $P_N$ and $Q_M$ are determined by the coefficients of the series (1.12) $t_l^{(i)}(z)$ and depend on energy $z$. The form (1.13) of the $t$-matrix has certain advantages over (1.12), namely, the expression (1.13) converges (with increasing $N$ and $M$) throughout the analyticity domain of the function $t_l(\lambda)$ in the $\lambda$-variable, whereas (1.12) converges usually only at small values of $\lambda$ [5]. Furthermore, the expression (1.13) contains the explicit form of the $t$-matrix (scattering amplitude) singularities. The zeros of the polynomial $Q_M(\lambda, z)$ correspond to these singularities. Thus, we can determine the pole trajectories of the scattering amplitude $k_l(\lambda)$ $(k^2 = 2mz/\hbar^2)$ from the equation:

$$Q_M(\lambda, k_l(\lambda)) = 0 . \tag{1.14}$$

To do this, it is necessary to find the coefficients $t_l^{(i)}(z)$ of the series (1.12) at each value of $z$, to transform the resultant series into the Padé approximant (1.13), and then, to use equation (1.14) for finding the value $\lambda$ corresponding to the given $k = \sqrt{(2mz/\hbar^2)}$ on the $k_l(\lambda)$ trajectory. Alternatively, we may use the well-known Hadamard method (see Chapter 1) which makes it possible to examine the positions of the analytic function poles on the basis of the information contained in some portion of the Taylor series of the function. Thus, even in the case of two particles the given approach involves a sufficiently complicated procedure for constructing the $k_l(\lambda)$ trajectory which is incomparable with the simple analytic expression (1.10).

### 5.1.3 Determination of Branch Points

As seen from the above, a knowledge of the branch-point $\lambda_0$ of the function $k_l(\lambda)$, as well as knowledge (at $l = 0$) of the function value $k_0 = i\varkappa = k_0(\lambda_0)$ at this point, is of fundamental importance in the approach discussed. Numerous calculations [1–4] have shown that the accuracy in calculating $\lambda_0$ is of decisive importance as regards the accuracy of the eventual results and the stability of the analytic continuation procedure. Several specific methods for finding $\lambda_0$ have been studied in [1–4]. In Section 1.2.4 we presented the universal method for finding $\lambda_0$ and $k_0$, which will be applied here (see formulae (1.2.37)–(1.2.44)).

Let the values $\{k_{ij}\}$ of the function $k(\lambda)$ be known at a finite number of points $\{\lambda_{ij}\}$, $i = 1, 2 ..., p$. Also, it is known that $k(\lambda)$ has a root branch point at some value of $\lambda = \lambda_0$. Since at $l \neq 0$ the problem is simplified by the known value $k_0 = 0$, we shall examine first how $\lambda_0$ is to be found in this case. The procedure is reduced to constructing the following Padé approximant through the sets of

$\{\lambda_i\}$ and $\{k_i\}$ (these are the same sets that were used to determine the trajectory (1.10) and to find the resonance parameters), $i = 1, 2, ...p$:

$$\lambda^{[N,M]}(k) = \frac{P_N(k)}{Q_M(k)}$$

where $N + M + 1 \leq p$. Since $\lambda(k)$ is analytic near $k = k_0$, the approximant $\lambda^{[N,M]}(k_0)$ converges to the exact value of $\lambda_0 = \lambda(k_0)$ at $N, M \to \infty$ and, therefore, $\lambda_0$ can be determined to within an arbitrary accuracy from the equality:

$$\lambda_0 \simeq \lambda_0^{[N,M]}(k_0) = \frac{P_N(k_0)}{Q_M(k_0)}$$

where $k_0 = 0$. At $l = 0$, $\lambda_0$ is determined by the same equality, but $k_0 \neq 0$ and it is necessary to find the value $k_0$ beforehand. To find $k_0$, simultaneously with $\{\lambda_i\}$ and $\{k_i\}$, we calculate the values of $\{z_i = 1/(\partial k/\partial \lambda)_i\}$ where $(\partial k/\partial \lambda)_i$ can readily be calculated using the Hellmann–Feynmann theorem [9]:

$$\left(\frac{\partial k}{\partial \lambda}\right)_i = \frac{m}{2\hbar^2 k_i}\left(\frac{\partial E}{\partial \lambda}\right)_i = \frac{m}{2\hbar^2 k_i}\langle \psi_i(k_i, r)|\,V\,|\psi_i(k_i, r)\rangle$$

where $i$ is the number of the reference point $(\lambda_i, k_i)$ and $\psi_i(k_i, r)$ are the normalized bound-state wave functions. Using the sets $\{k_i\}$ and $\{z_i\}$, we construct the Padé approximant

$$k^{[N,M]}(z) = \frac{P_N(z)}{Q_M(z)}.$$

Since the function $k(z)$ is analytic near the point $z = 0$, the approximate equality

$$k_0 \simeq k^{[N,M]}(0)$$

gets exact at $N, M \to \infty$.

Further, the resultant value of $k_0$ is substituted in $\lambda^{[N,M]}(k)$, whereupon $\lambda_0 \simeq \lambda^{[N,M]}(k_0)$ is obtained.

The method for determining $k_0$ and $\lambda_0$ was tested using some model functions (see Section 1.2.4). These examples are useful because they demonstrate the convergence of the method in its "pure" form, i.e. without the noise inevitable in realistic calculations. The results have been presented in Table 1.3. Column 1 shows the values of $\lambda_0$ at a given $k_0$. Column II gives the values of $\lambda_0$ and $k_0$ obtained self-consistently. Convergence to the exact values is rapid and stable. Table 5.1 compares the values of $\lambda_0$ for a number of states of the rectangular potential with the exact values. The convergence is good: usually, as many as 4–5 significant digits are reproduced reliably. More accurate determination becomes difficult due to the errors in calculating the values of $\{k_i\}$ obtainable by numerical integration of $(1.1)^2)$.

### 5.1.4 Wave Functions of the Gamow States: Orthogonality and Normalization Relations

The complex energy of the resonance state is just as obtained to be an analytic continuation of the bound state energy depending on the coupling constant as the complex wave function of the Gamow state is obtained to be an analytic continuation in the coupling constant of the wave function of the respective bound state.

It is evident that the continuation in the coupling constant can be replaced by the continuation *in k* along the $k_l(\lambda)$ trajectory determined by $(1.10)$ to the point $k_R$ corresponding to the wave vector of the Gamow state $k_R = k_l^{[N,M]} (\lambda = 1)$. To provide for a stable analytic continuation in $k$, it is necessary to allow for the analytic properties of the Gamow function $\Phi_{nl}(k, r)$ as a function of $k$. In Section 2.2 it was shown (see eq. $(2.2.35)$) that the function

$$\varphi(k, r) = \begin{cases} \Phi_{n0}(k, r)/\sqrt{k}, & l = 0 \\[2ex] \Phi_{nl}(k, r), & l \neq 0 \end{cases}$$

would be an analytic function of $k$ for $r < R_0$ in the Jost function analyticity domain.

Therefore, to find the resonance wave function $\varphi(r)$ we shall use the technique which was used to find the complex resonance energy $E$, i.e. we shall construct the Padé approximant

$$\varphi^{[N,M]}(k, r) = \frac{P_N(k, r)}{Q_M(k, r)} = \frac{a_0(r) + a_1(r)k + \ldots + a_N(r)k^N}{b_0(r) + b_1(r)k + \ldots + b_M(r)k^M}$$

whose coefficients are dependent on $r$.

The resonance function in the inner region is determined by the expression

$$\varphi_R(r) \simeq \varphi^{[N, M]}(k_R, r) = \frac{P_N(k_R, r)}{Q_M(k_R, r)}. \tag{1.15}$$

According to (2.34) of Chapter 2, we have at $r \geq R_0$ :

$$\Phi_{nl}(k_R, r) \underset{r \to \infty}{\approx} N_l h_l^{(+)}(k_R, r) \tag{1.16}$$

---

$^2$ To reduce the "noise" in the resultant values, it is necessary to use very accurate methods for calculating the energy eigenvalues $E_i = E(\lambda_i)$. One of such methods is described in Section 3.1.3.

where the asymptotic constant $N_l$ can be easily obtained also by analytic continuation from the appropriate values of the asymptotic constant for bound states. In particular, absolutely the same relationship [12] exist between the residue of the $S$-matrix $C_l$ at the *resonance pole* $k_R$ and the asymptotic constant $N_l$ from (1.16):

$$C_l = -i(-1)^l N_l^2$$

as in the case of *bound states* [12], i.e. the relation between $C_l$ and $N_l$ for a resonance is an analytically-continued relation for the respective bound state.

Moreover, the value of $N_l$ may easily be obtained also by smooth matching of (1.15) and (1.16).

In the actual calculations it is more convenient to carry out the analytic continuation of $\Phi_{nl}(k, r)$ in $k$ to the point $k_R$ only after the value of $k_R$ has been determined by analytic continuation in the coupling constant. Fig. 5.2 shows the results of calculating several resonance and virtual-state functions and compares these functions with the exact functions.

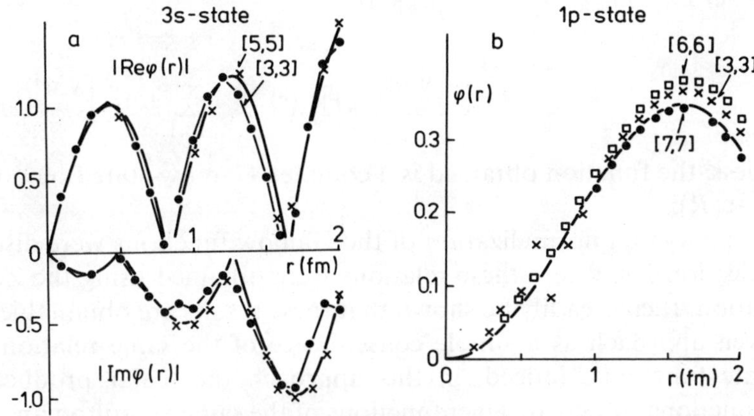


Fig. 5.2

Wave functions of the resonance (a) and antibound (b) states obtained by analytic continuation (1.15). The exact wave function is shown with the solid line. Approximate values are only shown where they do not coincide with the exact values. The Padé approximant order is indicated in square brackets.

It should be noted that straightforward numerical integration of the equation aimed at finding $\Phi_{nl}(k_R, r)$, especially in the case of broad or above-barrier resonances and for virtual states, is often associated with numerical instability of the calculation procedure because the solution increases sharply in the outer region.

In the general case, the calculations of the Gamow functions by (1.15) imply the continuation in $k$ of the bound-state function $\Phi_{nl}(k_B, r)$ at each point $r$ to the point $k_R$ corresponding to the resonance. If the bound state is calculated by a variational or some other method using the expansion of the function $\Phi_{nl}(k_B, r)$ in some complete basis $\{\varphi_i(r)\}$,

$$\Phi_{nl}(k, r) = \sum_i c_i(k)\varphi_i(r)$$

then the resonance-state function can be obtained as an expansion in the same basis. The expansion coefficients $c_i(k) = \langle \varphi_i(r)| \Phi_{nl}(k, r) \rangle$ are analytic functions of $k (\tilde{c}_i(k) = c_i(k)\sqrt{(k)}$ are the analytic functions at $l = 0$). This follows from the analytic properties of $\Phi_{nl}(k, r)$ . Thus, for $c_i(k)$ we may construct the Padé approximants

$$c_i^{[N,M]}(k) = \frac{P_N(k)}{Q_M(k)}$$

whereupon they may be used to continue $c_i(k)$ to the points $k_R$, i.e. we obtain the resonance function to be the expansion in the *same real basis* $\{\varphi_i(r)\}$:

$$\Phi_{nl}(k_R, r) \cong \sum_{i=1}^{n} c_i^{[N,M]}(k_R)\varphi_i(r) =$$

$$= \sum_{i=1}^{n} \mathrm{Re} \left[ c_i^{[N,M]}(k_R) \right]\varphi_i(r) + i \sum_{i=1}^{n} \mathrm{Im}[c_i^{[N,M]}(k_R)]\varphi_i(r) .$$

Nevertheless, the function obtained is a complex Gamow-state function (in the region $r < R$).

Orthogonality and normalization of the Gamow functions were discussed in detail in Section 2.2, where these relations were obtained using the Zel'dovich regularization. It can readily be shown that these results are obtainable in terms of the given approach as a simple consequence of the same relations for the bound-state functions. Indeed, in this approach the scalar product of two Gamow functions (which are eigenfunctions of the same Hamiltonian $H_0 + V$) is defined as

$$I_{ij} = (\Phi_i(k_R^{(i)}), \Phi_j(k_R^{(j)})) = \underset{\lambda \to 1}{\mathrm{Cont}} \langle \Phi_i(k_B^{(i)}) | \Phi_j(k_B^{(j)}) \rangle . \tag{1.17}$$

Here *"Cont"* denotes the analytic continuation in $\lambda$ to $\lambda = 1$ . The round brackets pertain to the scalar product with the Gamow functions, and the Dirac brackets to the ordinary scalar product including the bound-state functions; $k_B^{(i)} \underset{\lambda \to 1}{\to} k_R^{(i)}$ and $k_B^{(j)} \underset{\lambda \to 1}{\to} k_R^{(j)}$ , where $k_B^{(i)}$ and $k_B^{(j)}$ are the wave vectors of the bound states into which the resonance states $k_R^{(i)}$ and $k_R^{(j)}$ change with increasing the coupling constant. Further, since

$$\langle \Phi_i(k_B^{(i)}) | \Phi_j(k_B^{(j)}) \rangle = \delta_{ij}$$

and $k_B^{(i)}$ and $k_B^{(j)}$ are analytic functions of $x$ (going over to the variable $x$ is realized by uniformizing the functions $k^{(i)}(\lambda)$ and $k^{(j)}(\lambda)$ according to the known theorem of analytic continuation of a composite function [13] we get:

$$I_{ij} = \left(\Phi_i(k_R^{(i)}), \, \Phi_j(k_R^{(i)})\right) = \delta_{ij} \qquad (1.18)$$

in the analyticity domain of the functions $k^{(i)}(x)$ and $k^{(j)}(x)$. Here, the relation (1.18) is understood just in the sense of analytic continuation. It should be emphasized that such a determination works well when the Zel'dovich-type regularization fails, say, in the case of virtual states or broad resonances (see Section 1.1). In fact, the relation (1.18) means in terms of our approach that, if the initial bound states are chosen to be normalized and mutually orthogonal, the Gamow functions obtained from them will automatically exhibit these properties (in the sense of direct determinations (2.20) and (2.40) in Chapter 2).

### 5.1.5 Matrix Elements Comprising the Gamow Functions

If we proceed in the same manner as above (see also [14]), we shall find the matrix elements comprising the resonance and virtual states

$$\langle \psi | \, \tilde{O} \, | \Phi ) = \operatorname*{Cont}_{k \to k_R} \int_0^\infty \psi^*(r) \tilde{O} \Phi(k, r) \, \mathrm{d}r \qquad (1.19)$$

(where $\tilde{O}$ is an arbitrary one-particle operator) and the same expression for $(\Phi | \, \tilde{O} \, | \psi \rangle$ via the analytic continuation in $k$ of the corresponding matrix elements well-defined in the upper $k$ half-plane. Let us now describe the sequence of operations involved in using the method of analytic continuation in coupling constant. First, we shall examine the calculations of the one-particle matrix elements

$$\mathcal{M}_{ij} = (\Phi_i | \, \tilde{O} \, | \Phi_j) = \operatorname*{Cont}_{\lambda \to 1} \int_0^\infty \Phi_i[k_i(\lambda), r] \, \tilde{O} \Phi_j[k_j(\lambda), r] \, \mathrm{d}r \qquad (1.20)$$

where $\Phi_i(k, r)$ and $\Phi_j(k, r)$ are functions of bound states with energies $\hbar^2 k_i^2/2m$ and $\hbar^2 k_j^2/2m$, respectively. In the case of non-diagonal matrix elements there exist two trajectories $k_i(\lambda)$ and $k_j(\lambda)$ with two branch points $\lambda_0^{(i)}$ and $\lambda_0^{(j)}$, so that $\mathcal{M}_{ij}(\lambda)$ has also two branch points in the $\lambda$-plane. In this case, to make the analytic continuation, we must carry out uniformization in conformity with the general rules [15], i.e. we must go over to the variables

$$x^{(i)} = \left(\lambda - \lambda_0^{(i)}\right)^{\frac{1}{2}} ; \qquad x^{(j)} = \left(\lambda - \lambda_0^{(j)}\right)^{\frac{1}{2}}$$

and, then, develop the obtained Riemannian surface on the $t$-plane

$$x^{(i)} = \varDelta \frac{1 + t^2}{1 - t^2} \; ; \qquad x^{(j)} = \varDelta \frac{2t}{1 - t^2}$$

where

$$\varDelta = \lambda_0^{(j)} - \lambda_0^{(i)} .$$

After that, the matrix element $\mathcal{M}_{ij}(t)$ as a function of t becomes a single-value analytic function of $t$ and, given this function, the Padé approximant must converge rapidly.

If only one of the functions in (1.20) is a resonance function, the uniformization gets trivial.

In the general case of (1.20) we may also use the analytic continuation directly in $k_i$ and $k_j$ if the trajectories $k_i(\lambda)$ and $k_j(\lambda)$ have been found. This modus operandi is similar to that proposed by Romo [14], the only difference being that the continuation is along the $S$-matrix pole trajectories $k_i(\lambda)$ and $k_j(\lambda)$. Owing to the simplicity of the analytic properties with respect to $k$, the PA convergence is even better in this case, but we must use the generalized PA in two variables [16]. However, the simplest procedure of the analytic continuation of matrix elements is based on a certain ansazt involving a versatile analytic construction and allowing for the main analytic properties of the function to be continued (for details, see below).

Let us now discuss methods for calculating the two-particle matrix elements

$$\mathcal{M}_{rs,\,r's'} = \left( \varPhi_r(1)\varPhi_s(2) \right| V_{12} \left| \varPhi_{r'}(1)\varPhi_{s'}(2) \right) \tag{1.21}$$

which are necessary if two particles appear in the continuum [22]. In the general case where all four functions in (1.21) are resonance functions (or virtual-state functions) $\mathcal{M}_{rs,\,r's'}(\lambda)$ has four branch points corresponding to four trajectories $(i = r, s, r', s')$. To attain a mathematically correct continuation, it is necessary to uniformize the $n$-sheeted $(n > 3)$ Riemannian surface. It is known that such a Riemannian surface cannot be topologically mapped onto a plane (sphere), but can be mapped onto a torus [15]. However, we shall not go into details of the technique of such conformal mappings because one or more orbitals in (1.21) often prove to be bound states in the cases of practical importance. Moreover, there exists a method of analytic continuation of a many-sheeted function which makes it possible to avoid the uniformization procedure (see below). We may also make a continuation in several variables $k_i(\lambda)$ instead of one variable $\lambda$ (the general coupling constant of a selfconsistent field). Apparently, the following simple procedure of analytic continuation of the matrix elements $\mathcal{M}_{rs,\,r's'}$ is most effective in practice (see also Chapter 3). If all four resonance orbitals $r, s$ and $r', s'$ have angular momentum $l \neq 0$, then $\mathcal{M}_{rs,\,r's'}$ has, as a function of $\lambda$ (general coupling constant of the self-consistent field $V$ in which all four orbitals $r$, $s$ and $r'$ $s'$ are eigenstates) four root branch points $\lambda_r$, $\lambda_s$ and $\lambda_{r'}$, $\lambda_{s'}$ corresponding to the coupling-constant values at which each of the orbitals becomes a quasistationary state. In this case the analytic continuation can be carried out with the help of a simple ansatz allowing for the analytic properties of $\mathscr{M}_{rs,\,r's'}(\lambda)$, such as

$$\mathscr{M}_{rs,\,r's'}(\lambda) = \{P_r(\lambda)\,(\lambda - \lambda_r)^{\frac{1}{2}} + P_s(\lambda)\,(\lambda - \lambda_s)^{\frac{1}{2}} +$$

$$+\ P_{r'}(\lambda)\,(\lambda - \lambda_{r'})^{\frac{1}{2}} + P_{s'}(\lambda)\,(\lambda - \lambda_{s'})^{\frac{1}{2}}\} / Q_M(\lambda)$$

where $P_i(\lambda)$ $(i = r, s, r', s')$ and $Q_M(\lambda)$ are polynomials of low order with respect to $\lambda$ (for example, $P_i(\lambda) = a_i + b_i\lambda$) . Of course, the same procedure may also be used in the case where the number of branch points in $\lambda$ is smaller than four (i.e. for example, in the case of two-particle matrix elements).

The main difference (in the context of the present section) of the discussed approach from the others is that in making the analytic continuation in coupling constant the resonance matrix elements are *simultaneously* calculated and regularized. It is not necessary first to calculate the resonance state functions and then numerically formulate a tedious procedure for regularizing the matrix elements. Moreover, since the entire continuation is carried out on the basis of the calculations for the purely discrete spectrum, all the calculation procedures should be standard and easily accessible.

Concluding this section, it seems expedient to discuss the meaning of the Hellmann-Feynamn theorem for energies above the threshold. Zel'dovich [45] was the first to show that in the case of resonance states the correction for the resonance energy can be found from the conventional formula of first-order perturbation theory:

$$\Delta E_n = \frac{(\Phi_n|\,\Delta V\,|\Phi_n)}{(\Phi_n\,|\,\Phi_n)} = \frac{(\Phi_n|\,V\,|\Phi_n)}{(\Phi_n\,|\,\Phi_n)}\,\Delta\lambda \tag{1.22}$$

where it is assumed that $\Delta V = \Delta\lambda V$ if the divergent integrals are regularized by the common procedure (2.41) (Chapter 2), i.e. by introducing a truncation factor into the integrals and going over to the limit $\Delta\lambda \to 0$ . On the other hand, the material treated above shows that the integrals used here can also be interpreted as analytic continuation of the respective integrals in the case of a discrete spectrum. Since, however, (1.22) in the limit $\Delta\lambda \to 0$ coincides actually with the Hellmann–Feynman theorem, we may conclude that this theorem is valid both below and above the threshold if we interpret the appropriate matrix elements as continued analytically from the discrete spectrum domain.

### 5.1.6 Resonance in the Complex Potential

A number of physical problems require that the resonance and near-threshold states should be calculated for a non-Hermitian Hamiltonian including absorptive interaction. The resonance and near-threshold states in the $N + \bar{N}$ system, in $\bar{N} + 2N$ systems, etc. are among the most characteristic examples of this type. Since it is well-known [17] that the absorption due to an annihilation channel is short range (of the order of the core radius in the $NN$ interaction), the one boson exchange potential with phenomenologically fitted (for example, to the total annihilation cross section) absorptive short-range potential proves to be a good model for resonances in the $N + \bar{N}$ system. It is also well-known that the problem of the $N + \bar{N}$ resonance width is of paramount importance [17].

The optical model is another characteristic example. Determining discrete eigenstates in a complex optical potential is a non-trivial problem because, as shown below, all of them have complex energy. The imaginary part of the optical potential simulates the inelastic channels, i.e. the possibility for a particle to be ejected to other channels or to emerge from them, thereby imparting non-zero widths even to the bound states (in the present-day therminology, such states are called inelastic bound states). The ACCC method is very suitable when studying resonances in terms of such models. Indeed, the transition to the case of complex (or even imaginary) values of the coupling constant can easily be attained directly in the formulae (1.10) and (1.11) where all the coefficients and the value of $\lambda_0$ have been found by solving the purely Hermitian problem, i.e. by a simple transition to the complex $\lambda$ (if the radial form of the real and imaginary parts of potential are the same).

In case of "good" potentials, the $S$-matrix corresponding to the complex Hamiltonian is a meromorphic function of $\lambda$ for all $\lambda$ at which $\mathrm{Im}\,\lambda \neq 0$ [36] and shows the following symmetry properties [18, 36]:

$$S_l(k, \lambda)\, S_l(-k, \lambda) = 1 \; ; \tag{1.23a}$$

$$S_l^*(k, \lambda)\, S_l(k^*, \lambda^*) = 1 \; ; \tag{1.23b}$$

$$S_l^*(-k^*, \lambda^*) = S_l(k, \lambda) \; . \tag{1.23c}$$

Here, (1.23a) is the conventional relation between the $S$-matrix zeros and poles; (1.23c) means that, if $k_0$ is the $S$-matrix pole for the interaction $\lambda V$, $-k_0^*$ will be the pole of the interaction $\lambda^* V$. Therefore, it is sufficient to consider only the singularities corresponding to the absorptive potential, whereupon the emitting potential eigenvalues may be obtained from (1.23c). On the other hand, the approximants (1.10) and (1.11) satisfy the necessary condition (for real coefficients), i.e.

$$k_l^{[N,M]}(\lambda^*) = -\left(k_l^{[N,M]}(\lambda)\right)^* \; .$$

At the same time, the real coefficients are obtainable automatically when solving the bound-state problem with a real potential. Therefore, if the standard problem with a real potential has been solved, we need not make any new calculations, but may immediately use (1.10) or (1.11) to find the resonances and the near-threshold states for an arbitrary complex potential with the same radial dependence.

This applies both to the wave functions and to the matrix elements comprising these functions. If, however, the separation of the total Hamiltonian involves use of distorted waves, we may also continue in the partial coupling constant to its complex values. Here we obtain the pole trajectories (and the respective wave functions) for a general complex Hamiltonian with different radial forms of its real and imaginary parts. The transition in (1.10) and (1.11) to the complex values of $\lambda$ is justified due to the following. The function $k_l(x)$ is an analytic function of $x$ at real $\lambda$ and, in continuity, must be also an analytic function of $x$ in some region of the complex $\lambda$-plane including the positive real axis. This means that the Padé approximants (1.10) and (1.11) have to yield an adequate representation of $k_l(x)$ at such complex $\lambda$ too.

As a numerical illustration we shall again use a simple rectangular potential. Fig. 5.3 shows the trajectories of the 1s,2s and 3s states in the complex well $V(\lambda) = \lambda(1 + 0.1\,\text{i}) \times V_0\,\Theta(r - R_0)$ in the plane of the dimensionless variable $\beta = kR_0$.

Fig. 5.4 shows the trajectories of six $p$-levels for the potential $(1 + \text{i}\lambda)V_0\,\Theta(r - R_0)$. From the figures presented below it is clearly seen that at $\text{Im}\,\lambda \neq 0$ the symmetry of $S$-matrix poles with respect to the imaginary $k$-axis is violated.

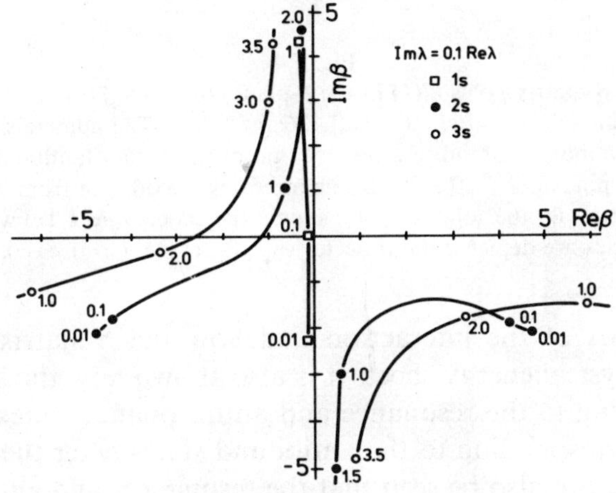


Fig. 5.3

Trajectories of the $S$-matrix poles $k_0(\lambda)$ in the $\beta$-plane $(\beta = kR_0)$ for the 1s, 2s, and 3s states of the complex rectangular well $\lambda(1 + 0.1\text{i})V_0\Theta(r - R_0)$.

So, in the upper and lower half-planes the trajectories can no longer follow the imaginary axis, but (see Fig. 5.3) move on the left and right sides of the axis, respectively (in the case of absorptive potentials). After that, the poles move apart from each other to the left and to the right without colliding with each other. The only exception is the odd pole (of the ground state) in the $S$-wave which goes on moving downwards on the left and along the imaginary $k$-axis. Fig. 5.4 shows also how the poles corresponding to the bound states and lying on the imaginary positive $k$-semi-axis enter the complex plane after switching on

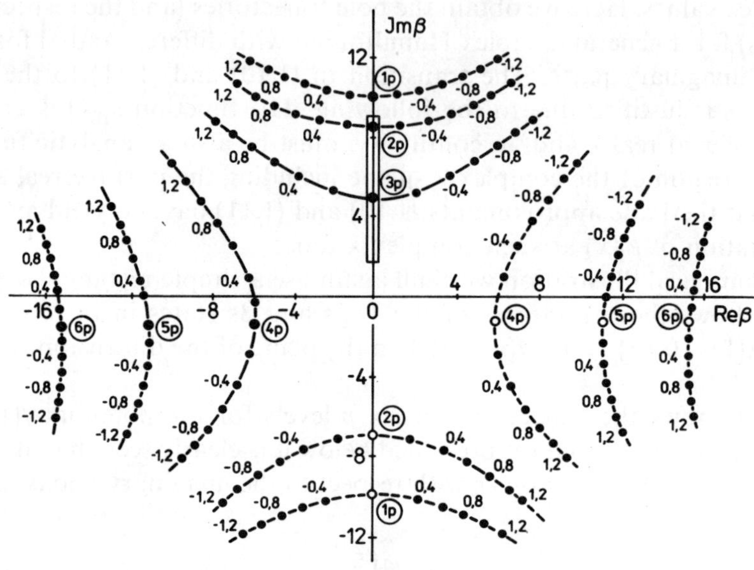

Fig. 5.4

Trajectories of the $S$-matrix poles $k_1(\lambda)$ in the $\beta$-plane $(\beta = kR_0)$ for six first p-states in the complex rectangular well potential $(1 + i\lambda) V_0 \Theta (r - R_0)$. The numerals at curves are the values of $\lambda$. The symbols ● (○) indicate the position of the bound (antibound) or resonance (antiresonance) state pole at $\lambda = 0$. The frame indicates the domain from which the analytical continuation was made. By the solid lines are shown the trajectories $k_l(x)$ while by the dashed lines are depicted the trajectories $k_l(-x)$ (i.e. $l = 1$ case).

the imaginary part of the interaction and how the $S$-matrix complex poles appear on the physical energy sheet. It is also shown how the symmetry of the poles corresponding to the resonance and antiresonance states is violated and how the poles corresponding to the antibound states enter the complex plane. From the figure it can also be seen that the resonances and antiresonances can enter the physical sheet smoothly in the case of a complex potential and that the positive-energy bound state (PEBS) appear when the positive $k$-semiaxis is intercepted (see also [36, 37]). The values of $\lambda$ corresponding to the intercept are the eigenvalues in the Hilbert–Schmidt problem. Such PEBS and resonance states lying in the vicinity of the kinematic cut can strongly influence the energy dependence of phase shifts found in terms of the optical model. Such effects were observed in the phase-shift calculations of proton scattering by medium nuclei. The effect of such $S$-matrix poles on phase shifts are examined in detail in [36,

46] where the generalized (to the case of complex interaction) Levinson theorem was also formulated:

$$\delta_l(0) - \delta_l(\infty) = \pi(n_b^l - n_p^l + n_i^l - \tfrac{1}{2} n_r^l)$$

Here, $n_b^l$ is the number of bound states in the $l$-th partial wave which is due to the real part of the interaction; $n_p^l$ is the number of PEBS; $n_i^l$ is the number of the $S$-matrix poles which moved from the 3rd to the 2nd quadrant of the $k$-plane because of the imaginary part of the interaction; $n_r^l$ is the number of poles located on the negative real $k$-semiaxis.

Thus, the inclusion of the imaginary part of the interaction (of at least the same radial form as the real potential) fails to result in any change in the number of $S$-matrix poles, but gives rise to their redistribution on the $k$-plane, thereby leading to substantial changes of the phase shifts.

## 5.1.7 The Resonance and Near-Threshold States in a Deformed Potential

In this section we shall discuss the near-threshold resonance states in a deformed potential. The development of a convenient and practically effective method for calculating such states and the corresponding transition matrix elements is very important in many problems of nuclear and atomic physics (in particular in the problems of $\alpha$-decay of deformed nuclei, and radiative capture of protons in them), in atomic problems concerned with the appearance of the molecular term in the continuum, etc. In nuclear problems it is convenient to proceed in the following way [3].

Let the initial deformed potential $V_0 f(r, \theta)^3)$

$$V(\lambda; r, \theta) = V_0 \{F(r) + \lambda[f(r, \theta) - F(r)]\} \tag{1.24}$$

be separated in such a way that the spherical component $F(r) \equiv f(r, \theta_f)$ be singled out and that the remaining deformed part $[F(r) - f(r, \theta)]$ be positive throughout the space. In particular, for the spheroidal Woods-Saxon potential, which is standard in nuclear physics,

$$V(r, \theta) = V_0 \{1 + \exp[(r - R(\theta))/a]\}^{-1} \tag{1.25}$$

(where $R(\theta) = R_0(1 + \sum_\mu \beta_\mu Y_{\mu 0}(\theta))$ and $\beta_\mu$ are the parameters of deformation) this can easily be done by selecting $\theta_f$ such that

$$F(r) \equiv \max_{0 \leq \theta \leq 2\pi} |f(r, \theta)|$$

---

$^3)$ For simplicity, we examine the case of spheroidal deformation.

so that we find $F(r) = \{1 + \exp((r - R_{\max})/a)\}^{-1}$. At $\lambda = 0$ the potential will be purely spherical, while at $\lambda = 1$ it turns into a given spheroidal potential. Under such a separation the function $E(\lambda)$ varies monotonically and the convergence of the PA sequence proves to be good. The total Hamiltonian[4]

$$H(\lambda) = H_0 + V_0\, F(r) + \lambda V_0\, g(r, \theta)$$

corresponding to the separation (1.24) leads to the eigenvalues (and to the eigenfunctions) which, having been classified by the quantum number $\Lambda$, i.e. the projection of the total orbital momentum on the symmetry axis of the spheroidal potential, exhibit the properties of analyticity in $\lambda$. These are generally the same as in the spherical case. The only difference being that the case of $\Lambda = 0$, to which the states with all $l$ contribute starting from $l = 0$, has to be examined separately because the branch point of trajectory $k_l(\lambda)$ at $l = 0$ lies in the lower $k$-half-plane on the imaginary semi-axis and at the threshold $k_l = 0$ at $l > 0$.

To understand the general situation with the motion of poles in the lower $k$ half-plane for a deformed potential, it is very useful to consider any exactly solvable model. As such a model we take a deformed separable potential

$$V(\boldsymbol{p}, \boldsymbol{p}') = -\lambda' g(\boldsymbol{p}) g(\boldsymbol{p}') \tag{1.26}$$

with the formfactor

$$g(\boldsymbol{p}) = \frac{1}{p^2 + \beta^2} + \mu\, \frac{p\, Y_0(\hat{\boldsymbol{p}})}{p^2 + \beta^2}$$

where $\mu$ is the relative intensity of the $p$-wave.

As a rule [5], the bound-state energy is defined by the condition

$$(\lambda')^{-1} = \int\limits_0^\infty \frac{p^2\, dp}{\varkappa^2 + p^2} \left[ (p^2 + \beta^2)^{-2} + \mu^2 p^2 (p^2 + \mu^2)^{-2} \right] =$$

$$= \pi^2 \beta^{-1}(\varkappa + \beta)^{-2} + \pi^2 \mu^2 (\beta^{-1} - \varkappa^2 \beta^{-1}(\varkappa + \beta)^{-2}) .$$

Redefining the coupling constant $\lambda$ as $\lambda' \pi^2 \beta^{-1}$ we find

$$\lambda^{-1} = (\varkappa + \beta)^{-2}[1 + \mu^2(2\varkappa\beta + \beta^2)] . \tag{1.27}$$

The value $\mu = 0$ corresponds to the pure $s$-wave, thereby yielding the pole trajectory in the following form ($k = i\varkappa$)

$$\varkappa = -\beta \pm \sqrt{\lambda} \tag{1.28}$$

---

[4] For simplicity, we neglect the spin-orbital interaction.

with the branch point at $\lambda = 0$ located in the lower $k$-half-plane at the point $\varkappa = -\beta$. The two poles (corresponding to the bound and virtual states) move as shown in Fig. 5.5a. It should be noted that after a collision of the poles the

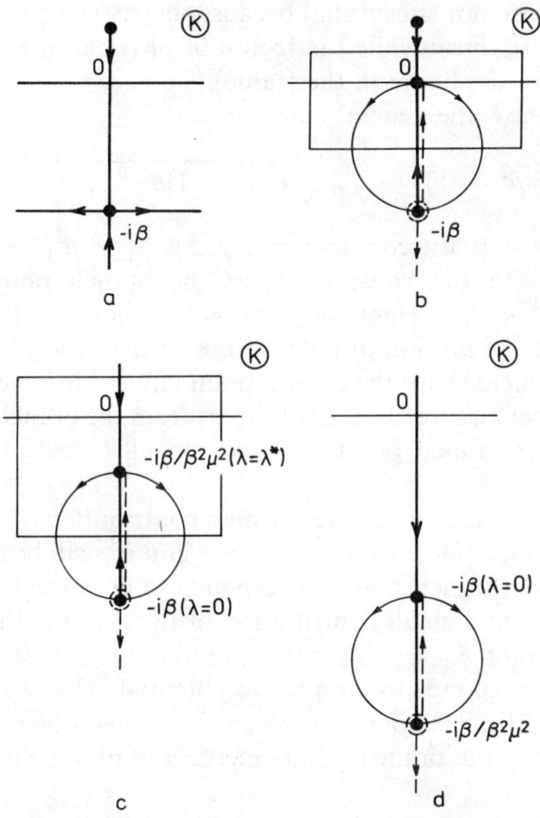

Fig. 5.5

The $S$-matrix pole trajectories for the case of the deformed (separable) potential. (a) The pure $s$-wave case, (b) the pure $p$-wave case, (c) the $p$-wave contribution is large, (d) the $p$-wave contribution is small (see explanations in the text) $\lambda^*$ denotes the branch point.

branches of the pole trajectory begin corresponding to the repulsive potential.

The value $\mu \to \infty$ corresponds to the pure $p$-wave, so $\lambda\mu^2$ plays the role of a coupling constant. In this case the trajectory of the pole is given by the equation

$$\varkappa = \beta\left[(\lambda\mu^2 - 1) \pm \sqrt{\lambda\mu^2}\,\sqrt{\lambda\mu^2 - 1}\,\right].$$

As a rule (i.e. for a short-range local potential), the threshold $\varkappa = 0$ coincides with the branch point where $\lambda\mu^2 = 1$. At the point $\varkappa = 0$ (as well as for the local potential) two poles, of which one moves from the uper $k$-semiaxis and another along the lower imaginary semi-axis, collide (see Fig. 5.5b).

Two resonance trajectories, however, do not go to infinity as in the case of the rectangular well, but collide again at $\lambda\mu^2 = 0$ at the same point $k = -i\beta$, as in the case of the $s$-wave. After that, one trajectory turns along the imaginary semiaxis to zero, while another goes to $-\infty$ along the same semiaxis. These branches however are not substantial because they correspond to the repulsive potential $(\lambda\mu^2 < 0)$, meanwhile the region of physical interest is close to the threshold (shown in the figure in the frame).

Let us examine now the general case

$$\varkappa = \beta(\lambda\mu^2 - 1) \pm \sqrt{\lambda}\sqrt{(\lambda\mu^2 - 1)\beta^2\mu^2 + 1}\;. \tag{1.29}$$

At the threshold $\varkappa = 0$ the coupling constant is $\bar{\lambda}_0 = \beta^2(1 + \beta^2\mu^2)^{-1}$ and no longer coincides with the value of $\lambda_0$ at the branch point where $\lambda_0 = (\mu^2\beta^2 - 1)/\beta^2\mu^4 < \bar{\lambda}_0$. Here $\varkappa(\lambda_0) \equiv \varkappa_0 = -\beta/\beta^2\mu^2$. In this respect the situation resembles the motion of poles in the $s$-wave case where the threshold $\varkappa = 0$ does not coincide with the branch point either. However, in our case the poles will move otherwise (see Fig. 5.5c). Apart from the branch point at $\lambda = \lambda_0$, there exists another branch point (common to all cases) at $\lambda = 0$ where $\varkappa = -\beta$.

Now, if $\mu^2\beta^2 > 1$ (i.e. at a sufficiently high contribution of the $p$-wave), then $\lambda_0 > 0$ and the branch point $k(\lambda_0) \equiv k_0 = -i\beta/\beta^2\mu^2$ will be the nearest to the threshold. The latter branch point corresponds to an attractive potential.

If $\mu^2\beta^2 < 1$ (i.e. at a small contribution of the $p$-wave) the nearest branch point will be the point $k_1 = -i\beta$ corresponding to $\lambda = 0$, thus leading to $\lambda_0 < 0$, which corresponds to a repulsive potential. The two cases are shown in Figs. 5.5c and 5.5d. The region of physical interest is shown in Fig. 5.5c in the frame. At $\beta^2\mu^2 = 1$ the branch points merge and disappear:

$$\varkappa = -\beta + \lambda\mu \pm \lambda\mu = -\beta + \frac{\lambda}{\beta} \pm \frac{\lambda}{\beta}\;.$$

Therefore, one pole at $\varkappa = -\beta$ always remains at any value of $\lambda$ while another

appears at $\varkappa = -\beta + \dfrac{2\lambda}{\beta}$.

The example discussed shows that at $\Lambda > 0$ the branch point coincides with the threshold. This case is the same as the case $l \neq 0$ and was examined fully in Section 5.1.3. At $\Lambda = 0$ the branch point location depends on deformation and we have $k(\lambda_0) \neq 0$, which is analogous with the case of $l = 0$ in a spherically symmetric potential. In these cases (i.e. at $l = 0$ and $\Lambda = 0$) the method for finding the branch point $\lambda_0$ described in Section 5.1.3 may be used.

On the whole, the motion of resonance poles in a realistic problem will be the same as in the separable model discussed above with only the unimportant (for our purposes) exclusion that the trajectory in the vicinity of the second branch point in the lower $k$-halfplane may behave differently at $\varkappa = -\beta$.

It should be emphasized again that the separation (1.24), together with allowance for the analytic properties of eigenvalues and eigenfunctions, permits continuation from the region of small $\lambda$ (i.e. from the region of bound states in a weakly deformed potential), so that the standard programs for calculating the bound states in deformed nuclei (based on the expansion of the potential in spherical harmonics) will be quite effective. The same technique is also convenient to use for calculating the bound states in a strongly deformed potential, where the conventional direct method for solving coupled equations may lead to considerable errors.

## 5.2. The Structure of the Threshold Singularities of Energy Eigenvalues in the Presence of Long-Range Interaction Potentials

Here we shall discuss the variations in the analytic properties of the $S$-matrix near the threshold which arise in the presence of long-range interaction potentials of the Coulomb and dipole types following [38–41]. In this case we shall use the method to study the near-threshold singularities of the $S$-matrix using the properties of the function $\Delta(z)$, the so-called complex level shift function (see the definition of $\Delta(z)$ in the previous chapter), near the point $z = 0$ and the direct method of studying the Jost function structure near the threshold by solving the wave equation.

The main change in the form of the $S$-matrix pole trajectories arising from the long-range interaction (of attractive character) is the disconnection of the resonance trajectories from the bound state branch. Physically this is due to the fact that not a single (non-singular) short-range perturbation (for example of Rydberg states) can push them into the continuous spectrum because their character is fully determined just by the long-range Coulomb tail of the interaction potential. The same situation arises also for a dipole potential $\sim 1/r^2$. The mathematical cause of this singularity is the appearance of a cut in the lower $k$-halfplane (on which the Jost function has a discontinuity) beginning just from the threshold at $k = 0$.

Our discussion of the $S$-matrix analytic structure will involve the physical problems in which it is important.

### 5.2.1. Coulombic Case. Attraction

Let us examine the physical problem of dissociative recombination of a molecular ion via near-threshold intermediate resonance due to vibrational excitations of the intermediate molecular complex, i.e. the process

$$ e \; + \; AB^+ \; \rightarrow \; (AB)^*_v \; \rightarrow \; A \; + \; B $$

or the inverse process, the so-called associative ionization:

$$A + B \rightarrow (AB)^*_v \rightarrow AB^+ + e \,.$$

Vibrationally-excited molecular resonances arising in the intermediate state provide another good example of the so-called Feshbach resonances (i.e. discrete levels embedded into continuum, QBSEC). QBSEC excitation by elastic and inelastic scattering of a nucleon by nuclei is examined in Section 4.6. The projection formalism described in the previous Chapter is very suitable when treating problems of this kind. Therefore we introduce a projector $P$ which is to include all the important vibrational states $\{\Phi_i\}$ of the system $(AB)^*_v$ i.e. $P = \sum_{i=1}^{N} |\Phi_i\rangle \langle \Phi_i|$ and a projector $Q = 1 - P$ complementary to the former.

Further, for simplicity, we shall limit ourselves to one intermediate state, i.e. $P = |\Phi\rangle \langle \Phi|$ because the basic conclusions are not affected by this limitation. In terms of the adiabatic model we assume also that the internuclear distance $R$ is a parameter of the problem.

As shown in Chapter 4, the poles of the complete $T$-matrix in the $E$-plane are defined by the equation

$$E - \epsilon_0 - d(E) = 0 \tag{2.1}$$

where

$$d(E) = \langle \Phi | H_{PQ} G_Q(E) H_{QP} | \Phi \rangle \tag{2.2}$$

and $\epsilon_0 = \epsilon_0(R) = \langle \Phi | H(R) | \Phi \rangle$. The total Hamiltonian $H(R)$ of the system $e + AB^+$ includes the interaction among all particles of the system. However, only the most long-range Coulombic component of the interaction $H_c$ between an electron and a molecular ion $AB^+$ is now important (because we are interested in the behaviour of the function $d(E)$ when $E \rightarrow 0$). In this case we can replace the projected many-particle resolvent $G_Q(E)$ in (2.2) by the purely Coulombic Green function:

$$G_c^{(+)}(k) = (k^2/2 - H_c + i\varepsilon)^{-1} \tag{2.3}$$

(for simplicity in this section we shall use the atomic system of units). The analytic properties of the Coulombic Green function in the co-ordinate representation are well known (see, for example, [42]). The strategy of the approach is evident [40, 41]. Since $G_c^{(+)}(k)$ has an essential singularity at $k = 0$, only the singular terms due to Coulombic potential are to be left in the expansion of $G_c^{(+)}(k)$ near $k = 0$. It is of interest that this modus operandi is quite universal.

For example, in the same way we may investigate the near-threshold singularities of the analytically continued $S$-matrix for a short-range potential [43], i.e. we can find all the properties of the Jost function $f(k)$ at $k \rightarrow 0$ described in the previous section. Since the effective Hamiltonian in our case is

$$\mathcal{H} = H_{QQ} + H_{QP}(E - H_{PP})^{-1} H_{PQ} \tag{2.3a}$$

while the additional term due to the coupling of the $P$ and $Q$ subspaces is a separable operator, we essentially examine the problem of scattering by a potential which is the superposition of a purely Coulombic interaction and a separable energy-dependent potential (singular at the threshold).

Let us decompose a function $d(E)$ in the usual way (see Chapter 4) into real and imaginary parts:

$$d(E) = \Delta(E) - i\,\Gamma(E)/2 \tag{2.2a}$$

where

$$\Gamma(E) = 2\pi\,|\langle\Phi|\,H_{PQ}\,|\psi_c(E)\rangle|^2 . \tag{2.2b}$$

The real shift function $\Delta(E)$ may be presented using the spectral decomposition of the Coulombic Green function in the form

$$\Delta(E) = \frac{1}{2\pi}\,P\int \frac{dE'\,\Gamma(E')}{E - E'} + \frac{1}{2\pi}\sum_{n=1}^{\infty}\frac{\Gamma_n}{E - E_n} \tag{2.4}$$

where

$$\Gamma_n = 2\pi\,|\langle\Phi|\,H_{PQ}\,|\psi_n\rangle|^2 . \tag{2.5}$$

In the formulae for $\Gamma(E)$ and $\Gamma_n$ the functions $\psi_c(E)$ and $\psi_n$ are Coulombic wave functions of the continum and of the discrete spectrum, respectively, (the functions $|\psi_n\rangle$ representing the bound states in a purely Coulombic field are contained in $Q$).

Using (2.2) and (2.2a), we may write the element of the $S$-matrix $S(E)$ corresponding to the purely resonance part of the $T$-matrix (Chapter 4)[5]) at a given internuclear distance $R$ as

$$S(k) = \frac{f(-k)}{f(k)} = \frac{k^2/2 - \epsilon_0 - \Delta(E) + i\,\Gamma(E)/2}{k^2/2 - \epsilon_0 - \Delta(E) - i\,\Gamma(E)/2} \tag{2.6}$$

and the Jost function $f(k)$ as

$$f(k) = 1 - \frac{d(E)}{k^2/2 - \epsilon_0} . \tag{2.7}$$

The $f(k)$ zeros determine the resonance poles of the $S$-matrix. It should be emphasized that $S(k)$ given by (2.6) is only the resonance part of the scattering matrix for the Hamiltonian which is the superposition of a purely Coulomb Hamiltonian $H_c$ and a short-range separable potential (the second term in

---

<sup>5</sup>) Because the $T$-matrix in Chapter 4, corresponding to the direct transition in the Coulomb field, does not contain resonance poles.

(2.3a)). It is this separable potential that is responsible for the appearance of resonances, because, as is well-known, resonances do not arise in a purely Coulombic field. A lot of publications are devoted to studying the analytic properties of the $S$-matrix for a combination of the Coulombic and short-range potentials (see [36, 40, 41] and the references therein). The key role in these studies is played by the Coulombic Green function $G_c^{(+)}(k; \boldsymbol{r}, \boldsymbol{r}')$. Expanding $G_c^{(+)}(k; \boldsymbol{r}, \boldsymbol{r}')$ and the shift function $d(k)$ in partial waves, we find [42]:

$$
G_c^{(+)}(k; \boldsymbol{r}, \boldsymbol{r}') =
$$

$$
= - \sum_{l, m} \frac{1}{rr'} \, Y_{lm}(\hat{\boldsymbol{r}}) \, Y_{lm}^*(\hat{\boldsymbol{r}}') \frac{f_l(k, r_<)}{k} \, [g_l(k, r_>) + i f_l(k, r_>)] = \quad (2.8)
$$

$$
= \sum_{l,m} \frac{1}{rr'} \, Y_{lm}(\hat{\boldsymbol{r}}) \, Y_{lm}(\hat{\boldsymbol{r}}') g_l^{(+)}(k; r, r') \qquad (2.8a)
$$

where $f_l$ and $g_l$ are the regular and irregular Coulombic wave functions, respectively, and

$$
d(k) = \sum_{l, m} d_{lm}(k) \qquad (2.9)
$$

where

$$
d_{lm}(k) = \int dr \int dr' \, \chi_{lm}(r) \, g_l^{(+)}(k; r, r') \, \chi_{lm}(r')
$$

and $\chi_{lm}(r) = \langle r | H_{QP} | \Phi \rangle$ are the localized short-range form factors. Since at $k \to 0$ the analytic properties of the Jost function do not depend on the form of $\chi_{lm}(r)$, for the sake of simplicity, we take $\chi_{lm}(r)$ to be a delta function $\chi_{lm} = \chi \delta(r - r_0)$ and find:

$$
d_{lm}(k) = -k^{-1} \, |\chi|^2 \, f_l(k, r_0) \, (g_l(k, r_0) + i \, f_l(k, r_0)) . \qquad (2.10)
$$

Thus, we obtain the result which was nearly evident beforehand, namely, that the analytic behaviour of the $S$-matrix and of the Jost function in our problem at $k \to 0$ is determined by the analytic properties of the Coulombic wave functions. In particular, the singularities of $d_{lm}(k)$ at small values of $k$ are determined by the singular structure of the irregular Coulombic function $g_l(k, r_0)$. Leaving only the basic terms in the expansion of $g_l(k, r_0)$ at $k \to 0$ (see the formulae in [44]), we find finally:

$$
d(k) = \frac{\beta}{2\pi} \, [m - ik/2Z + \ln(-ik) + \psi(1 - iZ/k)] \qquad (2.11)
$$

where $\psi$ is the digamma function [44], $Z$ is the molecular ion charge (in our case $Z = 1$); $\beta$ and $m$ are constants ($\beta > 0$, $m = \Delta(0)$ [40]).

Let us examine the properties of the complex shift function $d(E)$ at real positive values of $k$, i.e. at $E \to 0$. Using the well-known property of the digamma function

$$\operatorname{Im} \psi(1 + iy) = -\frac{1}{2y} + \frac{\pi}{2} \coth (\pi y)$$

we easily find from (2.2a) and (2.11):

$$\Gamma(E) = \frac{\beta}{2}\left(1 + \coth (\pi/\sqrt{2E})\right) \underset{E \to 0}{\simeq} \beta\left(1 + \exp(-2\pi/\sqrt{2E})\right).$$

$$(2.12)$$

In other words, the function $\Gamma(E)$ has a finite value at zero energy which is in agreement with the well-known threshold behaviour of cross section in the presence of an *atractive* Coulomb interaction (say, the cross section of the photoproduction of charged particles near threshold). Using the asymptotic expansion of the digamma function, we obtain at $E > 0$:

$$\Delta(E) = \frac{\beta}{2\pi}\left(m + \sum_{n=1}^{\infty} \frac{(-1)^{n-1} B_{2n}}{2n} (2E)^n\right) \tag{2.13}$$

where $B_{2n}$ are the Bernoulli numbers. This means that $\Delta(E)$ at $E \to +0$ is a smooth function of $E$.

Now, we shall examine $d(E)$ at negative energies $E < 0$. It is not difficult to show [40] in this case that

$$\Delta(E) = \frac{\beta}{2\pi}\left(m + \sum_{n=1}^{\infty} \frac{(-1)^{n-1} B_{2n}}{2n} (2E)^n\right) +$$

$$+ \pi \cot (\pi/\sqrt{(-2E)}) \tag{2.14}$$

i.e. the function $\Delta(E)$ has an infinite sequence of poles corresponding to the transformation of the argument of $\cot(y)$ into a number divisible by $\pi$:

$$\frac{\pi}{\sqrt{(-2E)}} = n\pi, \quad \text{i.e.} \quad E_n = -1/(2n^2).$$

In other words, $\Delta(E)$ contains the poles on the negative semiaxis $E$ corresponding to the purely Rydberg states. This is completely in agreement with the representation (2.4). The point $E = 0$ is the accumulation point of these poles. Further, as follows from (2.11), the complex shift function $d(k)$ has a logarithmic cut along the imaginary negative $k$-semiaxis. Using the asymptotic expansion of the digamma function again, we can write $d(k)$ on two rims of this cut as $(k = iu \pm \eta$ , $\eta \to +0$ , $u < 0)$:

$$d(iu \pm \eta) = \frac{\beta}{2\pi} \left( m - \sum_{n=1}^{\infty} \frac{B_{2n}}{2n} u^{2n} \mp i\pi \right) . \tag{2.15}$$

That is, when going through the cut along the imaginary negative $k$-semiaxis, the complex shift function has a discontinuity by a value equal to $\beta$.

Now, we shall examine the locations of the complex zeros of the Jost function $f(k)$. According to (2.7), the $f(k)$ zeros are defined by the condition

$$k^2/2 - \epsilon_0(R) - d(k) = 0 \tag{2.16}$$

where, because of the adiabaticity of the motion of nuclei, the energy of the term $\epsilon_0(R) = \langle \Phi | H(R) | \Phi \rangle$ depends on the internuclear distance $R$. The solutions to (2.16) on an imaginary positive $k$-semiaxis determine the bound states of the system, whereas the complex zeros located in the lower $k$-half-plane determine the resonances of the system. Using the asymptotic expansion of the digamma function at $k \to 0$ again, we find the following expression for the complex $k = k_1 - ik_2 (k_1 > 0$ , $k_2 > 0)$ :

$$d(k) = \frac{\beta}{2\pi} \left( m + \sum_{n=1}^{\infty} \frac{(-1)^{n-1} B_{2n}}{2n} k^{2n} + \pi \cot \left( \frac{i\pi}{k} \right) \right) \tag{2.17}$$

whence it is not difficult to ascertain that the known property $d(-k^*) = d^*(k)$ is valid. This condition, along with (2.16), guarantees that the resonance poles are located symmetrically with respect to the imaginary axis.

Let us examine now the trajectories of the resonance poles when the internuclear distance $R$ changes. For simplicity we shall neglect the dependence of $\beta$ and $m$ on the parameter $R$. Thus, $\epsilon_0(R)$ will be the only quantity depending on $R$. It decreases with increasing $R$. Equation (2.16) is equivalent to two conditions

$$\mathrm{Im}\,(k^2/2 - d(k)) = 0 \tag{2.18a}$$

and

$$\mathrm{Re}\,(k^2/2 - \epsilon_0(R) - d(k)) = 0 . \tag{2.18b}$$

On the imaginary $k$-semiaxis, $k^2/2 = -|E|$ is a real quantity and, because of (2.15), $\mathrm{Im}\,d(k) \neq 0$ at any value of $k$ on the negative imaginary semiaxis (here $|\mathrm{Im}\,d(k)| = \pi$ ). Therefore, the condition (2.18a) does not permit a pair of resonance poles to come close to the negative imaginary $k$-semiaxis. This means that in the examined problem the *merging of two conjugate resonance poles is not possible* as in the case of the short-range potential[6].

---

[6] This is evidently due to the fact that a logarithmic cut goes along the negative imaginary $k$-semiaxis, which prevents the resonance poles from merging.

On the positive imaginary $k$-semiaxis, besides an infinite number of zeros of the Jost function $f(k)$ which are determined by the condition (2.16), there exist an infinite number of the $f(k)$ poles which are due to the poles of $d(k)$ at points $k_n = i/n$, $n = 1, 2, 3 ...$, i.e. at energies of unperturbed Rydberg state (see also

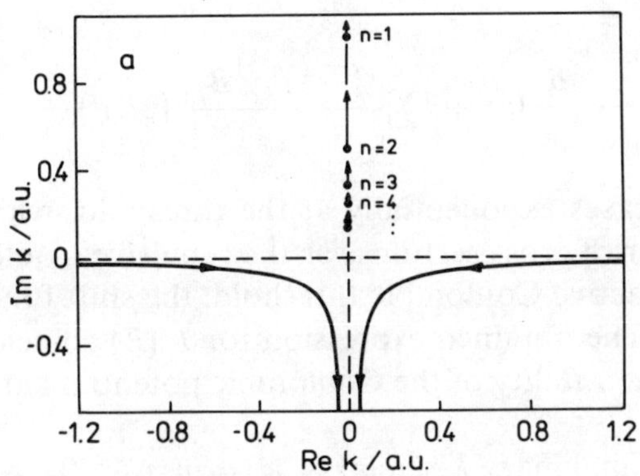


Fig. 5.6a

The $S$-matrix pole trajectories for short-range plus *attractive* Coulomb potential case at increasing short-range attraction. The numbers $n$ mean radial quantum numbers of the Rydberg states.

[38]). Each zero of $f(k)$ lies between two consecutive $f(k)$ poles and, as $R$ increases, such a zero moves upwards along the imaginary $k$-semiaxis. As is evident, it remains continually between two neighbouring $f(k)$ poles. The specific character of the trajectories of the resonance and bound-state zeros of $f(k)$ as the internuclear distance $R$ increases is shown in Fig. 5.6a [40]. From here it follows that the trajectories of the resonance poles are fully disconnected from the trajectories of the bound-state poles, as can be seen by comparing this figure with Fig. 5.1 presenting the resonance trajectories for the short-range potential. In other words, the resonance poles do not arise in the examined case from the corresponding bound states, but come to the vicinity of the Coulombic threshold from *inifinity*; hence, they *cannot* be found by analytic continuation of bound states.

### 5.2.2. Coulombic Case. Repulsion

The treatment of the repulsive Coulombic threshold follows directly from the above discussion via the replacement of the sign of the charge $Z$ by $-Z$. For $d(k)$ we find [40] (in case the Coulombic potential of unit strength):

$$d(k) = -\frac{\beta}{2\pi}\left(m + \frac{ik}{2} + \ln\left(-ik\right) + \psi(1 + i/k)\right). \qquad (2.19)$$

From here it follows that $d(k)$ has an essential singularity at $k = 0$ and a cut on the negative imaginary $k$-semiaxis.

At real positive energies $E > 0$ we find, by analogy with the previous section,

$$\Gamma(E) = \frac{\beta}{2} \left[\coth\left(\pi/\sqrt{2E}\right) - 1\right] \underset{E \to 0}{\simeq} \beta \exp\left(-2\pi/\sqrt{2E}\right), \qquad (2.20)$$

$$\Delta(E) = -\frac{\beta}{2\pi} \left(m + \sum_{n=1}^{\infty} \frac{(-1)^{n-1} B_{2n}}{2n} (2E)^n\right). \qquad (2.21)$$

That is, $\Gamma(E)$ decreases exponentially at the threshold, which agrees with the threshold behaviour of cross sections for the repulsive Coulombic potential. In contrast to the attractive Coulombic threshold, the shift function $\Delta(E)$ is continuous at $E = 0$. The obtained expression for $\Gamma(E)$ coincides essentially with the well-known penetrability of the Coulombic potential barrier at low energies $E$ [49].

On the negative imaginary $k$-semiaxis $k = iv, v < 0$) one has

$$d(iv + \eta)\big|_{\eta \to +0} = \frac{\beta}{2\pi} \left(m - \sum_{n=1}^{\infty} \frac{B_{2n}}{2n} v^{2n} - \pi \cot\left(\frac{\pi}{v}\right) \mp i\pi\right).$$

$$(2.22)$$

That is, the quantity $\mathrm{Im}\, d(k)$ exhibits a discontinuity on the negative imaginary $k$-semiaxis again and the function $\mathrm{Re}\, d(k)$ has poles at point $k_n = -i/n, n = 1, 2, \dots$ i.e. at the Rydberg state energies $E_n = -1/2n^2$, but on the non-physical sheet $E$. Owing to the continuity of the shift function $\Delta(E)$ at $E \to 0$ in this case, as the intensity of the short-range attraction decreases, the bound state pole is pushed *continuously* to the continuum and becomes a resonance. Since the Coulombic barrier can be treated as a limiting case of the centrifugal barrier at $l \to \infty$ (it is sufficient to compare (1.9b) and (2.20)), the pole motion pattern will be qualitatively similar to the case of the short-range potential (i.e. at the threshold $E = 0$ bound state pole will collide with the corresponding "virtual" state pole produced at the threshold from the "dynamic" cut; after that, they will separate at right angles with respect to the imaginary $k$-axis and yield two symmetric resonance poles). A difference from the purely short-range case is that owing to (2.20) the resonance branch of the trajectory presses itself more closely to the real axis (in the $k$-plane) than the trajectory for the purely short-range potential. Furthermore, since the "virtual" state poles are hidden on the cut, they do not manifest themselves here in scattering. Thus, in the case of the repulsive Coulombic threshold, in spite of the long-range character of the interaction, the pole trajectories for resonances and bound states join at the threshold again, so the method of continuation in the coupling constant discussed in this chapter is applicable again. In actual calculations, however, we have to take into consideration the fact that, because of the exponential smallness of the resonance width near the threshold, the error in calculating the reference energies $\{E_i\}$ must be very small (smaller than the width of the level $\Gamma$). In the opposite case, in the analytic continuation the error in the value of the width will be inadmissibly high.

### 5.2.3. Remarks

To conclude this subsection, we shall make some remarks concerning other physical processes where we are faced with the same problems.

1. The problem of the near-threshold singularities of the $S$-matrix in the presence of a Coulombic field and of the form of the corresponding resonance trajectories discussed in this subsection provides us with another interesting physical interpretation which is important in some applications, namely, the passage of the term into continuum when the repulsive Coulombic centre and the centre of a short-range potential $V(r)$ [39] approach each other. The corresponding Schrödinger equation is of the form (in the atomic system of units)

$$\left\{ -\frac{\Delta_r}{2} + V(r) + \frac{1}{|r - R|} \right\} \psi(r, R) = E(R)\, \psi(r, R)$$

where $R$ is the distance between the Coulombic centre and the short-range potential and $E(R)$ is the energy of the term. At a certain distance $R_0$ the term of the potential $V(r)$ intercepts the boundary of the continuous spectrum and becomes a resonance, i.e. $E(R_0) = 0$. At $R < R_0$ the repulsive Coulombic barrier will prevent a particle from leaving the field $V(r)$ and we arrive at the case discussed in Subsection 5.2.2 with the only difference that now the perturbation parameter is $\Delta R = R - R_0$. It may be shown [39] that in the first order in $\Delta R$ the energy shift (from the threshold) is

$$E^{(1)} = -\Delta R/R_0^2 , \tag{2.23a}$$

$$\Gamma^{(1)} = 1/8\ \varkappa^3 \exp\left\{ 8\varkappa^{-1} - 2\pi\varkappa^{-2}\ \sqrt{2/(-\Delta R)} \right\} \tag{2.23b}$$

where $\varepsilon_0 = -\varkappa^2/2$ is the energy of the unperturbed level in the potential $V(r)$. That is, again in the vicinity of the repulsive Coulombic threshold, the real part of the shift function $E^{(1)}(\Delta R)$ is continuous (with respect to the perturbation parameter $\Delta R$) and the width is exponentially small.

2. Problems similar to those discussed in Subsection 5.2.1 arise in the case of electron scattering by polar molecules near the threshold (see [41] and the references therein). A convenient model to study the near-threshold singularities and resonance trajectory shape is the interaction potential in the form of the superposition of a short-range attraction and a long-range attractive dipole term. Work [41] has analysed in detail the S-matrix analytic properties and the

behaviour of the pole trajectories with a change in the coupling constant of the short-range potential. The trajectories of the resonance poles are *disconnected* from the trajectories of bound states because of the long-range attraction. However, in contrast to the case of the Coulombic threshold, the resonance

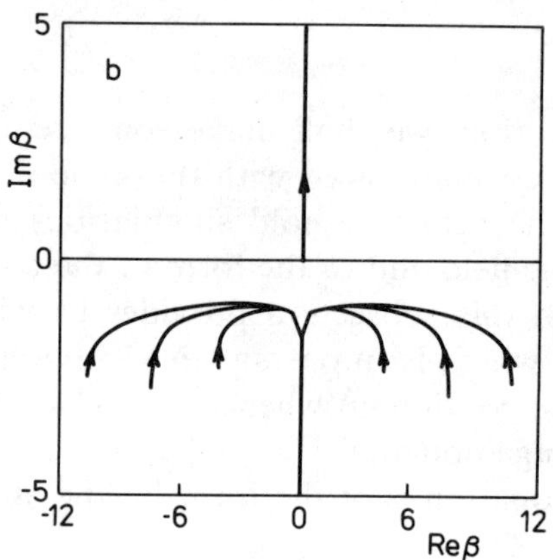


Fig. 5.6b

The $S$-matrix pole trajectories for short-range plus *attractive* long-range ($\varphi \sim \text{const}/r^2$) potential.

poles in the lower $k$-halfplane nearly collide on the negative imaginary $k$-semiaxis. Contrary to the case of the short-range potential, these poles after approaching each other go on moving downwards along the negative imaginary $k$-semiaxis (see Fig. 5.6b). Moreover, as shown in [41], it is possible to introduce an analogue of the virtual poles, but such poles will be *disconnected* from the resonance poles again. Thus, we see that the behaviour and the structure of the singularities in this case are (in one sense) of an intermediate character between the Coulombic attractive case and the case of short-range interaction.

## 5.3. Many-Particle Resonance and Near-Threshold States in the Systems with Real Potentials

In the case of resonances in a system of three and more particles, the scheme of the described ACCC method usually changes only a little. As in the two-particle system, we study the dependence of the bound-state evolution on the total, or some partial, coupling constant. Then, using the analytic properties of the eigenvalues and eigenfunctions of a multiparticle Hamiltonian and the analytic continuation technique described above (i.e. the construction of the Padé approximant in terms of a uniformizing variable), we find the required complex energies and wave functions of multiparticle resonance. However, in its practical aspects this situation gets considerably complicated. First, the multiparticle

Hamiltonian has many continua corresponding to the scattering in the subsystems of two or more particles. Secondly, the analytic behaviour of the energy of a many-particle bound state may get very complicated as the partial coupling constants change. For the time being we shall limit ourselves to a system of three identical particles. Solutions of the three-particle Schrödinger equation

$$\{H_0 + \lambda\,(V_{12} + V_{13} + V_{23})\}\,\psi_\lambda(E_\lambda) = E_\lambda\psi_\lambda(E_\lambda) \tag{3.1}$$

can be analytically continued from the region $\lambda > 1$ to the point $\lambda = 1$ which corresponds, according to the assumption, to the resonance. However, we cannot be quite sure that the obtained result gives, as in the two-particle system, a pole of the three-particle scattering amplitude. This amplitude is determined from the solution of the corresponding Faddeev equation for the components $W_{\alpha\beta}$ of the three-particle scattering matrix $T\,(z)$:

$$T\,(z) = \sum_{\alpha,\,\beta} W_{\alpha\beta}$$

and $T\,(z)$ is defined by the equation

$$T\,(z) = V_{123} + V_{123}G^{(+)}(z)V_{123}$$

where $V_{123} = V_{12} + V_{13} + V_{23}$; $G^{(+)}(z)$ is the three-particle Green function.

In [6] it is shown that the eigenvalues of (3.1) $\lambda_n(E)$ determine the singularities of the total Green function $G^{(+)}(E)$ as well as the poles of the three-particle $T$-matrix:

$$T\,(E) = V_{123} + \sum_n \frac{V_{123}\,|\psi_n\rangle\,\langle\psi_n|\,V_{123}}{1 - \lambda_n(E)} + T'(E)\,.$$

It is of importance here that at the energy values $E_R$ for which $\lambda_n(E_R) = 1$ the eigenvalues of the Hilbert-Schmidt problem $\xi_n(E_R)$ for the Faddeev equations are also equal to unity, i.e.

$$\lambda_n(E_R) = \xi_n\,(E_R) = 1\,. \tag{3.2}$$

On the other hand, the equality $\xi_n(E_R) = 1$ determines the three-particle resonances and, hence, the analytic continuation of the solutions for the three--particle Schrödinger equation (3.1) to the point $\lambda = 1$ also determines the true three-particle resonances. This means that the method of analytic continuation in the coupling constant permits the direct generalization to the case of three- and many-particle systems.

It should be noted, however, that the behaviour of the singularities of the three-particle amplitude when the coupling constant of two-particle interaction changes has not been studied sufficiently as yet. Therefore, we cannot exclude the possibility that on the nonphysical energy sheets of the three-particle system some singularities arise which are not an analytic continuation of three-particle bound states. Apparently, such singularities cannot be described in terms of the given approach.

Thus, the generalization to the case of few particles demands a study of the analytic behaviour of eigenenergy $E(\lambda)$ at the many-particle thresholds, which is a very difficult task. Leaving the complete solution of this problem to the future, we shall outline what can be done in this respect now, assuming that the analytic properties of the many-particle and two-particle eigenenergies are the same at the two-particle threshold.

### 5.3.1. Three-Particle Resonances with Two-Particle Decay

There are many cases where the resonance in a three-particle system is located either below or above the three-particle threshold, but the three-particle decay channel need not be taken into account. In particular, they include the numerous cases where a three-particle system decays into a particle and a long-lived two-particle resonance and the analytic behaviour of eigenenergies and wave functions at the threshold is the same as in the two-particle problem, i.e.

$$k_l(\lambda) \sim \begin{cases} \sqrt{\lambda - \lambda_0}, & l > 0, \\ (\lambda - \lambda_0), & l = 0 \end{cases} \tag{3.3}$$

with the only difference that the wave vector $k_l(\lambda)$ must be counted from the nearest two-particle threshold, i.e.

$$k_l(\lambda) = \left[ \frac{2\mu}{\hbar^2} (E_3(\lambda) - E_2(\lambda)) \right]^{1/2}$$

where $\mu$ is the reduced particle mass in the decay channel; $E_3(\lambda)$ is the energy of the three-particle state, $E_2(\lambda)$ is the energy of the nearest two-particle threshold. Generally, with a change in $\lambda$ in (3.1), the two-particle threshold will also move.

To make the analytic continuation in $\lambda$, we must calculate the trajectory $E_2(\lambda)$ beforehand. The calculation scheme is as follows. For a number of values $\{\lambda_i\}$ we find the three-particle energies $\{E_3(\lambda_i) \equiv E_{3,\,i}\}$ in one way or another (either by a variational method or by numerical solution of the Faddeev equations). After that, we find the values $\{k_l(\lambda_i) \equiv k_i\}$ using the corresponding values $\{E_{2,\,i}\}$, whereupon we construct the Padé approximant (1.10) or (1.11), find the value $\lambda_0$, and make analytic continuation to the point $\lambda = 1$. Then, from the numbers $k_l(1)$ we determine the energies of the three-particle resonances

$$E_3(\lambda = 1) = \frac{\hbar^2 k_l^2(1)}{2\mu} + E_2(\lambda = 1) = E_R - \frac{i}{2}\,\Gamma. \tag{3.4}$$

If necessary, the same procedure is applied to the wave functions of the Gamow states or to the transition matrix elements.

In terms of the shell-model approach [19], the above-described procedure is a method which allows for the configurations with one particle in the continuum.

### 5.3.2. Few-Particle Systems with Potentials Including Strong Repulsive Core

If we use a suitable separation of two-particle potentials and the analytic continuation in the coupling constant of repulsive core, then the ACCC approach may be applied effectively to the treatment of systems of few particles whose interactions contain a strong repulsive core. As will be shown below, in nuclear-physics applications we can effectively use the circumstance that most of the long-lived nuclear states lie relatively near to the threshold (with respect to the location which would be real if we switch on the NN repulsive core).

Let us examine the multiparticle Hamiltonian

$$H = \sum_{i=1}^{N} h_0(i) + \sum_{i<j}^{N} V_{ij} = \sum_{i=1}^{N} h_0(i) + \sum_{i<j}^{N} (V_{ij}^a + V_{ij}^r) \tag{3.5}$$

where the two-particle potentials $V_{ij}$ contain the strong repulsive core $V_{ij}^r$ (or another singularity; for example, strong velocity dependence, and so on) see Fig. 5.7a.

Let now the separation of two-particle potentials be made to single out the

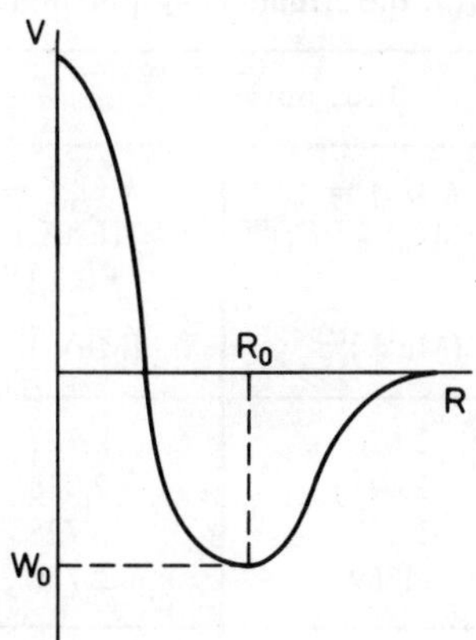


Fig. 5.7a
The forms of the initial and separated potentials.

positive operator $V_{ij}^p \neq V_{ij}^r$ in $V_{ij}$ (see Fig. 5.7a), whereupon an appropriate partial coupling constant $\lambda_p$ should be introduced:

$$H(\lambda_p) = H_0 + \lambda_p \sum_{i<j}^{N} V_{ij}^p \tag{3.6}$$

where

$$H_0 = \sum_{i=1}^{N} h_0(i) + \sum_{i<j}^{N} (V_{ij} - V_{ij}^p) . \tag{3.7}$$

In the model case with a local repulsive core, the separation may be defined as

$$V_{ij}^p = \begin{cases} V_{ij}(r) - W_0 , & r \leq R_0 \\ 0 , & r > R_0 \end{cases} \tag{3.8}$$

$$V_{ij}^a = V_{ij} - V_{ij}^p = \begin{cases} W_0 , & r \leq R_0 \\ V_{ij}(r) , & r > R_0 . \end{cases} \tag{3.9}$$

The meaning of the notation is clear from Fig. 5.7a.

The general calculation scheme is the same as before. We have to calculate the eigenenergies of Hamiltonian $H(\lambda_p)$ at $\lambda_p \ll 1$ using any method (say, the variational one). After that, the Padé technique must be used to extrapolate the function $E_N(\lambda_p)$ to the values $\lambda_p = 1$. In the context discussed here, all energies are in the purely discrete spectrum, so that even the direct extrapolation in $\lambda_p$ is possible. However, the actual calculations for the systems of 3 and 4 particles

Table 5.2

Three- and four-nucleon systems with the Afnan-Tang potential $S$ 1.

| Systems | Three-nucleon | | | Four-nucleon | |
|---|---|---|---|---|---|
| Approximated function | $E_3(\lambda_p)$ | $k_3(\lambda_p) =$ $= (E_3(\lambda_p) - E_d)^{1/2}$ | $k_3(\lambda_p) =$ $= (E_3(\lambda_p) -$ $- E_2(\lambda_p))^{1/2}$ | $E_4(\lambda_p)$ | $k_4(\lambda_p) =$ $= (E_4(\lambda_p) -$ $- E_3(\lambda_p))^{1/2}$ |
| PA order | MeV | $(\text{MeV})^{1/2}$ | $(\text{MeV})^{1/2}$ | (MeV) | $(\text{MeV})^{1/2}$ |
| [1, 1] | 0.60 | 2.626 | 2.727 | 10.6 | 4.40 |
| [2, 2] | 1.24 | 2.641 | 2.728 | 16.1 | 4.68 |
| [3, 3] | 3.57 | 2.650 | 2.728 | 19.2 | 4.74 |
| [4, 4] | 6.86 | 2.669 | 2.736 | 21.8 | 4.79 |
| Value obtained from the direct calculation | 7.76 | 2.738 | 2.738 | 31.1 | 4.83 |

(see Table 5.2) show that the convergence in the case of direct extrapolation (even if we use the Padé approximants) will be slow, being slower for 4 than for 3 particles, etc. On the other hand, even the simplest allowance for the analytic properties speeds up the convergence markedly. Thus, if we go over from the energy $E_N(\lambda_p)$ to the quantity $\bar{k}(\lambda_p) = [E_N(\lambda_p) - E_t]^{1/2}$ (where $E_t$ is the energy of the nearest threshold at $\lambda = 1$) and construct the Padé approximant for $\bar{k}(\lambda_p)$, the convergence improves appreciably. If we go over to a wave vector counted from the nearest threshold $E_t$ and take into account the evolution of $E_t(\lambda_p)$, i.e. if we construct the Padé approximant of the type

$$k(\lambda_p) = [E_N(\lambda_p) - E_t(\lambda_p)]^{1/2} \simeq \frac{P_N(\lambda_p)}{Q_M(\lambda_p)} \tag{3.10}$$

(to do this, the dependence $E_t(\lambda_p)$ should be found first, which reduces to studying a bound state in a system of less than $N$ particles), then the convergence (in the studied cases) becomes good. As examples may be cited the ground states of the $^3$H and $^4$He systems (i.e. three- and four-particle systems) calculated by the above-mentioned method with the $NN$ potential including a strong repulsive core (of a 1 GeV height), the Afran-Tang $S1$ potential [20]. In the two cases we limited ourselves to completely symmetrical states. Table 5.2 shows the convergence for $\lambda_p = 1$ in three different cases, namely, the direct extrapolation for energy (columns 2 and 5), the extrapolation for $\bar{k}(\lambda_p)$ (column 3), and, finally,

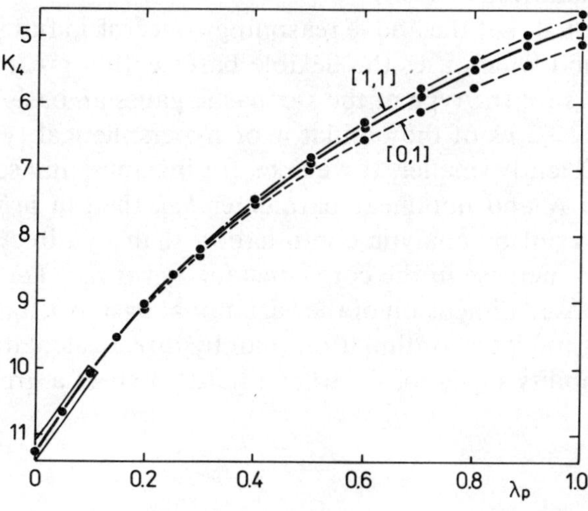


Fig. 5.7b

The convergence of $k_4(\lambda_p)$ for a four-particle system with the Afnan-Tang's potential $S1$ containing a repulsive core. The solid line is the exact value, the dotted line is the [0,1] PA; the dashed – dotted line is the [1,1] PA. The three points from which a continuation was performed (for the [1,1] PA) fall within the frame.

the analytic continuation for the wave vector counted from the mobile threshold $k(\lambda_p)$ (columns 4 and 6). Fig. 5.7b shows the convergence in the entire domain $\lambda_p$. From the figure it is clearly seen that to obtain the exact result at $\lambda_p = 1$ it is sufficient to use the [1,1] Padé approximant which was constructed using only three initial energies for $\lambda_p = 0$ (core is switched off at all), $\lambda_p = 0.05$, and $\lambda_p = 0.1$ . In other words, the simplest use of the analytic behaviour of extrapolated quantities makes it possible to decrease the intensity of the repulsive core by a factor of 10.

There is another advantage of such an approach. To calculate the initial energy of a many-particle system, it is very convenient to use the variational method. In this case the convergence rate and, hence, the basis dimension depend very strongly on the repulsive core intensity or on other singular parts of the Hamiltonian. Therefore, to calculate the initial energies where the core is reduced by a factor of 10–20, a much smaller dimension of the matrix than in the case of the core of total intensity is sufficient[7]). In particular, for the $3N$-system examined above with the core of total intensity $(\lambda_p = 1)$, the dimension of the matrix necessary for obtaining the final result with the same accuracy as in the case of the Padé extrapolation will be approximately $20 \times 20$ (the stochastic variational method [21] was applied), whereas an $8 \times 8$ matrix was used to calculate the initial energies. For the four-particle system the difference would be even more substantial. This means that even in the cases where the calculations with a full core can be made using a variational basis the procedure of the Padé extrapolation in the coupling constant makes the calculations much less laborious.

It should be noted that the above reasoning concerns in full measure only the variational method which uses the flexible bases with a great number of non--linear parameters (of the type of the stochastic gaussian basis [21] used by us). When the "rigid" bases of the oscillator or hyperspherical types are used the gain may be sufficiently smaller. If we take, for instance, an oscillator basis and fix its dimension $N$ and nonlinear parameter $\hbar\omega$, then in *principle we cannot* obtain a better result by analytic continuation than by direct calculations. If, however, with an increase in the core constant $\lambda_p$ we vary $\hbar\omega$ trying to get the best results for a fixed dimension of the variational basis, we may hope to obtain a better result by analytic continuation than by direct calculations, but in view of the weak flexibility of the basis we can hardly expect a great effect.

---

[7]) If the extrapolated function is very smooth, as in our case, then the error of the final result is of the same order as the error of the initial data.

## 5.4. Theory of Quasi-Stationary States with Two Particles in Continuum

In this section we shall study the resonances in a system of two light particles in the field of a heavy core. Such a system is a widely used model for very many processes in atomic and nuclear physics (for example, the autoionization states in atoms or the intermediate structure resonances in nuclear reactions).

The dilatation method (the method of complex scale transformations) discussed in detail in Chapter 7 is one of the effective and extensively used methods for finding the resonance states in such systems, especially for studying autoionization states in atoms. However, the dilatation method is not used extensively to study those nuclear systems where the interaction Hamiltonian does not exhibit dilatational invariance and the three-particle resonance may decay in channels with different compositions of fragments. Therefore, we shall describe two alternative approaches to studying such resonances. The first, simplified, approach works properly in those cases where the subsystem of light particles contain neither bound states nor near-threshold resonances. In other words, the first method does not permit any allowance for the channel of the decay of a three-particle quasi-stationary state into a bound pair and a core. The second method is based on examination of the solution for the appropriate Faddeev equation and is free of this limitation, although it appears to be more complicated. The first approach, is as follows.

Let the Hamiltonian of the system of two particles 1 and 2 in the field of a core 3 be

$$H = \mathcal{H}_0 + V_{12} \tag{4.1}$$

where

$$\mathcal{H}_0 = H_0 + \lambda(V_{13} + V_{23})$$

and $H_0$ is kinetic energy operator.

The system wave function $\psi_n$ can be series-expanded in the eigenfunctions of the Hamiltonian $\mathcal{H}_0$ which are in turn presented (for simplicity we examine the case of an infinitely heavy core) by the product of one-particle functions, i.e. $\psi_n$ may be written as

$$\psi_n = \sum_{ij} c_{ij} \, \Phi_i(1) \, \Phi_j(2) \tag{4.2}$$

where $\Phi_i$ are the one-particle wave functions of particles 1 or 2 in the field of the core. If the examined state $\psi_n$ is a resonance at $\lambda = 1$, we shall raise $\lambda$, so that $\psi_n$ becomes a bound state. Now, its energy may be calculated using the standard formula [5]:

$$E_n(\lambda) = \langle \psi_n | \, H \, | \psi_n \rangle = \langle \psi_n | \, \mathcal{H}_0 \, | \psi_n \rangle + \langle \psi_n | \, V_{12} \, | \psi_n \rangle \, .$$

Using the representation (4.2) for $\psi_n$, we obtain

$$\langle \psi_n | \mathscr{H}_0 | \psi_n \rangle = \sum_{ijkl} c_{ij}\, c_{kl} \langle \Phi_i(1)\, \Phi_j(2) | \mathscr{H}_0 | \Phi_k(1)\, \Phi_l(2) \rangle =$$

$$= \sum_{ij} c_{ij}^2 \left[ \varepsilon_{1i}(\lambda) + \varepsilon_{2j}(\lambda) \right]$$

where $\varepsilon_{1i}(\lambda)$ and $\varepsilon_{2j}(\lambda)$ are the energies of one-particle states in the field of the core of particles 1 and 2, respectively, at the chosen $\lambda$. And finally, for $E(\lambda)$ we obtain

$$E(\lambda) = \sum_{ij} c_{ij}^2 \left[ \varepsilon_{1i}(\lambda) + \varepsilon_{2j}(\lambda) \right] +$$

$$+ \sum_{ijkl} c_{ij} c_{kl} \langle \Phi_i(1)\, \Phi_j(2) | V_{12} | \Phi_k(1)\, \Phi_l(2) \rangle . \tag{4.3}$$

Using the analytic properties with respect to $k$ and $\lambda$ of the quantities entering (4.3) (see Sections 5.1.1 and 5.1.5) we can write the following representation for $E(\lambda)$:

$$E(\lambda) \simeq \sum_i \frac{\sqrt{\lambda - \lambda_0^{(i)}}\; P_{N_i}^{(i)}(\lambda)}{Q_{M_i}^{(i)}(\lambda)} \tag{4.4}$$

where each of the terms allows for the contribution to the three-particle energy from the respective one-particle orbital of the 1st and 2nd particle (the index $i$ ranges over the numbers of the orbitals of the two particles); $P_{N_i}$ end $Q_{M_i}$ are the polynomials of $\lambda$ of degree $N_i$ and $M_i$, respectively. The relation (4.4) may be transformed into a more convenient form[8]):

$$E(\lambda) = \frac{\sum_i \sqrt{\lambda - \lambda_0^{(i)}}\; P_{N_i}^{(i)}(\lambda)}{Q_M(\lambda)} . \tag{4.4'}$$

The relation (4.4') makes it possible to continue analytically from the region of three-particle bound states to the region where one or two particles above the core are in the continuum, i.e. the corresponding one-particle orbitals are the Gamow functions. The use of the relation (4.4') makes it possible to eliminate uniformization of the many-sheet Riemannian surface and to continue analytically without uniformization. The sheets adjoining each of the two-particle cuts are prescribed by the sign of the corresponding root. The calculation using eq. (4.4') is made in a standard way. For $\lambda_i > 1$ which are such that the examined state is bound, we calculate the energy of this state using the variational method or the Faddeev equations. Further, two reference sets $\{E_i\}$ and $\{\lambda_i\}$ and also the

---

[8]) Of course, the above calculations are not the "derivation" of (4.4'); they may be regarded only as hints.

values of $\{\lambda_0^{(i)}\}$ found earlier from the two-particle problem are used to determine the Padé-approximant (real) coefficients in (4.4′). Then, setting $\lambda = 1$ in the expression (4.4′), we obtain the complex energy of the examined resonance state.

For the sake of brevity, we have limited ourselves to a very schematic examination and omitted many questions essential for the realization, such as how many one-particle states are to be taken into account, etc. These questions must be answered separately in each individual case.

Let us examine now the second approach which makes it possible to calculate an arbitrary resonance in the model under consideration. We shall again use the decomposition of the total Hamiltonian of the type of (4.1).

In such a reduction we shall write down the set of Lippmann-Schwinger equations which is equivalent here to the set of Faddeev equations (see Section 1.4):

$$\Psi_{3,k}^{(+)} = \Phi_{3,k} + g_{12}^{(+)}(E)\, V_3 \Psi_{3,k}^{(+)},$$

$$\Psi_{3,k}^{(+)} = G_{(1+2)}^{(+)}(E)\, V_{12} \Psi_{3,k}^{(+)} \tag{4.5}$$

where

$$g_{12}^{(+)}(E) = (E + i\varepsilon - H_0 - V_{12})^{-1};$$

$$G_{(1+2)}^{(+)}(E) = (E + i\varepsilon - H_0 - V_1 - V_2)^{-1},$$

$\Phi_{3,k}$ is the wave function of the initial state, i.e. the function of the two-particle (1 and 2) bound state incident onto the core 3, multiplied by the function of free motion of the centre-of-mass of the pair (1 and 2). It can be shown [22] that the set (4.5) reduces to a single equation to find the operator $X(E)$:

$$X(E) = \mathscr{G}_{(1+2)}(E) + \mathscr{G}_{(1+2)}(E)\, t_{12}(E)\, X(E) \tag{4.6}$$

where

$$\mathscr{G}_{(1+2)}(E) = G_{(1+2)}(E) - G_0(E)$$

and

$$G_{(1+2)}(E) = 1/(2\pi i) \int g_1(E - \varepsilon)\, g_2(\varepsilon)\, d\varepsilon$$

is the convolution of two one-particle Green functions. The kernel of (4.6) $\mathscr{G}_{(1+2)} t_{12}$ contains only connected parts and, thus, the equation for $X(E)$ is of the Fredholm type [22]. Following [22] we separate from $\mathscr{G}_{(1+2)}(E)$ the completely connected part: $\mathscr{G}_{(1+2)}(E) = \hat{\mathscr{G}}_{(1+2)}(E) + \mathscr{G}_{(1+2)}^c(E)$ where $\mathscr{G}_{(1+2)}^c(E)$ is the completely connected part. Further, using the spectral decomposition for one-particle Green functions $g_{1(2)}$ which includes resonance poles [23], we can easily obtain the following expansion for $\mathscr{G}_{(1+2)}^c(E)$:

$$\mathscr{G}_{(1+2)}^c(E) = \sum_{ij} \frac{\Phi_{i1}\Phi_{j2}\tilde{\Phi}_{i1}^*\tilde{\Phi}_{j2}^*}{E - \varepsilon_{1i} - \varepsilon_{2j}} \tag{4.7}$$

where $\Phi_{i1}(\Phi_{j2})$ are the one-particle Gamow functions or the bound-state wave functions. (As above, the tilde means the function reversed in time). Fuller [22] has shown that the approximation (4.7) gives a very satisfactory description of the kernel of equation (4.6) if we limit ourselves to the nearest poles. We shall be interested in the quasistationary intermediate states of two particles in the field of a core. These states will be solutions for the homogeneous equation

$$X(E) = \mathscr{G}^c_{(1+2)}(E)\, t_{12}(E)\, X(E) \tag{4.8}$$

at complex energies $E_R - i\,\Gamma/2$ on the nonphysical sheet. Substituting the expansion (4.7) in (4.8), we arrive after simple algebra at an algebraic finite--dimensional problem for eigenvalues

$$\sum_{ij=1}^{N} \left[ T_{rs,\,ij}(E) - (E - \varepsilon_i - \varepsilon_j)\, \delta_{ir}\delta_{js} \right] Z_{ij} = 0 \tag{4.9}$$

where

$$T_{rs,\,ij}(E) = (\tilde{\Phi}^*_{r1}\tilde{\Phi}^*_{s2}|\, t_{12}(E)\, |\Phi_{i1}\Phi_{j2}) \,. \tag{4.10}$$

In the case where $i(j)$ are the indices of resonance states, the energies $\varepsilon_i(\varepsilon_j)$ are complex numbers. The matrix elements $T_{rs,\,ij}(E)$ can be conveniently calculated by analytic continuation in the coupling constant $\lambda$ of the operator $(V_1 + V_2) \Rightarrow \lambda(V_1 + V_2)$ from the values $\lambda > 1$ to the value $\lambda = 1$. If we limit ourselves to the state with only a single particle in the continuum, then all the matrix elements (4.10) have two root branch points at a maximum and can easily be found after uniformization (see Section 5.1.5). In the general case of two particles in the continuum some complications arise, so we can proceed as discussed in Section 5.1.5 or in Section 5.3.2. If we retain only one resonance orbital for both particles 1 and 2, as often happens [19], additional simplifications will arise.


## 5.5. Scattering and the Reactions Involving Unstable Particles

In this section we shall discuss the application of the method of analytic continuation in the coupling constant to a description of the reactions with unstable particles in final, intermediate, and initial states. As good examples of reactions of the first type we can quote direct nuclear reactions with production of excited nuclei, the subsequent decay of which leads to a many-particle final state; for example, the reactions of stripping and pick-up to unbound states. As examples of reactions of the second type we can cite intermediate resonances (the Feshbach-type resonances) for the scattering of nucleons or deuterons by nuclei with excitation of the 2p–1h or 2p–2h configurations. And finally, the processes of scattering of unstable particles, such as delta-isobar, by nucleons and nuclei are examples of the process with an unstable particle in the initial state. We shall examine them consecutively.

### 5.5.1. Stripping to Unbound State

For definiteness, we shall use the DWBA approach which is widely used to analyse direct stripping and pick-up nuclear reactions. Let us write the matrix element of the reaction of stripping (say, for the (d,p) reaction) to the unbound state of a final nucleus

$$M_{str} = \langle \chi_p^{(-)}(k_p)\chi_{l_n}^{(-)}(k_n)|\ V_{np}\ |\chi^{(+)}(k_d)\varphi_d\rangle \tag{5.1}$$

where $\chi_p^{(-)}$ and $\chi_{l_n}^{(-)}$ are distorted waves of the final proton and neutron, respectively; $\chi^{(+)}(k_d)$ is the distorted wave of the incident deuteron. Since $\chi_p^{(-)}$ and $\chi_{l_n}^{(-)}$ fall at infinity slowly the integral (5.1) converges very slowly (if it converges at all). To eliminate this difficulty, Vincent and Fortune [24] proposed a very convenient procedure to rotate a part of the integration contour in (5.1) into the complex $r$-plane (the dilatation transformation) where one of the functions $\chi^{(-)}$ becomes exponentially convergent and, thus, regularizes the integral. However, if we are interested in a stripping for a well-pronounced resonance state of a final nucleus, then, instead of the Vincent–Fortune procedure, we may well use the ACCC-approach. Following [25], we may write the reaction cross section as an analytic function of the neutron wave vector $k_n$:

$$\frac{d^3\sigma_{str}}{d\Omega_n\, d\Omega_p\, d\varepsilon_p} = \text{const.} \sum_{l_n} \langle \chi_p^{(-)}(k_p(k_n^*))\, \chi_{l_n}^{(-)}(k_n^*)|\ V_{np}\ |\chi_d^{(+)}(k_d)\varphi_d\rangle \times$$

$$\times \langle \varphi_d\chi_d^{(+)}(k_d)|\ V_{np}\ |\chi_p^{(-)}(k_p(k_n))\, \chi_{l_n}^{(-)}(k_n)\rangle \tag{5.2}$$

which coincides at real $k_n$ with the conventional definition and gives its analytic continuation at complex $k_n$. Near the resonance $k_n \sim k_R = k_1 - ik_2$ we can factorize the wave function $\chi_{l_n}^{(-)}(k_n, r_n)$ (for $r_n$ in the interaction region of nuclear forces) [26]:

$$\chi_{l_n}^{(-)}(k_n, r_n) \sim \sqrt{\Delta_n(\varepsilon)}\ \Phi_{G,\,l_n}(r_n) \tag{5.3}$$

where the energy dependent factor

$$\Delta_n(\varepsilon) = \frac{\gamma_0}{(\varepsilon - \varepsilon_0)^2 + \gamma_0^2/4}\ ;$$

the width $\gamma_0 = \gamma(k_i)$ is proportional to the barrier penetrability at an energy equalling the resonance energy; $\Phi_{G,\,l_n}$ is the Gamow wave function of the neutron. After the substitution of (5.3) in (5.2) and the subsequent integration over energy[9]) we obtain the contribution of a single Gamow state, i.e. the purely resonance-polar contribution, to the differential cross section

$$\frac{d\sigma}{d\Omega_p} = \text{const} \cdot \sum_{l_n} \langle \chi_p^{(-)}(k_p(k_R^*)) \, \tilde{\Phi}_{G,\,l_n}^* | \, V_{np} \, |\chi_d^{(+)}(k_d)\varphi_d \rangle \times$$

$$\times \; \langle \chi_d^{(+)}(k_d)\varphi_d | \, V_{np} \, |\chi_p^{(-)}(k_p(k_R)) \, \Phi_{G,\,l_n} \rangle . \tag{5.4}$$

So, it is evident that the necessary matrix element for stripping to the resonance (Gamow) state $\Phi_{G,\,l_n}(k_R, r_n)$ is obtained as a result of analytic continuation of the corresponding matrix element

$$\langle \chi_d^{(+)}(k_d)\varphi_d | \, V_{np} \, |\chi_p^{(-)}(k_p(k_b)) \, \Phi_b(k_b, r_n) \rangle$$

for the stripping to the *bound state* $\Phi_b(k_b, r_n)$ . Using the transition matrix elements calculated in this way, we may easily find the cross section of the process. It is convenient here to keep the total energy $E = T_d + \varepsilon_d$ constant ($T_d$ is the kinetic energy of the incident deuteron; $\varepsilon_d$ is its binding energy). Based on such hints, we can rigorously define the differential cross section of the stripping to the resonance state as an analytic continuation in the coupling constant of the corresponding cross section of the stripping to the bound state

$$\frac{d\sigma}{d\Omega_p} = \underset{\lambda \to 1}{\text{Cont}} \left( \frac{d\sigma}{d\Omega_p} \right)_{\substack{\text{bound,} \\ E = \text{fixed}}} \tag{5.4a}$$

whence the experimentally measured cross section may be inferred by taking the real part of (5.4a) [25]

$$\left( \frac{d\sigma}{d\Omega_p} \right)_{\text{exp}} = \text{Re} \left( \frac{d\sigma}{d\Omega_p} \right) .$$

It can easily be shown, however, that the ratio of the imaginary part of the cross section of (5.2) to the real part at a resonance pole, i.e.

$$\text{Im} \left( \frac{d\sigma}{d\Omega} \right) / \text{Re} \left( \frac{d\sigma}{d\Omega} \right) \sim \left( \frac{\Gamma}{E_R} \right)^2$$

is small and, therefore, at $\Gamma/E_R \ll 1$ we may directly continue the initial cross section (5.2).

Thus, the differential cross section of the stripping to the resonance or virtual state is obtained in the given approach by direct analytic continuation in the coupling constant (or in the energy) of the corresponding cross section of

---

[9]) This derivation must be treated rather as a suggestion than as a rigorous procedure.

stripping to the bound state which can readily be calculated using the variety of computer programs available at every nuclear research centre. The effectiveness of the discussed approach was demonstrated by calculating [27] the test version for the reaction $^{16}$O (d, p)$^{17}$O* (5.08 MeV) at the incident deuteron energy

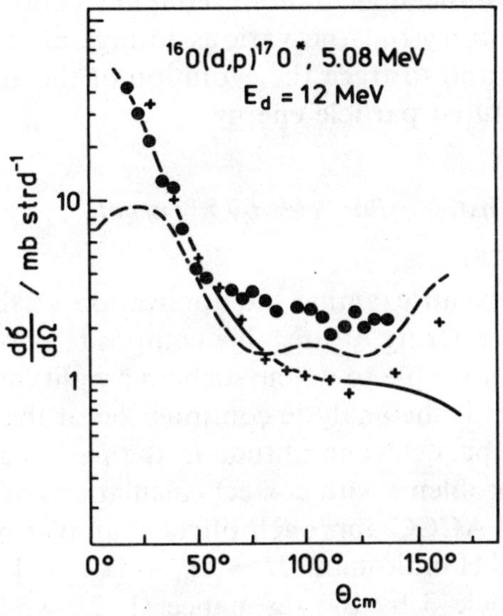



Fig. 5.8

Angular distribution for the $^{16}$0(d, p) $^{17}$0*. The dots are the experimental data, the solid line is the result of the Vincent-Fortuna calculations, the crosses are the results of our calculations; the dotted line is one of the initial distributions at $\hbar^2 k_n^2/2m < 0$ (stripping to a bound state) from which the continuation was made.

$E_d = 12$ MeV. The stripped neutron "sits down" on the one-particle resonance orbital $d_{3/2}$ in the field of the $^{16}$O nucleus with the excitation energy $E^* = 5.08$ MeV. Fig. 5.8 shows the results of these calculations and compares the results obtained on the basis of (5.4a) with the results of the Vincent-Fortune calculations, where the method of deformation of the integration contour in the $r$-plane was used, and with experimental data. Very good agreement between the two theoretical approaches can be seen. However, the approach treated here and based on analytical continuation in the coupling constant has some advantages, of which we shall mention the following:

(1) since the discussed method is based on the analytic properties of the amplitude of the process (as a function of the coupling constant), the ACCC method proves to be quite applicable to the cases of both zero- and finite-range interactions (this may be important for the theory of lithium stripping reactions of the type of ($^6$Li, t) or ($^6$Li, d));

(2) calculations of the stripping to bound and unbound states are made using the same computer program and to calculate the unbound final states it is sufficient merely to carry out several standard calculations for the conventional stripping (i.e. to the bound state);

(3) in this approach we obtain a convenient analytic dependences of the stripping-off neutron energy in the final nucleus and of the differential cross section at each value of the angle $\vartheta$ on the coupling constant $\lambda$, thereby making it readily possible to carry out the various fittings of the type of the known well-depth procedure and to trace the evolution of the angular distribution as a function of the captured particle energy.

### 5.5.2. Scattering of Unstable Particles by a Target

In this subsection we shall examine the application of the ACCC method to a description of the scattering of unstable composite particles (resonances) by a stable target. It is not simple to define such an amplitude directly [28]. One of the possible definitions is the analytic continuation of the $3 \rightarrow 3$ (i.e. three free particles to three free particles) amplitude in the pair energy to a two-particle resonance pole. The problems with correct calculations of the $3 \rightarrow 3$ amplitude arise in this case. The ACCC-approach offers a simpler method.

Let us introduce the Hamiltonian $H = H_0 + (V_{13} + V_{23}) + \lambda V_{12}$ describing the scattering of particle 3 by the resonance $(1, 2)$, where the value $\lambda = 1$ corresponds to the true resonance. We *define* the amplitude of scattering of a particle by a resonance (or vice-versa) as analytic continuation in $\lambda$ (at a fixed total energy $E$) of a well investigated scattering amplitude of particle 3 by the bound state of pair $(1,2)$ with energy $\varepsilon_{12} = \varepsilon_{12}(\lambda)$ and wave function $\varphi_{12}(\lambda)$. Let this scattering amplitude be defined via the appropriate transition operator $T_{12, 12}(E, \lambda)$:

$$\mathscr{A} \sim \langle k_3'(\lambda) , \; \varphi_{12}(\lambda)| \; T_{12, 12}(E, \lambda) \; |k_3(\lambda) , \; \varphi_{12}(\lambda) \rangle \tag{5.5}$$

where $|k_3(\lambda)\rangle$ denotes the plane wave of the free motion of the third particle with respect to the bound state of pair $(1, 2)$ with the wave vector $k_3$; here,

$$|k_3(\lambda)| = |k_3'(\lambda)| = \sqrt{2m[E - \varepsilon_{12}(\lambda)]/\hbar^2} \; .$$

The transition operator $T_{12, 12}$ is defined by the equality

$$T_{12, 12}(E, \lambda) = (V_{13} + V_{23}) + (V_{13} + V_{23}) \, G(E, \lambda) \, (V_{13} + V_{23})$$

where $G(E, \lambda)$ is the complete three-particle Green function. Now, the set of Faddeev equations can be used to find $G(E, \lambda)$. We can also express the operator $T_{12, 12}(E, \lambda)$ through the resolvent $R(E, \lambda)$ of the kernel of the Faddeev equations. In turn, the dependence on $\lambda$ enters $R(E,\lambda)$ through the Green function of the subsystem $(1, 2)$:

$$G_{12}(E, \lambda) = g_{12}\left(E - \frac{p^2}{2\mu}, \lambda\right)\delta(\boldsymbol{p} - \boldsymbol{p}')$$

where $g_{12}(\varepsilon, \lambda)$ is a meromorphic function of $\lambda$. Hence, each pole $\lambda_n(\varepsilon)$ of $g(\varepsilon, \lambda)$ leads to the cut in the $\lambda$-plane originating at the point $\lambda_n$ for the function $G_{12}(E, \lambda)$. Let us examine now the Hilbert-Schmidt expansion

$$G_{12}(E, \lambda) = \sum_n \frac{R_n}{\lambda^{-1} - \alpha_n(\sqrt{-E + p^2/2\mu})} . \tag{5.6}$$

At given total energy $E$ there exists a region $|\lambda - \lambda_n(E)|$ in which all the rest terms of (5.6) are regular except for the $n$-th term because the only point of accumulation is zero. This $n$-th term leads to the root cut in the $\lambda$-plane along the real axis. Examine the location of the branch point $\lambda_n(E)$ at different values of the total energy $E$. The positive values of $\alpha_n(\varepsilon)$ are known to rise monotonically with $\varepsilon$ at $\varepsilon < 0$ and become complex at $\varepsilon > 0$. The negative $\alpha_n(\varepsilon)$ do not lead to the poles in (5.6). Therefore, at $\lambda < \min\left(\alpha_n^{-1}\right)$ the denominator in (5.6) does not vanish and the kernel is regular. But $\min\left(\alpha_n^{-1}\right) = \max\left(\alpha_n\left(\sqrt{-E + p^2/2\mu}\right)\right)$, so at $E < 0$ the maximum is reached at $p^2 = 0$, i.e. the cut begins at the point $\alpha_n^{-1}(E)$. At $E > 0$, $\max\left(\alpha_n(\acute{u}(-E + p^2/2\mu))\right)$ is reached at $E = p^2/2\mu$ and, hence, the cut begins at the point $\alpha_n^{-1}(0)$. As a result, we arrive at the following conclusion.

At any fixed value of $E$ the kernel of the set of Faddeev equations is an analytic function of the variable $\sqrt{(\lambda - \lambda_n)}$ in the vicinity of the point $\lambda_n(E)$, where

$$\lambda_n(E) = \begin{cases} \alpha_n^{-1}(E) & \text{at } E < 0 , \\ \alpha_n^{-1}(0) & \text{at } E > 0 \end{cases}$$

if this vicinity does not contain the three-particle resolvent poles, i.e. the three-particle bound states. Therefore the three-particle resolvent proper and, hence, the amplitude of particle scattering by a bound state (5.5) will be analytic functions of the variable $\sqrt{(\lambda - \lambda_n(E))}$. From this it follows that by continuing the amplitude $A$ at a fixed value of $E$ from the values of $\lambda^{(i)} > \lambda_n$ (in this case a scattering by the bound state $\varphi_{12}^n$ with energy $\varepsilon_{12}(\lambda^{(i)})$ occurs) up to the value of $\bar{\lambda} < \lambda_n$ as an analytic function of $\sqrt{(\lambda - \lambda_n(E))}$ we can obtain the amplitude of particle scattering by a quasistationary state, or vice versa, of quasistationary state scattering by a target. Departures from the conventional unitarity condition arise in such a continuation [29] from the feasibility of the quasistationary state decay during the scattering.

The above-mentioned interpretation of the scattering of unstable particles goes very well with the dispersion N/D method for examining the same process (see the series of works [47]) in which the determination of the "potential" of particle interaction with a bound state is a more or less standard task [30].

## 5.6. Other Examples of the Application of the ACCC Method in Nuclear Physics

The optical model is extensively used in various areas of nuclear and atomic physics and in elementary particle physics. This model is attractive because it can reduce the many-particle problem to the effective two-particle problem with a complicated, nonlocal in the general case, and complex potential. However, the calculations of resonance states in terms of such potentials are faced with the same difficulties as those noted in Section 5.1. Evidently, the technique treated in Section 5.1 applies effectively in this case too. Let us now present some examples of its application.

(1) Low-energy $\alpha$–$\alpha$ scattering can properly be described by the optical potential. We have selected two $\alpha$–$\alpha$ potentials of different classes which are in common use.

(i) The microscopically-substantiated deep potential with forbidden states [31] which is of the Woods-Saxon form and describes very well the $\alpha$–$\alpha$ phases in a broad energy range. The observed $0^+$ and $2^+$ states of $^8$Be correspond to excited states in such a potential.

(ii) Ali-Bodmer potential with a substantial repulsive core of different forms in different partial waves [32].

Table 5.3

Values of the energy $E_R$ (MeV) and of the width $\Gamma$ (MeV) of some $\alpha-\alpha$ resonances obtained by the ACCC and dilatation method.

| Potential | | Ali-Bodmer | | Woods–Saxon |
|---|---|---|---|---|
| $J^\pi$ | | [5, 5] PA | Dilatation method*) | [5, 5] PA |
| $2^+$ | $E_R$ | 2.796 | 2.853 | 2.70 |
| | $\Gamma$ | 1.220 | 1.228 | 1.04 |
| $4^+$ | $E_R$ | 11.400 | 11.404 | 11.5 |
| | $\Gamma$ | 2.658 | 2.932 | 3.4 |
| $6^+$ | $E_R$ | 31.25 | – | 30.2 |
| | $\Gamma$ | 40.56 | – | 44.1 |

*)These values were obtained by T. Vertse using the dilatation transformation

This potential fits the $\alpha$–$\alpha$ phases in a narrower energy range compared with the potential (i). At low energies ($\leq 15$ MeV in the centre-of-mass system), however, the description of experimental phase shifts by the Ali–Bodmer potential is nearly as good as by the potential (i). Nevertheless, the values of the resonance parameters found for the interactions (i) and (ii) are different (see Table 5.3) thereby indicating their sensitivity to the fine details of the potential. The ACCC approach was used to find the positions of the $S$-matrix resonance poles in the two cases. The calculation results are shown in Table 5.3. For comparison, the table also gives the results obtained by T. Vertcse for the Ali-Bodmer potential by the dilatation method (for the relevant experimental data, see Table 6.5). It can be seen that the results of the two methods are in good agreement.

(2) The parameters of the one-particle neutron resonance and antibound states in the $^{209}$Pb and $^{17}$O nuclei, with the spin-orbit interaction allowed for. The selfconsistent field potential was chosen to be

$$V_{ws} = V(r) + V_{so}(r) \equiv f(r) - \frac{\varkappa}{2}\frac{\mathrm{d}f}{\mathrm{d}r}(\sigma l) ,$$

$$f(r) = -V_0 \left(1 + \exp \frac{r - R}{a}\right)^{-1} ,$$

$$R = r_0 A^{1/3} ,$$

$$\frac{2m}{\hbar^2} = \frac{0.048\,228\,A}{(A + 1)}\ \mathrm{MeV}^{-1}\mathrm{fm}^{-2}$$

with the following values of the parameters for

$$\text{n} + {}^{208}\text{Pb}: \quad V_0 = 46.232 \text{ MeV}; \quad r_0 = 1.24 \text{ fm}; \quad a^{-1} = 1.587 \text{ fm}^{-1};$$

$$\varkappa = 0.374 \text{ fm}^2$$

and for

$$\text{n} + {}^{16}\text{O}: \quad V_0 = 53.7 \text{ MeV}; \quad r_0 = 1.24 \text{ fm}; \quad a^{-1} = 1.54 \text{ fm}^{-1};$$

$$\varkappa = 0.216 \text{ fm}^2.$$

Tables 5.4 and 5.5 present some of the neutron antibound ($a$) and resonance ($r$) states in the $^{17}$O and $^{209}$Pb nuclei. It should be noted that the calculations of several antibound states from Table 5.5 involve a numerical instability, so that, as the approximant order $N, M$ increases, the convergence of the result may only be claimed tentatively; therefore, the relevant value in the table is marked by the symbol $\approx$[10]).

---

[10]) It should be emphasized, however, that such instabilities arise in the calculations of only a few antibound states. These instabilities are readily amenable to diagnosing as the PA order changes.

Table 5.4

Energies of some antibound ($a$) and resonance ($r$) states in $^{16}O$ . $\varepsilon_{nlj} = E_{nlj} - i\,\Gamma_{nlj}/2$ (MeV) .

| $nlj$ | 1 $d_{5/2}$ (a) | 1 $p_{1/2}$ (a) | 1 $d_{3/2}$ (r) | 1 $f_{7/2}$ (r) | 1 $f_{5/2}$ (r) |
|---|---|---|---|---|---|
| $E_{nlj}$ | $-3.71$ | $-12.3$ | 0.643 | 7.12 | 11.1 |
| $\Gamma_{nlj}$ | $-$ | $-$ | 0.058 | 2.62 | 12.1 |

Table 5.5

Energies of some antibound ($a$) and resonance ($r$) states for $^{209}$Pb. $\varepsilon_{nlj} = E_{nlj} - i\Gamma_{nlj}/2$ (MeV) .

| $nlj$ | $2h_{11/2}$ (r) | $1j_{13/2}$ (r) | $3s_{1/2}$ (a) | $5s_{1/2}$ (r) | $2d_{5/2}$ (a) | 4 $d_{3/2}$ (r) | 1 $m_{19/2}$ (r) |
|---|---|---|---|---|---|---|---|
| $E_{nlj}$ | 2.58 | 5.82 | $\approx -28.2$ | 13.2 | $-17.0$ | 11.2 | 33.2 |
| $\Gamma_{nlj}$ | 0.09 | 0.05 | $-$ | 37.3 | $-$ | 40.5 | 5.43 |

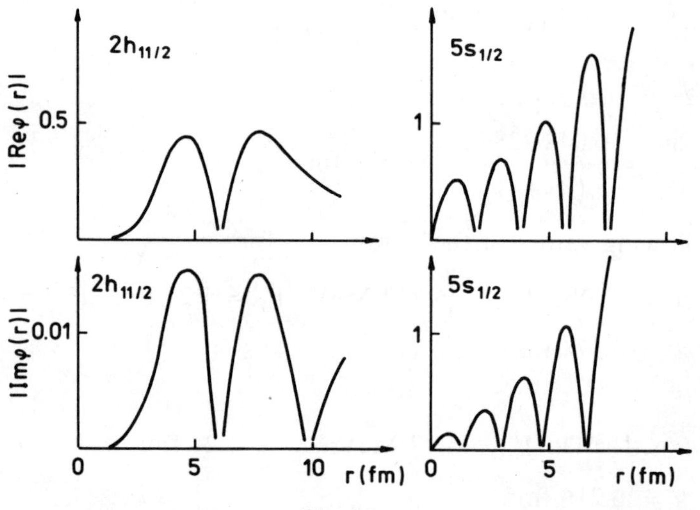


Fig. 5.9

The Gamow functions of the one-particle neutron resonances $5S_{1/2}$ and $2h_{11/2}$ in the Woods-
-Saxon potential, obtained by analytic continuation.

The respective studies have shown that in these cases the matrix of the set of
equations for finding the PA coefficients (1.10) proves to be ill-conditioned.
When the continued-fraction algorithm is used to find the PA coefficients (1.10),
the instability disappears. Fig. 5.9 shows the Gamow functions of the one-
-particle neutron resonance and of the antibound state in the $^{209}$Pb nucleus
calculated by the above-mentioned method.

(3) Since the optical potential of the Woods-Saxon form is very widely used in nuclear physics, we considered it expedient to discuss the $S$-matrix pole trajectories for the given potential in the present section. The $S$-wave trajectories

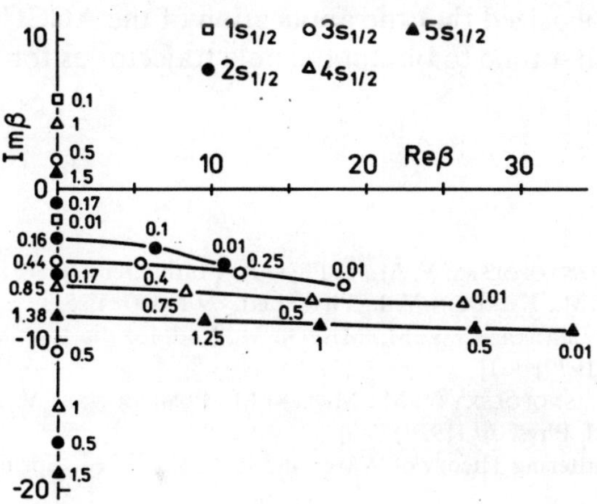

Fig. 5.10

Trajectories of the $S$-matrix poles $k_0(\lambda)$ for $s$-wave in the Woods-Saxon potential $(\beta = kR)$.

are shown in Fig. 5.10, and the trajectories for higher partial waves in Fig. 5.11. [33].

It should be noted that at $l = 0$ the position of the branch point $k_0 = k_0(\lambda)$ depends on the number of a particular level (radial quantum number $n$), contra-

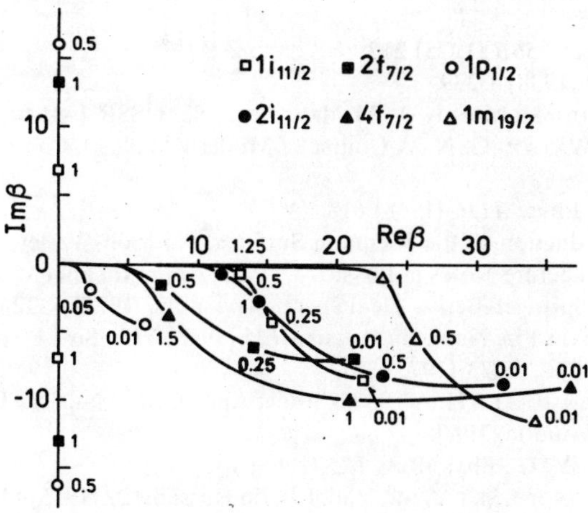

Fig. 5.11

Trajectories of the $S$-matrix poles $k_l(\lambda)$ $(l > 0)$ in the $\beta$-plane $(\beta = kR)$ for the Woods-Saxon potential.

ry to the case of the rectangular well [8]. As noted above, the trajectories obtained here at small values of $\lambda$ are very likely to differ from the real trajectories because the singularity of the function $k_l(\lambda)$ at $\lambda = 0$ was not allowed for in [33].

It should be emphasized that the application of the ACCC method has made it possible for the first time to obtain the pole trajectories for the Woods–Saxon potential [33].

*References*


1. KUKULIN V. I., KRASNOPOĽSKY V. M., J. Phys. A: Math. Gen. *10* (1977) L33.
2. KRASNOPOĽSKY V. M., KUKULIN V. I., Phys. Lett. *69A* (1978) 251.
3. KUKULIN V. I., KRASNOPOĽSKY V. M., MISELKI M., Yad. Fiz. (in Russian) *29* (1979) 818. [Sov. J. Nucl. Phys. *29* (1979) 421].
4. KUKULIN V. I., KRASNOPOĽSKY V. M., MISELKI M., POMERANTSEV V. N., Yad. Fiz. *30* (1979) 1 428. [Sov. J. Nucl. Phys. *30* (1979) 740].
5. NEWTON R. G., Scattering Theory of Waves and Particles, 2nd ed., Springer, Berlin, Heidelberg 1982.
6. BADALYAN A. M., SIMONOV Yu. A., The Variational Method for the 3-Body Resonances (Preprint Series):
   I. The Asymptotic Behaviour of the Resonance Wave Function, Preprint ITEP No. 966, Moscow 1972;
   II. The Equations for the Resonance Wave Function, Preprint ITEP No. 952, Moscow 1972; Yad. Fiz. (in Russian) *17* (1973). [Sov. J. Nucl. Phys. *17* (1973) 11].
7. OSTROVSKY V. N., Solovev E. A., ZhETF *62* (1972)   167. [Sov. Phys. JETP *35* (1972) 89].
8. NUSSENZVEIG H. M., Causality and Dispersion Relations, Academic Press, N. Y.–London 1972.
9. HELLMANN H., Einführung in die Quantenchemie, Deuticke, Leipzig and Vienna 1937.
   FEYNMAN R. P., Phys. Rev. *56* (1939) 340.
10. BATEMAN H., ERDÉLYI A., Higher Transcendental Functions, Vol. II, McGraw-Hill, N.Y.–Toronto–London 1955.
11. TJON J. A., Phys. Lett. *56B* (1975) 217;
    Phys. Rev. Lett. *40* (1978) 1 239.
12. DOLINSKY E. I., MUHAMEDJANOV A. M., Izv. Acad. Sc. USSR (ser. fiz.) *41* (1977) 2 055.
13. WHITTAKER E. T., WATSON G. N., A Course of Modern Analysis, Cambridge University Press, 1927.
14. ROMO W. J., Nucl. Phys. *A116* (1968) 618.
15. SPRINGER G., Introduction to the Riemann Surfaces, Addison-Wesley, Reading, Mass., 1957.
16. CHISHOLM J. S. R., Lecture Notes in Physics *47*, Padé Approximants Method and its Applications to Mechanics, Springer Berlin–Heidelberg–New York. 1976, p. 32.
17. SHAPIRO I. S., Uspekhi Fiz. Nauk (in Russian)*125* (1978) 577. [Sov. Phys. Usp. *21* (1978) 645].
18. JOFFILY S., Nucl. Phys., *A215* (1973) 301.
19. MAHAUX C., WEIDENMÜLLER H. A., Shell-Model Approach to Nuclear Reactions. North–Holland, Amsterdam–London, 1969.
20. AFNAN I. R., TANG Y. C., Phys. Rev. *175* (1968) 1337.
21. KUKULIN V. I., KRASNOPOĽSKY V. M., Yad. Fiz. (in Russian) *22* (1975) 1 110. [Sov. J. Nucl. Phys. *22* (1975) 578].
    KUKULIN V. I., KRASNOPOĽSKY V. M., J. Phys. G.: Nucl. Phys. *3* (1977) 795.
22. FULLER R. C., Phys. Rev. *C3* (1971) 1042.
23. ROMO W. J., Nucl. Phys. A302 (1978) 61.

24. VINCENT C. M., FORTUNE H. T., Phys. Rev. *C2* (1970) 782.
25. BERGGREN T., Phys. Lett. *73B* (1978) 389.
26. BANG J., ZIMANI I., Nucl. Phys. *A139* (1969) 534.
    COKER W. R., Phys. Rev. *C7* (1973) 2426.
27. KRASNOPOĽSKY V. M., KUKULIN V. I., Phys. Lett. *96B* (1980) 4.
28. KARMANOV V. A., KONDRATIUK L. A., SHAPIRO I. S., In: Problems of Modern Nuclear Physics, Proc. Second Problem Symp. on Nucl. Phys., Novosibirsk, 1970; Nauka, Moscow 1971, p. 496.
    KONDRATIUK L. A., SHAPIRO I. S., Yad. Fiz. (in Russian)*12* (1970) 77.
29. FREEDMAN D. Z., LOVELACE C., J. M. NAMYSLOVSKY, Nuovo Cim. *43A* (1966) 258.
30. BRAYSHAW D. D., Phys. Rev. Lett. *36* (1976) 73;
    BASDEVANT J. L., KREPS R. E., Phys. Rev. *141* (1966) 1398, 1404, 1409;
    MENESIER G., PASQIER R., PASQIER J. Y., Phys. Rev. *D6* (1972) 1 351.
31. NEUDATCHIN V. G., KUKULIN V. I., KORENNOY V. P., KOROTKICH V. L., Phys. Lett. *41B* (1972) 331.
32. ALI S., BODMER A. R., Nucl. Phys. *80* (1966) 99.
33. GAREEV F. A., GONCHAROV S. A., KUKULIN V. I., KRASNOPOĽSKY V. M., Preprints JINR P4-12004, P4-12002, Dubna 1978.
34. REED M. C., SIMON B., Methods of Modern Mathematical Physics. Academic Press, N. Y.–San Francisco–London 1979.
35. KLAUS M., SIMON B., Ann. Phys. (N. Y.) *130* (1980) 251.
36. KOK L. P., VAN HAERINGEN H., Ann. Phys. (N. Y.) *131* (1981) 426.
37. BEAM J. E., Phys. Lett. *30B* (1969) 67;
    BOLSTERLI M., Phys. Rev. *182* (1969) 1095.
38. DEMKOV Yu. N., KOMAROV V. I., ZhETF *50* (1966) 286. [Sov. Phys. JETP *23* (1966) 189].
39. KOMAROV V. I., SOLOVIEV E. A., TMF (in Russian) *32* (1977) 271.
40. DOMCKE W., J. Phys. B., At. Mol. Phys. *16* (1983) 359.
41. ESTRADA H., DOMCKE W., J. Phys. B.: At. Mol. Phys. *17* (1984) 279.
42. ULYANOV V. V., Integral Methods in Quantum Mechanics (in Russian), Ed. of Kharkov State University, Kharkov 1982.
43. DOMCKE W., J. Phys. B14 (1981) 4889.
44. ABRAMOWITZ M., STEGUN I., Handbook of Mathematical Functions, Chap. 6. New York--Dover 1965.
45. BAZ A. I., ZEL'DOVICH Ya. B., PERELOMOV A. M., Scattering, Reactions and Decay in Nonrelativistic Quantum Mechanics, Nauka, Moscow, 2nd ed. 1971 (in Russian). [English translation of lst ed.: Israel Program for Scientific Translations, Jerusalem, 1966];
    ZEL'DOVICH Ya. B., ZhETF *39* (1960) 776, 1483 (in Russian). [Sov. Phys. JETP *12* (1961) 542].
46. See Chap. III of Ref. 45.
47. SIMONOV Yu. A., GRACH I. L., SHAMTIKOV M. Zh., Nucl. Phys. A334 (1980) 80; BADALYAN A. M., SIMONOV Yu. A., ECHAYA *6* (1975) 299 (in Russian)
    SIMONOV Yu. A., The Resonances in the Three-Particle Systems: in: Proc. 1977 Europ. Symp. on Few Part. Problems in Nucl. Phys., p. 81.
48. HARTT K., YIDANA P. V. A., Phys. Rev. C31 (1985) 1105.
49. GAMOW G., Zeitschr. f. Phys. *5* (1928) 204; see also the classical textbook: Blatt J. M., Weisskopf V. F., Theoretical Nuclear Physics, New York, 1952.

# Chapter 6

# *S*-matrix Parametrization of Scattering Data. Extraction of Resonance Parameters from Experimental Data

In this chapter we shall examine the problem of extraction of spectroscopic information from scattering data. Basically, we shall talk about experimental data processing. However, it is clear that the whole discussed scheme will be conserved if the scattering amplitudes and phase shifts are not obtained from experimental data but are found by solving a dynamic (e.g. Schrödinger or Lippmann-Schwinger) equation. Those few cases where data from theoretical calculations will be used instead of the experimental ones and where it will be necessary to change the scheme will be discussed separately.

Experimental data in atomic, nuclear and elementary-particle physics are sets of total or differential cross sections and polarizations measured for some scattering process or reaction at a number of energies. Long-lived states (narrow resonances) manifest themselves already at a level of total cross sections (sharp increase in the cross section at an energy close to the resonance energy) and differential cross sections (at an energy close to the resonance energy the differential cross section takes the form characteristic of the decay of an isolated level, i.e. the form symmetrical with respect to the angle of 90 ° in the centre-of--mass system). However, to obtain more detailed microscopic information about the resonance state (determinations of its width, angular momentum, total spin, mechanism of formation and decay, etc.) it is necessary to analyse the experimental data at the level of scattering amplitudes and phase shifts. Such an analysis is necessary especially when the maxima in the cross sections are diffused and weakly expressed and can be associated either with the formation of a rapidly decaying (broad) resonance state or with effects of non-resonance character.

To extract scattering amplitudes and phaseshifts from experimental data, the powerful technique of phase shift analysis has been developed and this is excellently described in monograph [1]. We shall not examine this technique because problems of this kind fall outside the scope of this book (we refer interested readers to [1]). It is sufficient for us to know that from the total and differential cross sections, from polarizations, and from other similar data measured at certain values of energy we can obtain the scattering amplitudes and phase shifts either at the appropriate discrete values of energy (single energy phase shift analysis) or, having imposed supplementary conditions on the amplitude, to be a continuous function of energy interpolating the respective values between the experimental points (energy-dependent phase shift analysis). In other words, the experimental information is assumed to be of the form of partial phase shifts $\{\delta_{ij}^l\}$ with their experimental errors $\{\varepsilon_i^l\}$ at energies $\{E_i\}$ $(i = 1, ..., p)$. Theoretically, we calculate the phase shifts $\{\delta_{ij}^l\}$ (in this case the error $\{\varepsilon_{ij}^l\}$ is defined by the accuracy of the calculations and can always be made sufficiently small). In the general case, however, these data are insufficient to permit extraction of resonance state parameters from them. In order to find the parameters, additional processing of scattering amplitudes or phase shifts is necessary. The following methods of finding the resonance parameters on the basis of experimental data are the most popular:

(1) construction of the Argand diagram and its subsequent analysis which, in the general case, is rather complicated (see, e.g. [1,50]) and often gives very ambiguous results, especially in the case of sufficiently broad resonances (for more detailed discussion, see Section 2.1.1 and [1]);

(2) the Breit–Wigner parametrization of the scattering amplitude (see Section 2.1.1 and, in particular, the formula (2.1.17) and (2.1.27)) which also have a limited range of application. Complications arise here because of the necessity of allowing for the non-resonance background and for the energy dependence of the resonance width. This gives rise to some inadequacy and model dependence of such a parametrization in the case of the broader resonances [2];

(3) the $R$-matrix parametrization of the reaction amplitude or phase shift (see Section 2.5, formulae (2.5.14)–(2.5.16)). As noted above, the $R$-matrix formalism is widely used in atomic and nuclear physics, but it also has some shortcomings. Thus, the obtained resonance characteristics appear to depend on the model parameters of the $R$-matrix scheme ($R_c$ and $b$, see Section 2.5). This dependence is weak only in the case of narrow solitary resonances. At the same time, in case of the broader (but well pronounced) resonances we have to resort to a multilevel $R$-matrix parametrization in which a physical resonance state "spreads" over several $R$-matrix states which have little in common with the initial state. This necessity arises in this case because we make an attempt to simulate distant poles in the complex energy plane (with $\mathrm{Im}\, z_0 \sim \mathrm{Re}\, z_0$) by the $R$-matrix poles lying on the real $E$-axis.

In view of the above it is clear that it is desirable to get a parametrization which would be "model-free" to the greatest extent, thus allowing the scattering data to be used to restore the amplitude in the complex energy $z$-plane, or, which is even better, in the complex momentum $k$-plane, from the scattering data. The $S$-matrix Padé parametrization discussed below satifies these requirements to a high degree. Such a parametrization makes it possible to determine the resonance energy and width to a considerable extent in a model-independent way, namely through the locations of the poles of the partial-wave $S$-matrix (i.e. of scattering amplitude) in the complex energy (momentum) plane [2, 3].

If we know how the $S$-matrix is to be found in the physical region, from the scattering data, we can use it to obtain comprehensive spectroscopic information (bound state energies, resonance energies and widths, spectroscopic factors) by analytic continuation of the $S$-matrix to the region of negative energies and to the non-physical energy sheets and by finding there its poles and residues at the poles (see Chapter 2). For this purpose it is necessary to construct the $S$-matrix in a form that would allow a convenient and, which is most important, stable analytic continuation to the region of complex energies and to non-physical energy sheets. To carry out such a continuation the $S$-matrix must be parametrized to take the form of a certain sufficiently simple analytic function of energy. For the continuation to be stable, the parametrization must allow as completely as possible for the analytic properties of the $S$-matrix as a function of energy.

Numerous publications, especially in elementary-particle physics, are devoted to the stability and errors of analytic continuation of scattering amplitudes in energy, in cosine of scattering angle, and in angular momentum (see, for example [30–33], detailed review articles [24, 34, 35], the book of reviews [36], and the references therein). (The whole theory of dispersion relations is based essentially on one or other method of analytic continuation.) Since this field as a whole is not related directly to the problems discussed here, we shall not go deeply into the interesting general problems relevant to stability of analytic continuations (see Section 1.3.3 and the brief commentary at the end of this chapter), so the reader is referred to the above-mentioned reviews. Here we touch on only those concrete problems which are directly connected with the extraction of resonance parameters from the phase shift analysis results and closely concern our own investigations in the given field. Similar problems of the $S$-matrix parametrization are examined in [4, 5, 6, 10]. Here we shall briefly summarize the results of these works, including our own findings in the given field.

In Section 6.1 we shall discuss the basic problems relevant to $S$-matrix parametrization using the single-channel case as an example. In Section 6.2 we consider the possibility of the generalization of this formalism to the multichannel case.

## 6.1. Parametrization of Single-Channel $S$-matrix

The single-channel case corresponds to the physical situation when only the elastic scattering a + A → a + A at energies $0 \leq E \leq E_1$ takes place; here $E_1$ is the nearest threshold of inelastic scattering or of reaction channel. The $S$-matrix (in the one-channel case it has a dimension of $1 \times 1$) is a single-valued function of the variable $k = \sqrt{(2\mu E/\hbar^2)}$, where $\mu$ is the reduced mass of particles a and A; $E$ is the energy of relative motion in the centre-of-mass system. The analytic properties of the $S$-matrix, and hence the method of its parametrization, depend substantially on the character of the interaction between particles a and

A. We examine here the *S*-matrix parametrization for the following three types of interactions used most extensively in nuclear and atomic physics: the short--range, Coulomb, and polarization interactions.

### 6.1.1. The Case of Short-Range Interaction Potential

In the case of a finite-range potential the partial-wave *S*-matrix is an analytic function of $k$ in the entire $k$-plane [7]. Hence, in the $z$-plane it has a cut along the positive real axis from $z = 0$ to $z = \infty$ (kinematic cut); correspondingly, we have one physical energy sheet, and one non-physical energy sheet (see Section 2.1). The partial-wave *S*-matrix, parametrized through scattering phase shifts (see Section 2.1)

$$S_l(k) = \exp\{2i\,\delta_l(k)\}$$

can be unambiguously expressed in terms of the scattering function $X_l(k^2)$:

$$X_l(k^2) = k^{2l+1} \cot \delta_l, \tag{1.1}$$

$$S_l(k) = \frac{X_l(k^2) + ik^{2l+1}}{X_l(k^2) - ik^{2l+1}}. \tag{1.2}$$

Since

$$\delta_l(k) \xrightarrow[k \to 0]{} k^{2l+1},$$

the function $X_l(k^2)$ is regular at the origin and, moreover, may be shown to depend only on $k^2$, i.e. on $E$. In the case where the interaction is described by a Yukawa-type potential of range $\mu^{-1}$, the partial-wave *S*-matrix, along with the function $X_l(k^2)$, has a dynamic cut in the lower $k$-half plane separated from the origin (that is, from the threshold $k = 0$) by a distance $\mu/2$. Thus, for a short--range interaction, $X_l(k^2)$ is an analytic function of energy $z$ in some domain including the origin. With the help of (1.2) the problem of *S*-matrix parametrization can be reduced in this case to parametrization of $X_l(k^2)$. One of the well-known parametrizations is the effective-range expansion (ERE)

$$X_l(k^2) = \sum_i b_i^{(l)} k^{2i} = -\frac{1}{a_l} + \frac{1}{2} r_l\, k^2 - \frac{1}{4} P_l\, k^4 + \dots \tag{1.3}$$

where $a_l$ is the scattering length; $r_l$ is the effective range; $P_l$ is the form parameter, etc. However, as the convergence radius of the series (1.3) is in many cases rather small, the series (1.3) works well only in the near-threshold region. The convergence radius of the series (1.3) is limited both by poles of $X_l(k^2)$ and by "potential" cuts. Thus, when the scattering phase shift runs through the value $\delta_l(k_c) = n\pi$, the function $X_l(k^2)$ has a pole, and the value $k_c^2$ limits the convergence radius of the series (1.3). The passage of the phase shift through a value multiple of $\pi$ at small energies is a standard phenomenon for the majority of nuclear systems in lower partial waves. Here it will be sufficient to present the following examples: for instance, d + $\alpha$, d + t, $\alpha$ + $\alpha$, etc. (see also below). The near poles of $X_l(k^2)$ may be exist also at small negative energies: for example, in a doublet channel of n-d scattering for $l = 0$ [9]. To take into account such features, it was suggested [9] that polar terms of the type $ak^{2m}/(k^2 - k_c^2)$ were added to the truncated series (1.3), i.e. an approximation of the form:

$$X_l(k^2) \approx -\frac{1}{a_l} + \frac{r_l}{2} k^2 + \frac{pk^2}{1 - qk^2} \tag{1.4}$$

was proposed.

Such a procedure considerably improves the convergence of ERE [50]; nevertheless, it does not allow us to take universally into account other polar and non-polar singularities of the function $X_l(k^2)$ . In this connection, in a number of works [4, 5, 6, 10, 26] a parametrization of $X_l(k^2)$ with the help of Padé approximants was proposed:

$$X_l(k^2) \approx X_l^{[N,M]}(k^2) = \frac{P_N^{(l)}(k^2)}{Q_M^{(l)}(k^2)} . \tag{1.5}$$

Such a parametrization is very reliable and universal because it allows not only to reproduce the poles of $X_l(k^2)$ by zeros of $Q_M^{(l)}(k^2)$, but also to simulate the cuts of this function (see Sections 1.2 and 1.3). Furthermore, the PA convergence region (for $N$, $M \to \infty$) is much larger than the ERE convergence region.

Note that the representation (1.5) can be obtained from the theory of potential scattering. Indeed, let us examine the expression for the partial-wave $K$--matrix [7]:

$$K_l(k) = \mathrm{tg}\, \delta_l(k) = \int\limits_0^\infty j_l(kr) V(r) \psi_l^{(+)}(k, r)\, dr . \tag{1.6}$$

The scattering wave function $\psi_l^{(+)}$ can be expressed in terms of the regular solution for the Schrödinger equation $\psi_l(k, r)$ (see (2.2.24)–(2.2.27)):

$$\psi_l^{(+)}(k, r) = \frac{\psi_l(k, r)}{f_l(k)} \tag{1.7}$$

where $f_l(k)$ is the Jost function. The functions $\psi_l$ and $f_l(k)$ are known to be entire functions of $k$ (for finite potentials in the entire $k$-plane and for the Yukawa-type potentials in the region near the corresponding threshold) [8], i.e.

$$\psi_l(k, r) = \sum_{n=0} k^{l+n} C_{l,\,n}(r)$$

and

$$f_l(k) = A(k^2) + ik^{2l+1}B(k^2) \tag{1.8}$$

where $A(k^2)$ and $B(k^2)$ are the entire functions of $k^2$. Now, taking into account in (1.6) that

$$j_l(kr) = (kr)^l \sum_{m=0}^{\infty} b_m(kr)^{2m} \, ,$$

substituting (1.7) and (1.8) in (1.6), and representing $A(k^2)$ and $B(k^2)$ as a series, we find a fractional-rational representation of the (1.5) type for $X_l(k^2)$.

Thus, the $S$-matrix parametrization is now reduced to determining the coefficients of the polynomials $P_N^{(l)}$ and $Q_M^{(l)}$ in the PA (1.5).

Here we have three possibilities:

(1) If there is some reliable theoretical procedure allowing us to calculate, to within a high accuracy, the coefficients $\{b_{ij}\}_{i=1}^{p}$ of ERE (1.3) for sufficiently high $p$, we can use the formalism of type I PA (see Section 1.2) and calculate the coefficients $P_N^{(l)}$ and $Q_M^{(l)}$ in terms of given coefficients $\{b_{ij}\}_{i=1}^{N+M+1}$ using formulae (1.5) and (1.3).

(2) If the scattering phases $\{\delta_i^l\}$ are calculated or obtained from experimental data with very small errors $\{\varepsilon_i^l\}$, then, to construct the PA (1.5), we can use the technique of the type II PA (see Section 1.2). To do this, the values of $\{\delta_i^l\}$ and $\{E_i\}$ are recalculated to the values $\{X_i^l\} = \{X_l(k_i^2)\}$ $(k_i^2 = 2\mu E_i/\hbar^2)$ and then the coefficients of the polynomials $P_N^{(l)}$ and $Q_M^{(l)}$ are found by solving the system of linear equations

$$X_i^l Q_M^{(l)}(k_i^2) = P_N^{(l)}(k_i^2) \, , \qquad i = 1, 2, ..., N + M + 1 \, .$$

Here, of course, from the p values of the initial data, we use only $N + M + 1$ values.

(3) If it is desirable to use all the $p$ values of the initial data in the parametrization, or if the errors $\{\varepsilon_i^l\}$ are sufficiently high, then to find the coefficients of the PA (1.5) we use the technique of the type III PA described in detail in Section 1.2.3.

Parametrizing $X_l(k^2)$ in such a way, we obtain the following expression for the partial-wave $S$-matrix:

$$S_l(k) = \frac{P_N^{(l)}(k^2) + ik^{2l+1}Q_M^{(l)}(k^2)}{P_N^{(l)}(k^2) - ik^{2l+1}Q_M^{(l)}(k^2)} \, . \tag{1.9}$$

The $S$-matrix poles are determined by zeros of the denominator of the expression (1.9). Thus, in the case of short-range interaction the determination of the energy of $b$ and $a$-states and also of resonance energy and width reduces to the solution of a rather simple algebraic equation

$$P_N^{(l)}(k^2) - ik^{2l+1}Q_M^{(l)}(k^2) = 0 \, . \tag{1.10}$$

The only physically meaningful solutions for (1.10) are those that are stable as the PA order (i.e. $N$ and/or $M$) increases at least within certain intervals $N_1 \leq N \leq N_2$ and $M_1 \leq M \leq M_2$. The character of the state (i.e. whether it is bound, virtual, or of the resonance type) is unambiguously defined by the position of the root of this equation in the $k$-plane. After finding the poles, the $S$-matrix (1.9) residues are not difficult to calculate. Note that, in contrast to resonances, the energies of bound states are often known to within a high accuracy. In such cases equation (1.10) can be used as an additional condition (or a constraint) on the coefficients of the polynomial $P_N^{(l)}$ and $Q_M^{(l)}$. Introducing such additional conditions improves the accuracy and stability of the analytic continuation of the $S$-matrix in the complex $k$-plane and, which is of major importance, improves radically the accuracy of the determination of the $S$--matrix residue at the appropriate poles (see the example discussed below).

Since the nuclear interaction potential does not show any sharp cut-off, but behaves at $r > R$ $(R = \mu^{-1}$ is the radius of nuclear forces) in the same way as the Yukawa potential $\exp(-\mu r)/\mu r$), the Jost function $f(k)$ has a "dynamic" cut in the lower $k$-half-plane beginning at $k_c = -i\mu/2$. However, as the Padé approximants are able to represent the analytic function in the *entire* region of analyticity (at $N \to \infty$ and $M \to \infty$), the approximation (1.5) is still correct (we remind the reader that even if $\delta_l(k) \neq \pi n$, $n = 0, \pm 1, ...,$ the ERE covergence radius of (1.3) does not exceed $\mu/2$).

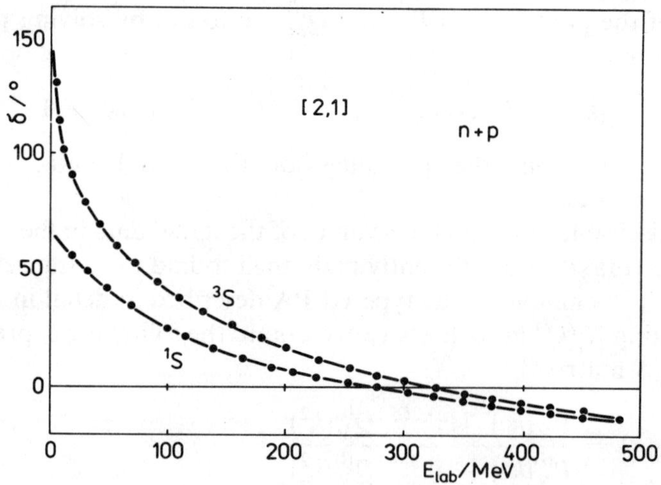


Fig. 6.1

Padé [2/1] fit of the singlet and triplet  n — p  phase shift.

As examples illustrating the approach described above for the case of short--range interaction we shall consider neutron–proton scattering in the $s$-wave and neutron–alpha-particle scattering in the $p$-wave. The singlet neutron–proton scattering (i.e. when the total spin of the system equals zero) is interesting because in this case the scattering phase shift is known from numerous phase shift analyses made in a broad energy range and behaves as follows. At near--zero energies the singlet $s$-wave phase shift of n-p scattering increases abruptly from zero to almost 60 °, then it decreases, goes through zero at $\sim 270$ MeV (in laboratory system), and becomes negative afterwards. Such behaviour of the phase–shift cannot be described by ERE (1.3), but must be expressed using either a modified formula of (1.4) type or the representation (1.5). This applies equally to the neutron–proton triplet (i.e. with the total spin $S = 1$) $^3S_1$ phase shift which behaves more regularly at low energies (decreases monotonically from $\pi$), but also runs through zero at $\sim 320$ MeV (in laboratory system). The phase-shift analysis results [11] were used here as basic input data.

Fig. 6.1 shows the description quality of the $^1S_0$ and $^3S_1$ phase shifts of n-p scattering obtained with the help of the parametrization (1.5) (the solid lines) together with the results of the appropriate phase shift analysis (the black points). It can clearly be seen that the approximant [2, 1] containing only 4 real parameters (the parameters of the PA were found using the technique of type III PA described in Section 1.1.2) gives a perfect description of the phase shifts in a broad energy range (0–500 MeV). Therefore, in this energy range the partial-wave $S$-matrix as a function of energy is described by a single simple analytic expression (1.9). The $S$-matrix poles were obtained from (1.10) (see Table 6.1). In the $^3S_1$ channel the nearest pole at energy $\varepsilon_d \sim -2.2$ MeV corresponds to the deuteron. Considering the smallness of the PA order used

Table 6.1

The effective-range expansion parameters $a$, $r_0$, and $P$; the positions of the partial-wave $S$-matrix poles $E_0$; the pole residues $C$ for the $^3S_1$ and $^1S_0$ n-p scattering states.

| Method used | $a$ (fm) scattering length | $r_0$ (fm) effective range | $P$ (fm$^3$) form parameter | $E_0$ (MeV) the pole position | Residue of $S$-matrix pole |
|---|---|---|---|---|---|
| | | $^3S_1$ channel | | | |
| Ref. [2] | $5.399 \pm 0.011$ | $1.732 \pm 0.012$ | – | $-2.22$ | $-i \, . \, 0.8[20]$ |
| Present method Ref. [10] | 5.43 | 1.68 | $-0.806$ | $-2.14$ | $-i \, . \, 0.72$ |
| | | $^1S_0$ channel | | | |
| Ref. [2] | $-23.68 \pm 0.028$ | $2.46 \pm 0.12$ | – | $-0.066$ | – |
| Present method Ref. [10] | $-23.52$ | 2.46 | $-1.26$ | $-0.068$ | $i \, . \, 0.074$ |

here, we must admit that the deuteron energy and the $S$-matrix residue at the deuteron pole are reproduced sufficiently well. In the ${}^1S_0$ channel the $S$-matrix pole lies in the $k$-plane on the negative imaginary semi-axis and corresponds to

Table 6.2

The convergence of the partial-wave $S$-matrix pole positions as the PA order increases for the $p_{3/2}$ adn $p_{1/2}$ n $-$ $\alpha$ scattering. Three lower lines are the pole positions predicted by other methods [14, 15] and the values presented in the standard compilation [13].

| PA order | $p_{3/2}$ | | $p_{1/2}$ | |
|---|---|---|---|---|
| | Re $z_0$ , MeV | $-$ Im $z_0$ , MeV | Re $z_0$ , MeV | $-$ Im $z_0$ , MeV |
| 1.1 | 0.772 55 | 0.322 27 | 1.831 0 | 3.135 7 |
| 2.2 | 0.772 59 | 0.322 46 | 1.987 4 | 2.598 8 |
| 3.3 | 0.772 73 | 0.322 33 | 1.988 2 | 2.605 7 |
| 4.4 | 0.772 71 | 0.322 53 | 1.971 0 | 2.610 3 |
| 5.5 | 0.772 73 | 0.322 47 | 1.975 7 | 2.610 2 |
| Ref. [13] | 0.89 $\pm$ 0.05 | 0.3 $\pm$ 0.02 | 4.9 $\pm$ 1.5 | 1.0 $\pm$ 0.4 |
| Ref. [14] | 0.777 8 $\pm$ 0.002 | 0.319 6 $\pm$ 0.006 | 1.999 $\pm$ 0.04 | 2.267 $\pm$ 0.08 |
| Ref. [15] | 0.771 4 $\pm$ 0.002 | 0.321 9 $\pm$ 0.003 | 1.969 5 $\pm$ 0.016 | 2.609 $\pm$ 0.08 |

the $a$-state. This state (the virtual deuteron) is located just near the threshold (see Table 6.1) and causes a sharp change of the ${}^1S_0$ phase shift near the threshold. Table 6.1 lists also the parameters of the effective-range expansion obtained from the expression (1.5) and compares them with the results of other works.

As a second model example we shall examine the n–$\alpha$ elastic scattering with angular momentum $j = 1/2$ and $j = 3/2$. This case is of interest because the $p$-wave channels $p_{1/2}$ and, especially, $p_{3/2}$ have pronounced resonances. The resonance is narrow $\left(E_R > \Gamma\right)$ in the $p_{3/2}$ channel and broad in the $p_{1/2}$ channel $\left(\Gamma > E_R\right)$. Moreover, this case was subjected to sufficiently high-accuracy phase shift analyses [12] and to single energy phase shift analysis aimed, among other things, at deriving the ${}^5$He resonance state parameters. The results obtained are summarized in Table 6.2 which demonstrates that the stability and convergence of the resonance parameters is preserved with increasing PA order in (1.5) and gives a comparison with the resonance parameters obtained in other works [13–15]. From Table 6.2 it may be seen that the results of the standard (that is, single-level) $R$-matrix analysis [13] may differ from the true resonance parameters even in the case of rather narrow resonances, such as the $p_{3/2}$ level of the ${}^5$He nucleus.

### 6.1.2. Long-Range Forces

As mentioned above, the scattering function $X_l(k^2) = k^{2l+1} \cot \delta_l$ is an analytic function of energy near the point $k = 0$ in the case of short-range interactions only. At the same time, in the case of long-range interaction between the particles a and A which shows a power-law behaviour at great distances, for example, $r^{-1}$ (Coulombic interaction) or $r^{-4}$ (polarization interaction of charge--induced dipole type), the approach explained above must be modified because in this case the function $X_l(k^2)$ exhibits a singularity at the point $k = 0$.

First, we shall examine Coulombic interaction, which is the most important of the long-range interactions. We shall assume that the interaction between two charged particles with charges $z_1$ and $z_2$ is not only Coulombic but also nuclear. In this case the modified scattering function $X_l^c(k^2)$ will be an analytic function of energy instead of $X_l(k^2)$:

$$X_l^c(k^2) = \frac{k^{2l+1}C_l^2\left[C_0^2(\cot\delta_l - i) + 2\eta H(\eta)\right]}{C_0^2} \tag{1.11}$$

where $\delta_l$ is now the Coulomb modified nuclear phase shift corresponding to the standard decomposition of the total phase $\delta_l^{\text{total}} = \delta_l + \tau_l^c$; ($\tau_l^c$ is the purely Coulombic phase shift); $\eta = z_1z_2e^2\mu/\hbar^2k$ is the Coulombic parameter; $C_0$, $C_l$ and $H(\eta)$ are the following standard functions [1, 17, 56]:

$$C_0^2 = \frac{2\pi\eta}{e^{2\pi\eta} - 1},$$

$$C_l^2 = C_{l-1}^2\left(1 + \frac{\eta^2}{l^2}\right),$$

and

$$H(\eta) = \psi(i\eta) + \frac{1}{2i\eta} - \ln\left[-i\eta\,\text{sign}\,(-z_1z_2)\right];$$

$$\psi(z) = -C - z^{-1} + z \sum_{m=1}^{\infty} \left[m(m + z)\right]^{-1} \tag{1.12}$$

is the digamma function; $C$ is the Euler constant.

Indeed, from the generalized (to the Coulombic case) theory of effective range expansion [17, 56] it is known that the function $X_l^{(c)}(k^2)$ is an analytic function of $k^2$ in some region including the point $k^2 = 0$ and can be series-expanded at its neighbourhood to take the form of a modified ERE:

$$X_l^c(k^2) = -\frac{1}{a_l^c} + \frac{1}{2}r_l^ck^2 - \frac{1}{4}P_l^ck^4 + \ldots . \tag{1.13}$$

For this function we can utilize the PA representation (1.5). Furthermore, the parametrization technique coincides with that described above in the case of a short-range potential where $X_l$ is replaced by $X_l^c$. Thus, the $S$-matrix poles are now determined by the zeros of the expression

$$C_l^2 k^{2l+1} \cot \delta_l - iC_l^2 k^{2l+1} = X_l^c(k^2) - 2\eta k^{2l+1} \frac{C_l^2 H(\eta)}{C_0^2} \qquad (1.14)$$

and, after substituting (1.5) in (1.14), we obtain the following transcendental equation to search for the $S$-matrix poles:

$$P_N^{(l)}(k^2) - 2\eta k^{2l+1} C_l^2 H(\eta) Q_M^{(l)}(k^2)/C_0^2 = 0 \qquad (1.15)$$

which can, nevertheless, be solved rather simply using, for example, a convenient algorithm based also on the PA technique (see formula (1.2.36)).

All of the expressions (1.11)–(1.15) are correct in some region of the complex $k$-plane, i.e. on the physical and nonphysical energy sheets. However, to find the coefficients of the PA (1.5) it is sufficient to know $X_l$ or $X_l^c$ at positive energies for which the scattering data are available. At real energies the expression for $X_l^c$ is simplified:

$$X_l^c(k^2) = k^{2l+1} C_l^2 \left( \cot \delta_l + 2\eta h(\eta)/C_0^2 \right) \qquad (1.16)$$

because [17]

$$\lim_{Im\, k \to 0} Im \left[ 2\eta H(\eta) \right] = i\, C_0$$

and

$$\lim_{Im\, k \to 0} Re \left[ H(\eta) \right] =$$

$$= -C - \ln \eta + \eta^2 \sum_{m=1}^{\infty} \left[ m(m^2 + \eta^2) \right]^{-1} \equiv h(\eta) .$$

Let us now explore some examples illustrating the efficiency of the above parametrization. First we shall describe proton-proton scattering in the singlet $^1S_0$ channel. The results of the phase shift analysis are also used here as input data. Just as in the case of the p-n system, the [2, 1] PA yields a high-quality description of experimental p-p phase shifts in a broad energy range $E = 0$–400 MeV (see Fig. 6.2). Table 6.3 gives the parameters of effective-range expansion found using the [2, 1] PA. In the singlet $^1S_0$ channel, for the proton-proton system, as well as in the singlet $s$-wave channel for the n-p system, we find the $S$-matrix pole on the nonphysical sheet. In the given case, however, the addition of the Coulombic interaction will shift the virtual $a$-state of the n-p pair (in the singlet $s$-wave channel) to the complex energy plane. The energy of this state, which was $E_0^{np} \approx -0.067$ MeV for the n-p system, now becomes (i.e. for the diproton case) $E_0^{pp} = (-0.146{-}i0.477)$ MeV. The energy of this $S$-wave singlet diproton state was also found by Kok [18] who used a much more complicated

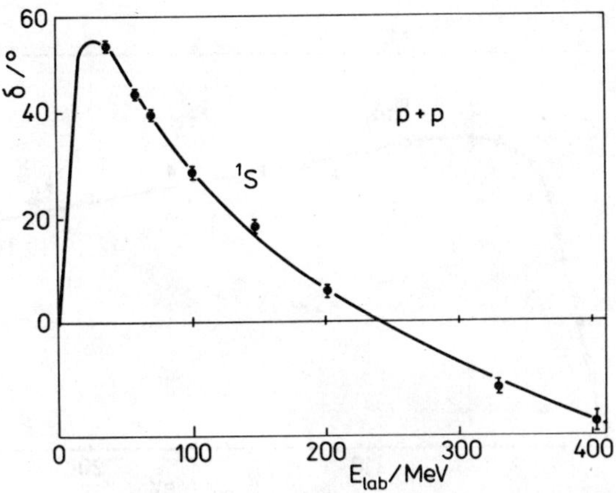



Fig. 6.2

Padé [2/1] fit of the $^1S_0$ p − p phase shift.

method and obtained a similar value of $E_0^{pp} = (-0.140{-}i0.467)$ MeV. We note that in the case of p-p scattering the singularity of the $S$-matrix lies farther from the threshold $k = 0$ than in the case of the singlet $^1S_0$ "deuteron" pole. In accordance with this, although the $^1S_0$ p–p phase shift is qualitatively similar to the $^1S_0$ n–p shift, it increases less markedly at low energies (see Fig. 6.2).

Table 6.3

The parameters of effective range expansion for p–p scattering.

| Method | $a$ (fm) scattering length | $r_0$ (fm) effective range | $P$ (fm$^3$) the form parameter |
|---|---|---|---|
| Ref. [2] | $-7.826 \pm 0.05$ | $2.786 \pm 0.01$ | – |
| Present method, Ref. [10] | $-7.72$ | $2.68$ | $0.191$ |

As a second illustrative example we shall examine the scattering of protons by an alpha particle in the $p_{3/2}$ channel. The scattering phase shifts obtained in [11, 12] were used here as input data. Fig. 6.3 shows the description quality of the experimental p–α phase shifts gained with the approximation (1.5) and the representation $(1.11)$ using the [2, 1] PA. Table 6.4 gives the parameters of the effective range and the resonance characteristics obtained by different methods.

The procedure described here was used in the same manner to parametrize the $\alpha$–$\alpha$ scattering phase shifts in partial waves $l = 0,2,4$. Fig. 6.4 depicts three phase

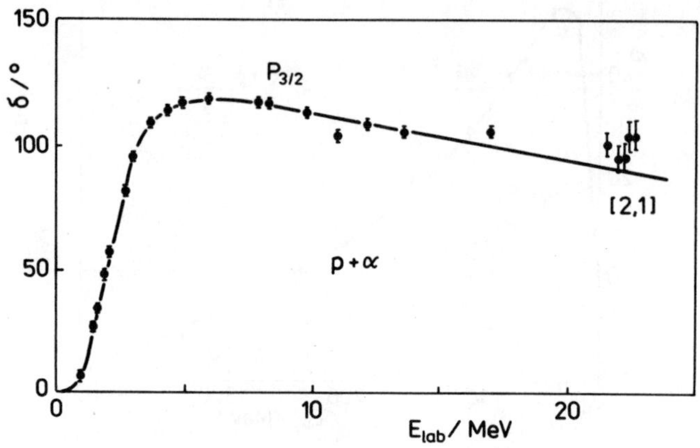


Fig. 6.3
Padé [2/1] fit of the $p_{3/2}$ $p - \alpha$ phase shift.

shifts of the $\alpha$–$\alpha$ scattering with $l = 0,2,4$ approximated by the method described above (the [2,1] PA was used again) in the 0–28 MeV energy range. The results of the phase shift analyses presented in the review [19] were used as input data (see also [37]). Table 6.5 gives the parameters of the known $2^{+}$ and

Table 6.4

The parameters of effective range expansion and the resonance energy for $p - {}^{4}He$ scattering

| Method | $a$ (fm) scattering length | $r_0$ (fm) effective range | $P$ (fm³) the form parameter | Re $z_0$ (MeV) | $-$Im $z_0$ (MeV) |
|---|---|---|---|---|---|
| Ref. [16] | $-44.83 \pm 0.51$ | $-0.365 \pm 0.013$ | $-2.39 \pm 0.15$ | – | – |
| Ref. [13] | – | – | – | $1.97 \pm 0.05$ | 0.4 |
| Ref. [14] | | | | $1.637 \pm 0.005$ | 0.646 $\pm$ 0.003 |
| Present method Ref. [10] | $-44.84$ | $-0.363$ | $-2.02$ | $1.64 \pm 0.01$ | 0.65 $\pm$ 0.01 |

$4^+$ $\alpha$–$\alpha$ resonances found by the above-mentioned method and compares them with the results of the $R$-matrix parametrization [13]. Since the $\alpha$–$\alpha$ resonances are highly pronounced, we could expect that the $R$-matrix parametrization

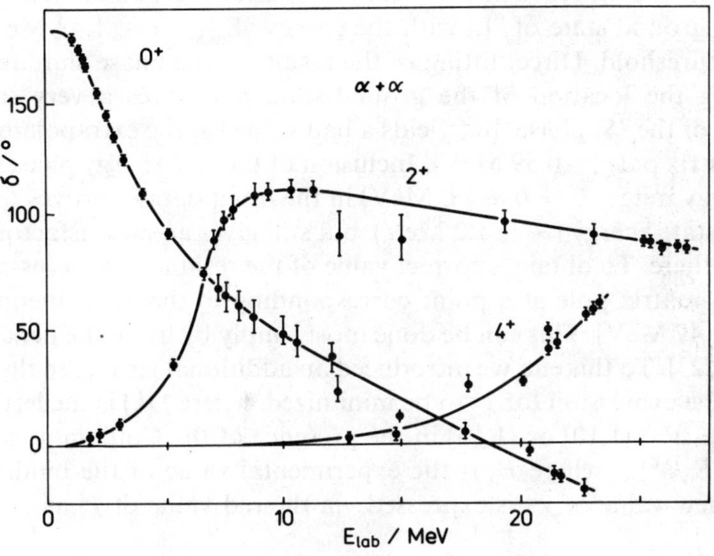


Fig. 6.4
Padé [2/1] fit of the $0^+$, $2^+$ and $4^+$ $\alpha - \alpha$ phase shifts.

would prove to be quite effective. In the case of $4^+$ resonance, however, the single-level $R$-matrix approximation turns out to be already quite unsatisfactory, so the result in Table 6.5 corresponds to the two-level approximation.

Table 6.5

The parameters of $2^+$ and $4^+$ $\alpha - \alpha$ resonances.

| Method | $2^+$ | | $4^+$ | |
|---|---|---|---|---|
| | Re $z_0$ (MeV) | $-$Im $z_0$ (MeV) | Re $z_0$ (MeV) | $-$Im $z_0$ (MeV) |
| Ref. [13] | $2.94 \pm 0.03$ | $0.78 \pm 0.05$ | $11.4 \pm 0.3$ | 1.75 |
| Present method, Ref. [10] | 2.92 | 0.62 | 11.4 | 1.7 |

Now we shall try to realize the important possibility offered by this approach to use independent information about the bound-state energies for improving $S$-matrix approximation. For this purpose we shall examine the deuteron-alpha scattering data. To stick to the single-channel problem, we have neglected an unimportant (everywhere, except a small interval $E \sim 5$–$7$ MeV) tensor mixing of $^3S_1$ and $^3D_1$ channels and also the deuteron break-up channel. The results of the phase shift analyses [21] (in the energy range $E = 3$–$12$ MeV) were used as input data. In the $^3S_1$ wave in the $\alpha$–d system there is a bound state corresponding to the ground state of $^6$Li with the energy $E_{\text{gr.st.}} = -1.47$ MeV relative to the $\alpha$–d threshold. Direct fitting of the results of the phase shift analysis [21] disregarding the location of the ground-state pole gives a very satisfactory description of the $^3S_1$ phase, but yields a bad value for the extrapolated location of the $S$-matrix pole $(-0.59$ MeV$)$. Inclusion of the low-energy phases from [22] (in the energy range $E = 0.8$–$1.6$ MeV) in the input data improves the value of the bound-state energy $(\sim -1.2$ MeV$)$, but still gives an unsatisfactory value of the residue there. To obtain a correct value of the residue, it is necessary at least to fix the $S$-matrix pole at a point corresponding to the experimental binding energy $(-1.47$ MeV$)$. This can be done most simply by using the penalty-function method [23]. To this end we introduced an additional term (penalty function) $|f_l(k_0)|^2$ to the expression for $\chi^2$ to be minimized, where $f_l(k)$ is the left-hand side of the expression $(1.10)$ or $(1.15)$ in the presence of the Coulombic interaction, $k_0 = \sqrt{(2\mu E_0/\hbar^2)}$ , where $E_0$ is the experimental value of the binding energy. Thus, the new value of $\tilde{\chi}^2$ is expressed via the old value of $\chi^2$ as

$$\tilde{\chi}^2 = \chi^2 + \xi \, |f_l(k_0)|^2 \, ,$$

where $\xi$ is a constant controlling the magnitude of the penalty. At a sufficiently high value of $\xi$ the minimization of $\tilde{\chi}^2$ provides a sufficiently accurate fulfilment of the equality $(1.10)$ or $(1.15)$ at the point $k_0$, i.e. for the required location of the $S$-matrix pole. After minimizing the new function $\tilde{\chi}^2$ for the squared $S$-matrix residue at the pole corresponding to the ground state of $^6$Li, a very resonable value of the squared asymptotic normalization $C = 3.6$ was obtained. The value of the squared residue $C^2$ is proportional to the vertex constant $G^2$ for the virtual decay of $^6$Li $\rightarrow \alpha + $d. The value of the residue obtained here corresponds to $G^2 \cong 0.18$ and agrees with the results obtained by other authors (see the tabulated data in [24]) in their analyses of other nuclear reaction data[1]).

In essentially the same way, i.e. by introducing an appropriate penalty function, we can carry out the necessary fitting allowing for other restrictions imposed on the behaviour of the $S$-matrix (or of the Jost function) by experimental data or by additional analyticity requirements.

---

[1]) It is interesting that the very accurate theoretical calculations of the vertex constant carried out in terms of the three-particle $\alpha + $n$ + $p model for the $^6$Li nucleus [37–39] have yielded a somewhat different value of the squared residue, $G^2_{\text{Li}\rightarrow\alpha d} = 0.84$. The reason for the difference is not yet known.

### 6.1.3. Polarization-Type Long-Range Interaction

Now, let the peripheral part of the interaction potential exhibit a general exponential dependence $r^{-n}$. In the case of dipole interaction $(n = 2)$, the analytic behaviour of the Jost function $f_l(k)$ and of the $S$-matrix $S_l(k)$ near $k = 0$ was analysed in many works (see, for example, [40]). A number of the cases important in practice (including the case of a nonspherically-symmetric short-range interaction [58] and the case where such a potential is superposed on a long-range Coulombic interaction [59]) have been examined elsewhere. The general case of a superposition of a spherically-symmetric short-range interaction, a long-range polarization potential, and a Coulomb interaction (i.e. $V = V_s + ar^{-1} + br^{-4}$) was studied recently [57, 60, 61].

Here we dwell in more detail on the so-called polarization potential $(n = 4)$ arising in case of interaction between a charged particle and a neutral atom (i.e. in the charge-induced dipole interaction) at low energies and used extensively in atomic physics. The interaction between an atom and a charged particle consists of two parts. At small distances $r < d$ (roughly speaking, $d$ is the "radius" of the atom) the interaction is very complicated because the nonlocal exchange forces play an important role. At greater distances $r > d$, the interaction may be represented, to a high accuracy, by a simple local interaction $V_p$

$$V_p = -\frac{\beta^2}{r^4}, \qquad r > d \tag{1.17}$$

($\beta$ is called the polarizability of the atom). If the polarization interaction is eliminated, i.e. $\beta = 0$, the effective-range expansion is of the standard form

$$X_0(k) = k \cot \delta_0 = \frac{1}{a_0} + \frac{1}{2} r_0 k^2 + \ldots \tag{1.18}$$

because the underlying interaction is a short-range one. If, however, the polarization interaction is included, the situation becomes much more complicated. In this case the effective-range expansion is [25]

$$X_0^p(k) = k \cot \delta_0 = -\frac{1}{a} + Bk + Ck^2 \ln \frac{\beta k}{4} + r_0'^2 k^2 + \ldots \tag{1.19}$$

Unlike the case of short-range interactions, this is not an analytic function of $k^2$ near $k = 0$, as it contains a logarithmic singularity. As shown in [10, 26], the function $X_0^p(k)$ should be replaced by the function $\psi_0^p(k)$ defined as

$$\psi_0^p(k) = \frac{\left(1 - \dfrac{c(k)}{k}\right) X_0^p(k) - c(k)}{1 + \dfrac{b(k)}{k} + X_0^p(k) \dfrac{a(k)}{k^2}} \tag{1.20}$$

where

$$a(k) = -\beta^2 \int_d^\infty \frac{\sin^2 kr}{r^4} \, dr \, ,$$

$$b(k) = -\beta^2 \int_d^\infty \frac{\sin kr \cos kr}{r^4} \, dr \, ,$$

$$c(k) = -\beta^2 \int_d^\infty \frac{\cos^2 kr}{r^4} \, dr \, . \tag{1.21}$$

The function $\psi_0^p(k)$ is analytic near $k = 0$ and may be continued analytically in order to seek the $S$-matrix resonance poles. In deriving (1.20) we made use of the fact that the polarization is very weak (for many sorts of atoms) compared with nonlocal short-range interaction and that, at a sufficiently large value of d, it may be treated as a perturbation. In the expressions (1.19) and (1.20) the polarization interaction is treated to the lowest order in $\beta^2$.

As an example we shall consider the $^1S$ resonance of $H^-$. The phase shifts have been published [27] and are very accurate. We set $d = 8a_0$ ($a_0$ is the Bohr radius) and $\beta^2 = 4.5a_0^2$ . The resonance position found is [10]

$$E_r = 9.5573 \text{ eV}, \qquad \Gamma = 0.0472 \text{ eV}.$$

This agrees well with the result obtained by the Feshbach projection technique [27]:

$$E_r = 9.55735 \text{ eV}, \qquad \Gamma = 0.04717 \text{ eV}.$$

A similar approach is also applicable to other long-range interactions (van der Waals forces, etc.).

Thus, in this case also, the Padé approximant technique of analytic continuation of the partial-wave $S$-matrix proves to be very effective. It should be emphasized, however, that the stability and error of the $S$-matrix analytic continuation depend decisively on the accuracy and consistency of the input data for phase shift analyses. When the errors in these data are high, the convergence of the approximation is destroyed, so some additional *a priori* information about the character of the sought solution is necessary for the stability to be restored (i.e. for the results to be stable) [31–35].

### 6.2. Multichannel $S$-matrix parametrization

Going from the single-channel to multi-channel problem always involves considerable complications of the formalism. In the latter case, complications arise in two directions.

(1) The analytic properties of the $S$-matrix as a function of energy get complicated in that the cuts and the non-physical sheets get more numerous and the relationships between the sheets (hence the classification of the $S$-matrix poles and their relevance to the physical states of the system) also get complicated. All these problems are discussed in the excellent review [28] where exhaustive information may be found.

(2) Because of the above mentioned complications of the analytic properties of the $S$-matrix, we fail to reduce its parametrization to a single real analytic function of energy. Instead, we have to introduce either several real functions or a complex function. This leads to an abrupt increase in the number of parameters necessary for the scattering data to be fitted and, simultaneously, to a build--up of numerical difficulties.

We do not seek here to expound comprehensively the generalization of the above described formalism to the case of many channels or, in particular, to discuss the almost uncountable problems relevant to many-channel resonances. The interested reader is referred to the sufficiently detailed discussion in the review [28].

In this section we shall only outline some potentialities for generalizing the formalism described above to the case of several channels.

Above, when examining the single-channel case, we discussed the resonances in elastic scattering, i.e. the case when a resonance state decays to the same channel from which it arose (the nature of the final state remains the same as that of the initial state). Now we shall examine the case where the final and initial states may be different, $a + A \rightarrow b + B$. Thus, for example, in the $p + {}^4He$ reaction at a relative-motion energy above 18.4 MeV the process $p + {}^4He \rightarrow d + {}^3He$ may take place, in the $d + {}^6Li$ collision an inelastic scattering with excitation of resonance levels of ${}^6Li$ nucleus, $d + {}^6Li \rightarrow d + {}^6Li^*$, is possible, etc.

Let energy $E_1$ correspond to the threshold of the inelastic channel. Then, at the energy $0 \leq E \leq E_1$ of the relative motion of particles a and A, we shall observe the elastic scattering $a + A \rightarrow a + A$ only, while at $E > E_1$ two processes

$$a + A \rightarrow a + A$$
$$\searrow$$
$$b + B$$

will proceed simultaneously (in the particular case of inelastic scattering $b = a$ and $B = A^*$). For the sake of simplicity we restrict ourselves to the case of two channels (elastic and inelastic).

### 6.2.1. Complex Phase-Shift Method

The simplest way of describing an elastic scattering in the energy region where inelastic processes are possible, i.e. at $E > E_1$, is to introduce complex phase shifts. We shall briefly explain the meaning of this procedure. A scattering wave function in the partial-wave representation can be written (see Chapter 2, eq. (2.26)) as

$$\psi_l^{(+)}(k, r) \sim [f_l^{(-)}(k, r) - S_l(k) f_l^{(+)}(k, r)]$$

where $f_l^{(-)}$ and $f_l^{(+)}$ are the Jost solutions turning, in the asymptotic region, into ingoing and outgoing spherical waves, respectively. In the case of elastic scattering the incoming particle flux equals the outgoing particle flux (in the elastic scattering channel) and, therefore, $|S_l(k)| = 1$. If the inelastic channel is open, some of the particles are "absorbed" by, and "flow" into, this channel, so the outgoing particle flux in the elastic channel proves to be smaller than the ingoing particle flux, i.e. $|S_l(k)| < 1$. The flux attenuation in the entrance channel may be allowed for by introducing the complex phase shift $\Delta_l(k) = \delta_l(k) + i\varepsilon_l(k)$, thereby giving rise to the following $S$-matrix parametrization:

$$S_l(k) = e^{2i\Delta_l(k)} = \eta_l(k) e^{2i\delta_l(k)} \tag{2.1}$$

where $\delta_l(k)$ and

$$\eta_l(k) = e^{-2\varepsilon_l(k)} \tag{2.2}$$

are called the (real) phase shift and inelasticity parameter, respectively. Thus, the open inelastic channel influences not only the phase but also the amplitude (through the coefficient $\eta_l$) of the outgoing wave in the elastic channel. The case of $\eta_l = 1$ coresponds to the elastic scattering, and $\eta_l = 0$ to the maximum absorption when the reaction cross section is maximum. The behaviour of $\eta_l$ and $\delta_l$ in various physical situations is described in detail in the monograph [1]. It is of importance that in our case we can also write the $S$-matrix formula (2.1) which is formally analogous to the corresponding single-channel representation. Remaining within the formal analogy, we can reduce the $S$-matrix parametrization to the parametrization of the function $\tilde{X}_l(k)$

$$\tilde{X}_l(k) = k^{2l+1} \cot \Delta_l(k) \tag{2.3}$$

which proves, however, to be complex and has to be approximated by formula (1.5) using the Padé approximants with complex coefficients. The $S$-matrix will be of the form of (1.9), but the coefficients of the polynomials $P_N^{(l)}$ and $Q_M^{(l)}$ will also prove to be complex. Moreover, the second part of (2.1) may also be used to reduce the $S$-matrix parametrization to the parametrization of two real functions $\eta_l$ and $X_l(k) = k^{2l+1} \cot \delta_l(k)$. In this case $X_l(k)$ and, hence $e^{2i\delta_l}$, are approximated in a standard way and we can write an expression of the type of (1.5) for $\eta(k)$. Thus, the $S$-matrix takes the form

$$S_l(k) = \frac{A_K^{(l)}(k) \, P_N^{(l)}(k^2) + ik^{2l+1} Q_M^{(l)}(k^2)}{B_L^{(l)}(k) \, P_N^{(l)}(k^2) - ik^{2l+1} Q_M^{(l)}(k^2)} . \tag{2.4}$$

Such a method for parametrizing the multichannel problem is relatively simple
(evidently, it may be generalized without changes to the case where several
inelastic channels are open), so the multichannel problem reduces to an effective
single-channel problem. The shortcomings of the approach ensue from this
simplicity; in particular, the expression for the *S*-matrix obtained above can
hardly be continued to the energy region $E < E_1$ where the phase shift becomes
real, because the analytic properties of the phase (of the *S*-matrix) at the
threshold $E = E_1$ are not contained in the approximation (2.1). Therefore,
although we identify the *S*-matrix poles found in terms of such an approach with
the resonance states, it is often difficult to understand in which channel these
resonances can be found. To clarify such problems, additional investigations
have to be carried out [1].

### 6.2.2. *Multichannel S-matrix*

As mentioned above, the difficulties with the complex phase method arise from
insufficient allowance for the analytic properties of the multichannel *S*-matrix.
First, in the two-channel case (for simplicity we limit ourselves to this case
again) the *S*-matrix is a matrix of dimension $2 \times 2$, i.e.

$$S(E) = \begin{pmatrix} S_{11}(E) & S_{12}(E) \\ S_{21}(E) & S_{22}(E) \end{pmatrix}. \tag{2.5}$$

Second, in the $E$ plane the matrix elements $S_{ij}(E)$ have two cuts instead of one in
the single-channel case (see Fig. 6.5) and are functions of two momenta $k_1$ and $k_2$

$$\left(S_{ij} = S_{ij}(k_1, k_2)\right), \quad k_1 = \sqrt{2\mu_1 E/\hbar^2}$$

and

$$k_2 = \sqrt{2\mu_2(E - E_1)/\hbar^2},$$

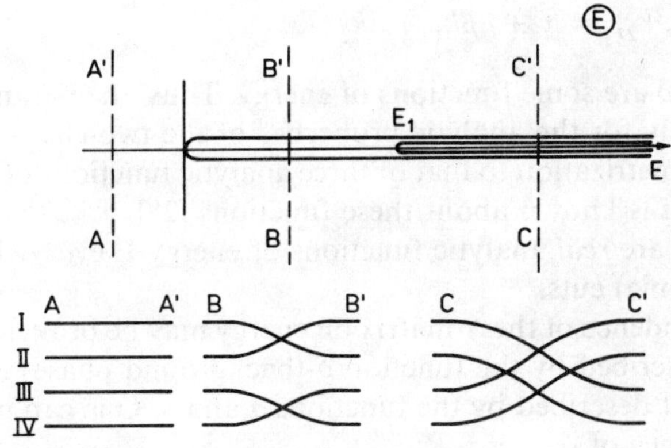


Fig. 6.5

The analytic structure of the two-channel *S*-matrix.

corresponding to the motion in the first and second channel, respectively. In accordance with this, the Riemannian energy surface in the two-channel case has four, rather than two, sheets:

$$
\begin{array}{llll}
\text{1st sheet} & \operatorname{Im} k_1 > 0\,, & \operatorname{Im} k_2 > 0\,, \\
\text{2nd sheet} & \operatorname{Im} k_1 < 0\,, & \operatorname{Im} k_2 > 0\,, \\
\text{3rd sheet} & \operatorname{Im} k_1 < 0\,, & \operatorname{Im} k_2 < 0\,, \\
\text{4th sheet} & \operatorname{Im} k_1 > 0\,, & \operatorname{Im} k_2 < 0\,.
\end{array}
\tag{2.6}
$$

Joining of the sheets in different places of the Riemannian surface is shown in Fig. 6.5. Thus, in the two-channel case, as many as three nonphysical sheets (2,3, and 4) arise in the problem instead of a single nonphysical sheet and the resonance $S$-matrix poles may be located on each of them (a detailed classification of singularities see in [28]). In the complex phase-shift method the difference between the nonphysical sheets is completely ignored.

For the two-channel $S$-matrix we may write the representation which alow completely for its analytic properties and for the structure of the Riemannian surface thus, for example, the Dalitz-Tuan representation (under the assumption that the angular momentum $l_2$ of the second channel is zero) is of the form:

$$
S_{11} = \frac{\alpha - i\beta k_1 + ik_2}{D(k_1, k_2)}\,,
$$

$$
S_{12} = S_{21} = -2i \exp(i\delta)\,\frac{\sqrt{\beta k_1 k_2}}{D(k_1, k_2)}\,,
$$

$$
S_{22} = \exp(2i\delta)\,\frac{\alpha + i\beta k_1 - ik_2}{D(k_1, k_2)}\,,
$$

$$
D(k_1, k_2) = \alpha + i\beta k_1 + ik_2
\tag{2.7}
$$

where $\alpha$, $\beta$, and $\delta$ are some functions of energy. Thus, the parametrization (2.7) allows completely for the analytic properties of the two-channel $S$-matrix and reduces its parametrization to that of three analytic functions of energy $\alpha$, $\beta$, and $\delta$. The following is known about these functions [28]:

(i) $\alpha$, $\beta$ and $\delta$ are *real* analytic functions of energy E everywhere outside the left-hand (dynamic) cuts.

(ii) The dependence of the $S$-matrix on energy may be broken into a slow and smooth part described by the function $\delta$ (background phase) and a rapid and pronounced part described by the functions $\alpha$ and $\beta$. One can often neglect the energy dependence of $\delta$.

(iii) The function $\beta$ is positive and at $k_1 \to 0$ we get $\beta \sim k_1^{2l_1}$ (we remind that $l_2 = 0$).

The simplest way of approximating the functions $\alpha$, $\beta$, and $\delta$ is to replace them by a truncated series. However, although this approximation is good for the very smooth function $\delta$, it proves to be effective for $\alpha$ and $\beta$ only in a narrow range near the energy at which the function is series-expanded. The authors of [28], for example, suceeded in describing a number of near-thrreshold resonances in nuclear and elementary-particle systems by series-expanding $\alpha$ and $\beta$ near the threshold of the inelastic channel. At the same time, the approximation cannot be used to describe the $S$-matrix in a sufficiently broad energy range or to continue it to a complex plane on nonphysical sheets sufficiently far (in case of broad resonances). In this case we again resort to the Padé approximation which can be effective in a much broader energy range compared with the truncated series. Thus, we can write the representation

$$\alpha(E) = \frac{P_N(E)}{Q_M(E)},$$

$$\beta(E) = \frac{k_1^{2l_1} R_L(E)}{S_K(E)} \tag{2.8}$$

for $\alpha$ and $\beta$ and a polynomial approximation

$$\delta(E) = \sum_{i=0}^{m} \alpha_i E^i \tag{2.9}$$

for $\delta(E)$. Now, the coefficients of polynomials $P_N$, $Q_M$, $R_L$ and $S_K$, as well as the coefficients of the series (2.9), are parameters used in fitting the scattering data. From the scattering data we obtain the values of the matrix elements $S_{kl}(E)$ at points $\{E_i\}_{i=1}^p$. As can be seen from (2.7), the total fitting of the $S$-matrix can no longer be reduced to a problem which is linear with respect to the PA coefficients. If we are interested, however, in the elastic channel only, we may limit ourselves to fitting the element $S_{11}$ only:

$$S_{11}(E) = \frac{P_N(E)S_K(E) - ik_1^{2l_1+1}R_L(E)Q_M(E) + ik_2 S_K(E)Q_M(E)}{P_N(E)S_K(E) + ik_1^{2l_1+1}R_L(E)Q_M(E) + ik_2 S_K(E)Q_M(E)} =$$

$$= \frac{p_n(E) - ik_1^{2l_1+1}q_m(E) + ik_2 r_s(E)}{p_m(E) + ik_1^{2l_1+1}q_m(E) + ik_2 r_s(E)} \tag{2.10}$$

where $p_n$, $q_m$ and $r_s$ are the polynomials of powers $n$, $m$, and $s$, respectively. As in the single-channel case, the fitting of this matrix element may be reduced to a problem which is linear with respect to the coefficients of the polynomials.

In such a parametrization, the determination of resonance $S$-matrix poles is reduced to solving the transcendental equation

$$\mathcal{D}(k_1, k_2) = p_n(E) + ik_1^{2l+1}q_m(E) + ik_2 r_s(E) = 0 . \tag{2.11}$$

The solutions for this equation on the corresponding nonphysical sheets (see eq. (2.6)) are identified with resonance states of various types (the classification in [28]).

A simplified version of the procedure discussed, namely the approximation of the multichannel $S$-matrix through the multichannel generalization of the effective-range theory [28, 41–43], was widely used for describing and deriving the parameters of multichannel resonances in the systems $\Lambda + N \rightleftarrows \Sigma + N$, $T + D \rightarrow {}^4He + n$, etc. (see [28, 49] and the references therein). Unfortnunately, because of the convergence limitations inherent to the effective-range approximation, the parametrization of the multichannel $S$-matrix is possible only in rather narrow energy intervals near certain thresholds. The general approximation of the multichannel $S$-matrix in a broad energy range implies a generalization similar to that described briefly above.

In contrast to the multichannel $S$-matrix parametrizations which are based almost entirely on a knowledge of the analytic properties of the multichannel $S$-matrix and, therefore, are model-independent to the maximum possible extent, the multichannel $R$-matrix parametrizations based on the canonical Wigner -Eisenbund ideology [44] have been developed to a very high degree, especially in nuclear physics (see also the exhaustive classical review [45]). Generally, the $R$-matrix approach is more model-dependent than the $S$-matrix approach because it involves the fairly arbitrary parameters, namely, the channel radius and the logarithmic derivative of the inner wave functions at the boundary. In many cases, however, the $R$-matrix approach yields a more practical scheme for parametrizing the multichannel amplitudes of the resonance nuclear reactions compared with the $S$-matrix method. In particular, the $R$-matrix method was extensively used for the purposes of phase shift analyses of few-nucleon systems by the well-known Hale-Dodder group in Los Alamos [46–48]. The basic problems arising in this case reduce to the question of how the substantial non-resonance background and the contribution of direct nuclear reactions can be correctly approximated without involving any excessive increase in the number of variable parameters of the $R$-matrix. The relevant studies have shown that these difficulties can well be overcome in many cases.

It seems expedient to make some concluding remarks concerning the stability and reliability of the $S$-matrix parametrizations discussed above. The $S$-matrix parametrization and continuation to the complex plane examined here are a particular case of a more general mathematical problem of analytic extrapolation of the analytic function. There exist two basic methods for making the analytical continuation of a function prescribed on a finite set of points. The first method is to prescribe a convenient ansatz for the function to be approximated. The second method is to continue directly from the data. Generally speaking, extrapolation of this kind is unstable. This problem was briefly discussed in Section 1.3.3 (the detailed discussion of this problem as applied to the extraction of the resonance parameters from experimental data treated here see also in

[51–55] where the references to earlier works may be found). The main efforts in the field are aimed at finding the criteria and conditions for stabilizing the analytical continuation directly from the data [51–55]. For example, Ref. [29] gives an expression for the error of the extrapolation to some point $z_0$ of the analytic function $f(z)$ limited by the value $\varepsilon$ on the curve $\Gamma$ within the analyticity domain $D$ and by a constant $M$ in the rest of the domain and at its boundary $G$:

$$\Delta f(z_0) \sim \varepsilon^{\omega(z_0)} M^{1-\omega(z_0)}$$

where $\omega(z)$ is the real function which satisfies the following equation in the analyticity domain:

$$\frac{\partial^2 \omega(z)}{\partial x^2} + \frac{\partial^2 \omega(z)}{\partial y^2} = 0$$

with the values

$$\omega(z) = 1 \qquad \text{for} \quad z \in \Gamma,$$

$$\omega(z) = 0 \qquad \text{for} \quad z \in G.$$

From here it follows that any estimate of the extrapolation error requires either a knowledge of the limitation $M$ or the introduction of appropriate hypotheses concerning the properties of the function $f(z)$. Therefore, the particular extrapolation methods are usually constructed on the basis of the most comprehensive examination of the analytic properties of the function to be continued. This applies also to the direct method of optimal extrapolation proposed by Pišút and Prešnajder [30] and based on weighted dispersion relations and to the methods based on conformal mapping [31, 32, 33]. The more copious the information taken into account, the more stable becomes the extrapolation. In the light of the above reasoning, the stability of the results obtained in Section 6.1 becomes more understandable. Here, the greatest possible allowance is made for the analytic properties of the $S$-matrix in the low-energy range. Also, for analytic continuation we use the PA which was shown in Section 1.2.1 to automatically make a wide class of conformal transformations, thereby enlarging the extrapolation domain and increasing the extrapolation stability.


### References

1. NICHITIU F., *Analiza de fază în fizica interactiolor nucleare*, Editura Academiei RSR, Bucuresti, 1980.
2. Particle Data Group, Rev. of Particle Properties, Rev. Mod. Phys. *52* (1980), part 2.
3. BALL J. S., LEE P. S., SHAW G. L., Rev. *D7* (1973) 2789.
4. NICHITIU F., Rev. Roumaine Phys. *27* (1982) 15.

5. HARTT K., Phys. Rev. *C22*(1980) 1377, *C23* (1981) 2399, *C29* (1984) 695.
6. MALYAROV V. V., POPUSHOY M. N., Yad. Fiz. (in Russian) *18* (1973) 1140. [Sov. J. Nucl. Phys. *18* (1974) 586].
7. NEWTON R. G., Scattering Theory of Waves and Particles, Mc Graw-Hill, N. Y. 1982.
8. DE ALFARO V., REGGE T., Potential Scattering. North-Holland Amsterdam 1965.
9. ADHIKARI S. K., TORREÃO J. R. A., Phys. Lett. *119B* (1982) 245.
   ZANKEL H., MATCHELITSCH L., Phys. Lett. *132B* (1983) 27.
10. KRASNOPOĽSKY V. M., KUKULIN V. I., HORÁČEK J., Czech J. Phys. *B35* (1985) 805.
11. ARNDT R. A., HACKMAN R. H., ROPER L. D., Phys. Rev. *C15* (1977) 1002.
12. ARNDT R. A., LONG D. D., ROPER L. D., Nucl. Phys. *A209* (1973) 429.
13. AJZENBERG-SELOVE F., LAURITSEN T., Nucl. Phys. *A227* (1974) 1.
14. AHMED M. V., SHANLEY P. E., Phys. Rev. Lett. *36* (1976) 25.
15. BOND J. E., FIRK F. W. K., Nucl. Phys. *A287* (1977) 317.
16. ARNDT R. A., ROPER L. D., SHOTWELL R. L., Phys. Rev. *C3* (1971) 2100.
17. KOK L. P., HAERINGEN H., Two-Range Potential and Effective-Range Theory. Int. report 164, Inst. for Theor. Phys., University of Groningen, Groningen 1981;
    VAN HAERINGEN H., Charged Particle Interactions, Theory and Formulas, Coulomb Press, Leyden 1985.
18. KOK L. P., Phys. Rev. Lett. *45* (1980) 427.
19. AFZAL S. A., AHMED A., ALI S., Rev. Mod. Phys. *41* (1969) 247.
20. KERMODE M. W., ALLEN L. J., MC TAWISH J. P., J. Phys. G: Nucl. Phys. *8* (1982) 71.
21. RISLER R., BOERMA D., Nucl. Phys. *A242* (1972) 265.
22. BARIT I. Ya., DULKOVA L. S., KUSNETSOVA E. V., SOBOLEVSKY N. M., Izvestiya Acad. Nauk SSSR, ser. Phys. (Proc. Acad., Sci. of USSR) *48* (1984) No2, 380; see also KELLER L. G. Nucl. Phys. *A156* (1970) 465;
    SCHMELZBACH P. A. et al., Nucl. Phys. *A185* (1972) 193–213.
23. POLAK E., Computational Methods in Optimization (A Unified Approach), Series: Mathematics in Science and Engineering Vol.77, Academic Press, N. Y.–London 1977.
24. BLOKHINTSEV L. D., BORBÉLY I., DOLINSKY E. I., EChAYA (in Russian) *8* (1977) 1189. [Sov. J. Part. Nucl. 8 (1977) 485].
25. O'MALLEY T. F., ROSENBERG L., SPRUCH L., Phys. Rev. *125* (1962) 1300.
26. SASAKAWA T., HORÁČEK J., J. Phys. *B15* (1982) L169.
27. HO Y. K., BHATIA A. K., TEMKIN A., Phys. Rev. *A15* (1977) 1423.
28. BADALYAN A. M., KOK L. P., POLIKARPOV M. I., SIMONOV Y. A., Phys. Rep. *82* (1982) 31.
29. BERTERO M., VIANO G. A., Nuovo Cim. *30* (1965) 1915.
30. PIŠÚT J., PREŠNAJDER P., Nucl. Phys. *B12* (1969) 110.
31. CIULLI S., FISCHER J., Nucl. Phys. *24* (1961) 465.
32. FRAZER W. R., Phys. Rev. *123* (1961) 2180.
33. CUTKOSKY R. E., DEO B. A., Phys. Rev. Lett. *22* (1968) 1272; Phys. Rev. Lett. *174* (1968) 1 859.
34. PIŠÚT J., EChAYa (in Russian) *9* (1978), 602. [Sov. J. Part. Nucl. *9* (1978) 246].
35. DUMBRAJS O. V., EChAYA *9* (1978) 602 (in Russian) [Sov. J. Part. Nucl. *6* (1975) 53];
    CIULLI S., POMPONIU C., SABBA-STEFANESKU I., EChAYA *6* (1975) 72 (in Russian) [Sov. J. Part. Nucl. *6* (1975) 29];
    NICHITIU F., EChAYA *12* (1981) 805 (in Russian) [Sov. J. Part. Nucl. *12* (1981) 321].
36. Low- and Intermediate Energy Kaon-Nucleon Physics (E. Reidel), ed. R. Ferrari and G. Violini, (Proceeds. Workshop, Univ. Roma, March 1980) 1981.
37. BATTER A. D. et al., Phys. Rev. Lett. *29* (1972) 1 331.
38. KUKULIN V. I., KRASNOPOĽSKY V. M., VORONCHEV V. I., SAZONOV P. B., Nucl. Phys. *A417* (1984) 128.
39. GHOVANLOU A., LEHMAN D. R., Phys. Rev. *C9* (1974) 1 730;
    LEHMAN D. R., RAI M., GHOVANLOU A., Phys. Rev. *C17* (1978) 744;
    BANG J., GIGNOUX C., Nucl. Phys. *A313* (1979) 119.

40. ESTRADA H., DOMCKE W., Journ. Phys. B.: At. Molec. Phys. *17* (1984) 279.
41. ROSS M. H., SHAW G. L., Ann. Phys. (N. Y.) *9* (1960) 391; *13* (1961) 147.
42. SHAW G. L., ROSS M. H., Phys. Rev. *126* (1962) 806.
43. NAGELS M. M. et al., Phys. Rev. *B109* (1976) 1.
44. WIGNER E. P., Phys. Rev. *70* (1946) 15; *70* (1946) 606;
    WIGNER E. P., EISENBUD L., Phys. Rev. *72* (1947) 29.
45. LANE A. M., THOMAS R. G., Theory of Nuclear Reactions at Low Energies, Rev. Mod. Phys. *30* (1958) 257.
46. HALE G. M., DEVANEY J. J., DODDER D. C., WITTE C., Bull. Am. Phys. Soc. *19* (1974) 506.
47. DODDER D. C., HALE G. M., JARMIE N., JEFF J. H., KLATON P. W. Jr., NISLEY R. A., WITTE K., Phys. Rev. *C15* (1977) 518.
48. HALE G. M., DODDER D. C., Proceeds IX Few-Body Int. Conf., Volume of Papers, p. 11.
49. BARIT I. Ya., SERGEEV V. V., Proc. Lebedev Phys. Inst., Vol. XLIV (in Russian), Nauka, Moscow 1969, p. 3.
50. ARVIEUX J., Nucl. Phys. *A221* (1974) 253.
51. CIULLI S., SPEARMAN T. D., Phys. Rev. *D27* (1983) 1580.
52. PETERSEN J. L., PIŠUT J., Nucl. Phys. *38B* (1972) 207.
53. NOGOVA A., PIŠUT J. PREŠNAJDER P., Nucl. Phys. *61B* (1973) 438; 445.
54. LOPEZ C., YNDURAIN F. J., Phys. Lett. *41B* (1972) 183.
55. LOPEZ C., YNDURAIN F. J., Nucl. Phys *64B* (1973) 315.
56. JACKSON T. D., BLATT J. M., Rev. Mod. Phys. *22* (1950) 77.
57. KVITSINSKY A. A., Theor. Math. Phys. (in Russian) *65* (1985) 226.
58. ALBEVERIO S. et al., Ann. Phys. *148* (1983) 308.
59. BOLLÉ D., GESZTESY F., SCHWEIGER W., J. Math. Phys. *26* (1985) 1661.
60. GIBSON A. G., In: Proceedings of the X European Symposium on the Dynamics of Few-Body Systems, Vol. II, CRIP, Budapest 1986, p. 335.
61. BENCZE Gy., CHANDLER C., FRIAR J. L., GIBSON A. G., PAYNE G. L., Low Energy Scattering Theory for Coulomb plus Long-Range Potentials, Preprint, Los Alamos Nat. Lab. 1986.

Chapter 7

# Resonances in Atomic Physics

## 7.1. Introduction

In the previous chapters we have discussed a number of general methods for calculating the resonance energies and the corresponding widths of resonance levels. We have also presented several examples illustrating the application of these methods which dealt mostly with resonance processes in nuclear physics. This chapter will be devoted to resonances in atomic physics. Basically, the few-particle systems treated in atomic physics differ from the nuclear-physics few-particle systems in two aspects. First, long-range forces act between charged particles and, secondly, the interactions in atomic physics, compared with nuclear physics, are known to a high accuracy. Of course, long-range forces do act in nuclear systems too, where they are, however, of minor importance, whereas they are the main forces in atomic systems. The long-range nature of these forces leads to complications even in the case of two-particle systems. For example, in nuclear physics the effective-range expansion is of a simple form [1]

$$k \cot \delta_0(k) = -\frac{1}{a} + \frac{1}{2} r_0 k^2 + \dots .$$

$$(1.1)$$

The function $k \cot \delta_0(k)$ is an analytic function of $k^2$ and the first two terms of this expansion describe the $N$–$N$ interaction in a rather broad energy range (0–10 MeV) quite accurately. For the polarization interaction $\beta^2 r^{-4}$, for example, which a slow electron "experiences" at large distances from the neutral atom, we get [2]

$$k \cot \delta_0(k) = -\frac{1}{a_0} + \frac{\pi \beta^2}{3 a_0^2} k + \frac{r_0}{2} k^2 + \frac{4 \beta^2}{3 a_0} k^2 \ln \left( \frac{\beta k}{4} \right) + \dots$$

i.e. the function which is much more complicated than (1.1), which is not analytic at the origin and contains also odd powers of $k$. Moreover, its range of application is much narrower because we may use it mostly at energies much below 1 eV.

When we go over to systems of three and more charged particles, we are faced with some fundamental problems. For example, the Faddeev equations have not yet been generalized to the case of three charged particles at energies above the break-up threshold. Also, the problem of the threshold ionization of a hydrogen atom by electron impact has never been resolved fully or correctly. The classical Wannier theory [3] shows that the ionization cross section just above the threshold is of the form

$$\sigma \sim (E - E_i)^{1.1268...} . \qquad (1.2)$$

However, neither this theory nor the quantum theory of this effect [4] have been fitted to experimental data [5].

Interactions in atomic physics are known to a much higher accuracy than nuclear interactions for which, in reality, only the behaviour at large distances due to exchange of one boson is known reliably. At small distances nuclear forces get much more complicated and they have not been studied yet to within a sufficient accuracy. This fact markedly limits the choice of methods for calculating resonances in atomic physics. Only very precise methods may be considered. For example, the energy of the lowest $^1S$ resonance in the system $e^-$ $+ e^- + p$ can be calculated accurately to within 6 significant digits [6], whereas the accuracy of the standard calculations of nuclear resonances is near one percent.

In principle, resonances can be observed in experiments of two types. Resonance states of a target atom $T$ may be excited in inelastic collisions as

$$P + T \rightarrow P + T^*$$
$$\searrow$$
$$A + B \qquad (1.3)$$

where $A$ and $B$ are the decay products of the resonance state $T^*$. In experiments of this type, the cross section of the reaction (1.3) is generally determined by the excitation cross section for formation of the resonance state $T^*$, rather than by its decay. It is possible to excite the state $T^*$ in a relatively broad energy range of the projectile particle $P$ and, therefore, resonances of this type may also be observed when the energy resolution of the incident beam of particles is bad. On the other hand, experiments in which an intermediate compound state $C$ arises

$$P + T \rightarrow C \rightarrow A + B \qquad (1.4)$$

require that the energy of incident particles should lie in a very narrow energy range, i.e. the energy resolution of the incident beam has to be very good, or else the resonance may easily be missed. This is why most of the earlier experiments were carried out with the reaction (1.3) and it is only since the late sixties, when it became possible to obtain low-energy electron beams with energy resolution in the millivolt range, that many low lying resonances have been discovered in experiments of the type of (1.4).

As in nuclear physics, we deal here also with two types of resonance. The "shape resonances" arise when the interaction between an incident particle and an atom (ion, molecule) takes a certain form. As a rule, the interaction is composed of an attractive well surrounded by a repulsive barrier (the target may be in an excited state). These resonances lie energetically above the state to which they are coupled most strongly and are usually broader. On the other hand, the "closed–channel", or "Feshbach", resonances may arise when the interaction between an incident particle and an atom in the excited state, which is prevented from returning to the ground state, may form a bound state. If the atom can be de-excited (a coupling exists between the two channels), the bound state changes into a resonance. The formation of the latter necessitates at least two channels, whereas one channel will suffice for the formation of a shape resonance. The closed-channel resonances lie energetically below the channel to which they are most strongly coupled and are typically much narrower than the shape resonances because their most important decay mode is energetically forbidden.

Furthermore, we may classify the resonances into (1) extra-particle resonances where an incident particle is held for some time near a target atom so that the two form a resonance state and (2) target-atom resonances which survive even if the particle which excited them is far from the excited target atom. Autoionizing states are a typical example of target-atom resonances. These states already exist in two-electron atoms; for example, the $He(2s^2)$ state when both electrons are excited from the $1s^2$ ground level to the $2s^2$ level. This state lies above the ionization threshold of helium. The ionization energy of a He atom is 24.58 eV, whereas the energy of the $1S(2s^2)$ state is 57.82 eV and its lifetime is $5 \times 10^{-15}$ s. This state can be de-excited in two ways, namely, radiative and non-radiative (so called Auger's transition).

$$He^{**}(2s^2) = \begin{cases} He(1s^2) + \hbar\nu + \hbar\nu \\ He^+(1s) + e^- \, . \end{cases}$$

The probability of the non-radiative process (autoionization) is several orders higher than that of the radiative process. These states are called autoionizing

Table 7.1

Energies and widths of two resonances and four autoionization levels of helium

| | State | Energy (eV) | $\Gamma$ (eV) | $\tau$(s) |
|---|---|---|---|---|
| $He^-$ | $(2s^22p)^2P$ | 57.22 | 0.09 | $7 \times 10^{-15}$ |
| | $(2s2p^2)^2D$ | 58.30 | 0.05 | $1 \times 10^{-14}$ |
| $He^{**}$ | $(2s^2)^1S$ | 57.82 | 0.138 | $5 \times 10^{-15}$ |
| | $(2s2p)^3P$ | 58.30 | $<0.015$ | $>4 \times 10^{-14}$ |
| | $(2p^2)^1D$ | 59.90 | 0.072 | $9 \times 10^{-15}$ |
| | $(2s2p)^1P$ | 60.132 | 0.042 | $2 \times 10^{-14}$ |

states; several lowest states of helium are listed in Table 7.1 which also presents two resonance levels of He$^-$ [7–9] lying in the neighbourhood of these.

In the atoms with three and more electrons the excitation of even a single electron is sufficient for an autoionization level to be formed. In the case of neon, for example, the $(1s^2 2s 2p^6 3s)^1 S$ level with 43.64 eV excitation energy which arises from excitation of a 2s electron to a 3s state is an autoionization level. The neon ionization energy is 21.56 eV. Studying the excitations of these states by charged particles is of great interest. If the energy of an incident charged particle lies just above the threshold energy of an autoionizing state, the incident particle will have a very low energy after excitation. Since the lifetime of the autoionizing states is very small ($\sim 10^{-14}$ s), the exciting particle will still be in close proximity to the atom at the moment of autoionization and its presence will strongly affect the process. The ionization cross section will not be described by the simple Breit–Wigner formula, but appears to be a more complicated function of the energy and charge of incident particles. We shall treat this problem later.

The range of atomic and molecular physics effects in which resonance processes play an important role is very broad (for example, the Auger effect, predissociation, charge exchange, etc.), so we shall be able to discuss very few of these problems. We shall omit resonances in molecules and deal only with some aspects of atomic resonances. For more detailed information we refer the reader to several reviews [10–17] devoted to resonances both in atomic and molecular physics.


## 7.2. Methods for Calculating the Resonance States

From the general standpoint, the existing methods for calculating the resonance energies and widths fall roughly into four categories.

(i) *Scattering method.* This method is to calculate a scattering wave function (phase shifts) in the region where a resonance is expected, i.e. where the phase shift increases sharply, whereupon the resonance location and width are found through fitting the cross section by the Breit–Wigner formula. This method has some disadvantages. If a resonance is narrow (which most of the closed-channel resonances are in atomic physics), we can omit it from the calculation because the phase shift is determined by modulo $\pi$. Therefore, the calculations have to be carried out at many energies by very small energy increments, thereby entailing excessive calculation time. The application of this method is also difficult in the case of resonances with several open channels. If a resonance is broad, the Breit–Wigner formula yields inaccurate results and must be corrected. Nevertheless, the initial calculations of resonances in atomic physics were carried out using this method. In the early sixties Burke and Schey [18] predicted the existence of the resonance in the e$^-$ + H system lying some 0.6 eV below the excitation threshold of the $n = 2$ state, which was confirmed shortly afterwards by experimental data [19]. This resonance is due to the potential produced by the dipole coupling between the degenerate 2s and 2p states of a hydrogen atom [20]. The potential is sufficiently attractive to support an infinite number of resonances below the $n = 2$ threshold (at great distances the potential behaves as $r^{-2}$ and supports an infinite number of bound states for the total angular momentum of 0, 1 and 2 in the absence of coupling to the 1s state of hydrogen).

(ii) *Quasibound-state methods*. These include, first of all, the Feshbach projecting method which was discussed in detail in Chapter 4 and which has yielded numerous results. In the two methods outlined above, we must operate with the scattering wave functions.

(iii) *Time dependent approach methods*. These methods are based on the solution of the time dependent Schrödinger equation [21] and have attracted relatively less attention and the advantages and disadvantages of such methods from the computational point of view are not yet clear.

(iv) *Complex eigenvalue of the Hamiltonian*. These methods are based on the calculation of complex eigenvalues of the Hamiltonian. (Eigenvalues of the Hamiltonian may be complex because the boundary condition which is to be imposed on the resonance wave function is complex, and the resultant Hamiltonian is non-Hermitian.) This method was originally rejected because the eigenfunctions corresponding to the complex energy (on the second sheet) increase exponentially and cannot be normalized in the standard way (see Chapter 2 above).

Nowadays, however, it is this method that makes it possible, after the divergences are appropriately removed, to obtain the most accurate values of the resonance energies and widths. We shall now deal with the methods based on calculating the complex eigenvalues of the Hamiltonian. It seems expedient, however, to begin by finding out why the wave function divergence arises. In Chapter 2 it was shown that to correspond to a resonance a wave function must contain only the outgoing wave in the asymptotic region and be a solution of the equation

$$H\psi = W\psi \tag{2.1}$$

where

$$H = -\frac{d^2}{dr^2} + V(r).$$

(for simplicity we discuss only $S$-wave scattering by a short-range spherically symmetric potential); the interaction potential is assumed to be truncated, i.e.

$$V(r) = 0; \quad \text{for} \quad r > R \tag{2.2}$$

and the following conditions must be satisfied

$$\psi(0) \; = 0 \,,$$

$$\psi'(R) = ik \; \psi(R) \,, \tag{2.3}$$

$$W \quad = k^2 \,. \tag{2.4}$$

In atomic physics, a wave function of this type is called the Siegert state [22] and behaves at $r > R$ as

$$\psi(r) = \text{const} \,.\, e^{ikr} \,.$$

In order that this state may correspond to a resonance Im $k$ must be negative, i.e.

$$k = \alpha - i\beta, \qquad \alpha, \beta > 0 \,.$$

Then, the Siegert (Gamow) state increases exponentially at $r > R$

$$\psi(r) = \text{const} \,.\, e^{i\alpha r} \,.\, e^{\beta r} \,.$$

The cause of such behaviour of this wave function may be properly illustrated using the following analogy [23]. Let us consider a star whose luminous emittance decreases exponentially with time. If the star shone all the time with the same intensity, the radiation intensity would decrease as $r^{-2}$ with distance. However, since the luminous emittance of the star decreases exponentially with time, the radiation intensity at a given time will be higher at great distances from the star than in its close vicinity because the distant radiation was emitted at some moment earlier when the luminous intensity was higher. At great distances the exponential increase will utterly prevail over the decrease $r^{-2}$. This behaviour is clearly due to the fact that the star shines eternally. If the star began to radiate for a finite time, the divergence would not occur. The same is valid in the case of the Siegert state, and its divergence is the price to be paid for applying a stationary theory to an apparently non-stationary process.


## 7.3. Variational Methods in the Theory of Resonances

The most important modern methods for calculating resonance energies and widths are always based on variational principles. It should be remembered how variational calculations are to be made when calculating bound-state energies. The wave function $\psi_0$ and the energy $E_0$ are the solution of the Schrödinger equation

$$H\psi_0 = E_0\psi_0 \,, \qquad \|\psi_0\| < \infty$$

if and only if the functional $E[\psi]$

$$E[\psi] = \frac{\langle \psi | H | \psi \rangle}{\langle \psi | \psi \rangle}$$

is stationary at $\psi = \psi_0$, i.e. its variation is zero:

$$\delta E[\psi_0] = 0 .$$

Then,

$$E_0 = E[\psi_0] .$$

To be applicable to calculating the resonance states, this variational principle has to be generalized [24, 25]. In this case the Hamiltonian is not Hermitian because of the complex-valued boundary condition (2.3) imposed on the resonance wave function.

The variational principle may be generalized, for example, in the following way. Let us define a functional $Z$

$$Z[\hat{\psi}, \psi] = \frac{\langle \hat{\psi} | H | \psi \rangle}{\langle \hat{\psi} | \psi \rangle}$$

where $\hat{\psi}$ and $\psi$ may be treated in this case as independent functions. The condition that the variation of this functional should vanish

$$\delta Z[\hat{\psi}_0, \psi_0] = \frac{\langle \delta \hat{\psi}_0 | H - z_0 | \psi_0 \rangle + \langle \hat{\psi}_0 | H - z_0 | \delta \psi_0 \rangle}{\langle \hat{\psi}_0 | \psi_0 \rangle} = 0 ,$$

where $z_0 = Z[\hat{\psi}_0, \psi_0]$, leads to the equations

$$H | \psi_0 \rangle = z_0 | \psi_0 \rangle , \tag{3.1}$$

$$\langle \hat{\psi}_0 | H = \langle \hat{\psi} | z_0 . \tag{3.2}$$

Since $z_0$ is complex, we cannot put $\hat{\psi}_0 = \psi_0$, i.e. proceed as in the case of the bound state calculations. In order that the equations (3.1) and (3.2) be satisfied simultaneously, it is necessary to use the functions $\Phi$ and $\tilde{\Phi}$ from the biorthogonal set defined in Section 2.2.2 (eq. (2.2.37)) for $\psi$ and $\hat{\psi}$, respectively. In this case the functional $Z$ takes the form

$$Z[\Phi] = \frac{(\tilde{\Phi} | H | \Phi)}{(\tilde{\Phi} | \Phi)} . \tag{3.3}$$

Instead of the integrals $(\tilde{\Phi} \mid \Phi)$, $(\tilde{\Phi} \mid H \mid \Phi)$ we must use either their regularized values (see Chapter 2, eq. (2.40)) or the values obtained by analytic continuation (see Chapter 5). Since $\tilde{\Phi}^* = \Phi$ (see (2.2.40)), the equations (3.1) and (3.2) prove to be satisfied and the condition that the variation of the functional $Z$ should vanish may be written in the form

$$(\delta\tilde{\Phi}_0 \mid H - z_0 \mid \Phi_0) = 0 . \tag{3.4}$$

This is a generalization of the Kohn variational principle to the case of complex energies [27]. It should be noted that this generalization is not unique. There exist several versions of such variational principles [28], but to study them is beyond the scope of this book.

Let us present two examples of calculations by the variational method. According to [27], the trial function $\Phi_0$ may be written as

$$\Phi_0(k_0, r) = \sum_{i=1}^{N-1} c_i\varphi_i(r) + c_N\varphi_N(k_0, r) \tag{3.5}$$

where $\varphi_i$ are real quadratically integrable functions; the function $\varphi_N$ provides for correct asymptotic behaviour, i.e.

$$\varphi_N(k_0, r) \xrightarrow[r \to \infty]{} e^{ik_0 r}$$

where

$$k_0 = \sqrt{\frac{2\mu z_0}{\hbar^2}} .$$

It should be emphasized that $k_0$ depends on $z_0$, which is what we actually sought. From (3.4) and (3.5) we obtain the following system of linear equations in the standard way:

$$\sum_{j=1}^{N-1} \mathcal{H}_{ij} c_j + \mathcal{H}_{iN} c_N = 0, \qquad i = 1, 2, ..., N - 1 ,$$

$$\sum_{j=1}^{N-1} \mathcal{H}_{Nj} c_j + \mathcal{H}_{NN} c_N = 0 ,$$

where

$$\mathcal{H}_{ij} = \mathcal{H}_{ji} = \langle \varphi_i \mid H - z_0 \mid \varphi_j \rangle ,$$

$$\mathcal{H}_{iN} = \mathcal{H}_{Ni} = \langle \varphi_i \mid H - z_0 \mid \varphi_N \rangle ,$$

$$\mathcal{H}_{NN} = \langle \varphi_N \mid H - z_0 \mid \varphi_N \rangle .$$

Here $\varphi_N$ depends on $z_0$, hence $\mathscr{H}_{iN}$ and $\mathscr{H}_{NN}$ also depend nonlinearly on $z_0$; the matrix $\mathscr{H}$ is non-Hermitian. The resonance energy $z_0$ may be found from the solvability condition for the above algebraic system, i.e.

$$\det\left(\mathscr{H}_{ij}(z_0)\right) = 0 . \tag{3.6}$$

The iteration method for solving (3.6) was proposed in [27]. This method proves to converge rapidly and yield very accurate results. Fig. 7.1 shows the wave

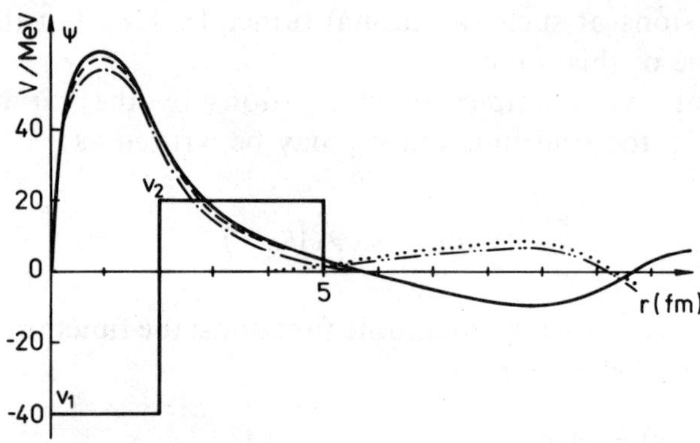


Fig. 7.1

The resonance wave function for a potential consisting of two rectangular wells. The solid line denotes the real part of the exact wave function, the dotted line the imaginary part. The real part of the approximate wave function (3.5) is denoted by the dashed line and its imaginary part by the double dot-dashed line. The dot-dashed line shows the wave function without $\varphi_N$.

functions obtained by this method for the potential consisting of two rectangular wells ($V_1 = -40$ MeV, $V_2 = 20$ MeV). The accurate value of the resonance energy is $z_0 = 4.941 - 0.48$ i MeV; calculation with the trial function (3.5) (which exhibits the correct asymptotic behaviour of the exact wave function) leads to $z_0 = 4.946 - 0.480$ i MeV, and the variational calculations in which the function $\varphi_N$ is missing (i.e. the trial wave function is approximated by quadratically integrable functions) yield the real value $E_0 = 5.36$ MeV.

Fig. 7.1 illustrates another general property of the resonance wave function which was mentioned above, namely that it is localized inside the interaction region, i.e. its amplitude inside the well is larger than behind the barrier.

In the above-mentioned method only one function $\varphi_N$ implies the outgoing-wave asymptotic behaviour of the resonance wave function, whereas on other functions $\varphi_i$ is imposed only the condition of their quadratic integrability. Another approach is possible where all the basis functions $\varphi_i$ satisfy exactly the condition (2.3). This method is realized in the following way [29]. Assume that the potential is zero at $r > R$. Introduce further the linear space $X_k$ as a space of continuous functions $\{\varphi_n\}$ defined on $(0, R)$ with continous derivative satisfying the conditions

$$\varphi_n(0) = 0 \,,$$

$$\varphi'_n(R - 0) = ik\varphi_n(R) \,.$$

The trial wave function will be taken to be of the form

$$\psi = \sum_{n=1}^{N} c_n \varphi_n$$

where $c_n$ are the variational parameters.

The calculation itself is carried out so that for a given $k$ we diagonalize the Hamiltonian in the space $X_k$. Eigenvalues obtained in this way depend, however, on $k$. The Siegert eigenvalues are those for which the realtion (2.4) holds. Thus, we must repeat the calculations at another value of $k$ until one of the eigenvalues merges with $k^2$. If the condition (2.2) is not satisfied, i.e. the potential is not of a finite range, this method cannot be used and must be generalized. Such a generalization, which consists in shifting the boundary condition, was proposed in [29]. The basic idea of the method is that in this case also the boundary condition at some finite point $R$ is written to take the form coinciding formally with (2.3):

$$\psi'(R) = ik'\psi(R) \tag{3.7}$$

where $k'$ depends, however, on the part of the interaction at $r > R$ and may be found in the following way. Let us write the Jost solution $f(k, r)$ as

$$f(k, r) = \exp\left(ikr - \int_r^{\infty} g(k, x)\, \mathrm{d}x\right) \tag{3.8}$$

(The Jost solution satisfies the condition $f(k, r) \sim e^{ikr}$ at $r \to \infty$ and differs from the Seigert state $\psi$ only by normalization). Substituting (3.8) in (2.1), we obtain the nonlinear equation [26]

$$g' + 2ikg + g^2 = V \tag{3.9}$$

for the function $g$ which must satisfy the boundary condition

$$\lim_{r \to \infty} r\, g(r) = 0 \,.$$

The solution for (3.9) may formally be written as the series

$$g(k, r) = \sum_{n=1}^{\infty} g^{(n)}(k, r) \tag{3.10}$$

where

$$g^{(1)}(k, r) = - \exp\left(-2ikr\right) \int_r^\infty V(x) \exp\left(2ikx\right) dx$$

and

$$g^{(n)}(k, r) = \exp\left(-2ikr\right) \sum_{j=1}^{n-1} \int_r^\infty g^{(j)}(k, x)\, g^{(n-j)}(k, x) \exp\left(2ikx\right) dx \ .$$

After finding $g(k, r)$ we may determine the logarithmic derivative of the Jost solution

$$ik' = \frac{f'(k, R)}{f(k, R)} = ik + g(k, R) \ .$$

Thus, we have determined the value of $k'$ which is necessary in (3.7), so we may use the previous method without alterations. If we choose $R$ to be sufficiently large, it is sufficient to use only the first few terms in the series (3.10) to achieve a high accuracy. With the potential

$$V(r) = 15\, r^2\, e^{-r}$$

$R = 10$ and considering the first two terms in (3.10) this method was used in [29] to find the sharp resonance $k = 2.617\,786\,17 - 0.004\,879\,88\,i$.


## 7.4. Stabilization Method


A common feature of the two methods examined below is that we may determine the resonance energy and width without making allowance for the asymptotic behaviour of the wave function. In other words, the resonance energy and width may be found using only the quadratically integrable functions. The first of these methods, the stabilization method of Hazi and Taylor [30], is only an approximate method and does not allow us to determine directly the resonance width. However, this method is often used as a first step in the complex scaling method (see the next section) which makes it possible to determine the resonance energy and width to within a high accuracy using only quadratically integrable functions. The use of the latter has two substantial advantages. As mentioned above we need not be afraid of the divergences inherent to the Siegert state and may use (which is very important in practice) the minutely developed present--day technique of calculating the bound states, because then the calculation of a resonant state is formally identical with the calculation of a bound state.

Let us consider first the stabilization method in the simplest case of a one--channel shape resonance which may arise when the interaction is of the form of an attractive well surrounded by a repulsive barrier.

It is well known (see Chapter 2) that if there exists a resonance with energy $E_0$ and a small width $\Gamma$, then for the energies in the $E_0 - \Gamma$, $E_0 + \Gamma$ range, the wave function will be localized inside, and will have a small amplitude outside the attractive well (see Fig. 7.1). The idea of the stabilization method is to approximate such a wave function using the quadratically integrable functions irrespective of the fact that the true wave function is not quadratically integrable. At first sight it is clear that we can properly approximate the wave function in only a finite interval inside the well. The series-expansion of the wave function in some quadratically integrable basis functions

$$\psi \approx \sum_{n=1}^{N} c_n \varphi_n \tag{4.1}$$

will surely diverge with increasing $N$. Nevertheless, using (4.1) we may obtain very accurate values of the resonance energies if the resonance width is sufficiently small. Let the Hamiltonian $H$ be diagonalized in the basis of quadratically integrable real functions $\varphi_n$. From the equation

$$\det \left| \langle \varphi_i | H | \varphi_j \rangle - \varepsilon_i^{(N)} \delta_{ij} \right| = 0 \tag{4.2}$$

we obtain the real eigenenergies $\varepsilon_i^{(N)}$ which depend on the number of terms $N$ in the expansion (4.1). According to this dependence, the eigenvalues $\varepsilon_i^{(N)}$ may be broken into two groups. The first group includes the eigenvalues which vary strongly with changing $N \to N + 1$, whereas the eigenvalues in the second group are stabilized, i.e. vary but slightly over a broad range of values of $N$. Hazi and Taylor [30] propose to interpret the eigenenergies which have been stabilized as resonance-state energies, whence the name of the method originates. From the very beginning it is clear that this method is an approximation and can yield reasonable results only in the case of narrow resonances. In the case of broader resonances, the wave function is less localized and the accuracy of the approximation decreases. Nevertheless, this method has two substantial advantages: (1) it uses only quadratically integrable functions thereby eliminating the difficulties arising from the divergence of the Siegert state and (2) the calculations of the resonance energies reduce to the calculations of the eigenvalues of a real symmetric matrix which is a standard problem. Let us illustrate the effectiveness of the stabilization method using the following example [30]. The potential of the one--dimensional harmonic oscillator with an exponential barrier

$$V(x) = \begin{cases} x^2 ; & x < 0 \\ x^2 e^{-\lambda x} ; & x > 0 \end{cases}$$

(where $\lambda$ is a positive constant) is taken to form a shape resonance. At $\lambda = 0$ this potential reduces to the potential of a harmonic oscillator which has only bound states. At $\lambda > 0$ these states turn into resonances. In this model it is immediately apparent that the harmonic oscillator wave functions can well serve as the functions $\varphi_n$ in which the expansion (4.1) is made. Thus, we assume that

$$\varphi_n(x) = \left(2^n n! \; \pi^{1/2}\right) H_n(x) \, e^{-x^2} \, .$$

The calculations of the matrix elements $\langle \varphi_n | H | \varphi_m \rangle$ are not difficult. We have to calculate only the perturbed eigenvalues $\varepsilon_i^{(N)}$. Table 7.2 presents the dependence of one of the stabilized eigenvalues on the number of functions $N$ in the

Table 7.2

The dependence of the stabilized eigenvalue on the number of terms $N$ in the expansion (4.1). $j$ is the eigenvalue number, $C_1$ is the first coefficient in (4.1)

| $N$ | $j$ | $\varepsilon_j^{(N)}$ | $C_1$ |
|---|---|---|---|
| 15 | 1 | 0.452 846 | 0.982 7 |
| 20 | 1 | 0.449 656 | 0.952 7 |
| 25 | 2 | 0.456 984 | 0.973 8 |
| 30 | 2 | 0.454 416 | 0.984 5 |
| 35 | 2 | 0.453 508 | 0.983 0 |
| 40 | 2 | 0.452 659 | 0.978 4 |
| 45 | 2 | 0.451 045 | 0.959 7 |

expansion (4.1) at $\lambda = 0.19$. We can see that the eigenvalue with energy of about 0.45 is stabilized over a large range of values $N$. At the beginning the first eigenvalue $(j = 1)$ stabilizes but at a certain value of $N$ this eigenvalue becomes destabilized and the eigenvalue with $j = 2$ stabilizes at approximately the same energy. $c_1$ is the first coefficient of the expansion (4.1) and it can be seen that $\psi$ is described very accurately already by the first function $\varphi_1$ because the resonance is very narrow. Its energy and width are

$$E_r = 0.453 \; 536 \, ,$$

$$\Gamma = 0.002 \; 805 \, .$$

The eigenvalue $\varepsilon_1(\varepsilon_2)$ determines the resonance energy very precisely and is almost constant invariable over a broad range of $N$. The degree of stability of such an eigenvalue is a measure of the resonance width. The calculation of the widths, however, requires that the continuum wave functions should be used (when only the quadratically integrable functions are used, certain information about the width may be obtained by the Stieltjes-moment-theory technique [31]).

To stabilize the eigenvalue we must apply not only the parameter $N$ but also other parameters; for example, the variational parameters in the trial function $\psi$ which is used to minimize the functional (3.3). A typical example of such a stabilization is the calculation of the resonance in the Ps-system $e^- + e^- + e^+$ [32]. The wave function for representing Ps was chosen to be of the Hyleraas-type

$$\psi = \sum_{klm} c_{klm} \exp\left[-\alpha(r_{1p} + r_{2p})\right] r_{12}^k \left(r_{1p}^l r_{2p}^m + r_{1p}^m r_{2p}^l\right)$$

where the indices 1,2, and $p$ denote electron 1, electron 2, and positron, respectively. The dependence of the eigenvalues on the scale parameter $\alpha$ is shown in Fig. 7.2. The stable behaviour of the eigenvalues is clearly seen. Here, as in the previous case, the stable behaviour goes over from one eigenvalue to another at certain values of the parameter $\alpha$.

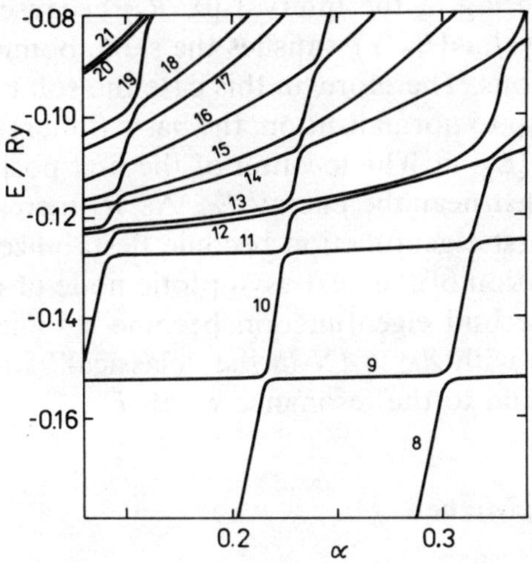


Fig. 7.2

The dependence of the energy eigenvalues on the variational parameter $\alpha$. The stable behaviour of several eigenvalues is clearly seen.

The stabilization method may be generalized to the case where the coefficients $c_m$ in the sum (4.1) or the function $\varphi_n$ are themselves complex. Then the eigenvalues which may be inferred from (4.2) are also complex [33, 34].

Apart from the "classical" stabilization procedure described above, there exist several associated approaches. Thus, proceeding from the similarity between a resonance (especially a narrow one) and a bound state, we may assume that placing a system into a box with infinite walls will not disturb the resonance spectrum significantly. The approach proposed in [35] is based on this idea. In this case also, the Hamiltonian of the system is diagonalized in a quadratically integrable basis. However, the following boundary conditions should be imposed on the basis functions:

$$\varphi_n(0) = \varphi_n(R_0) = 0 \, , \tag{4.3}$$

where $R_0$ is the radius of the box.

Let us examine further the behaviour of the eigenvalues and the eigenfunctions depending on $R_0$. It may be shown that, with increasing $R_0$, exactly the same stabilization of positive eigenvalues of the Hamiltonian may be observed near $E = E_R$, as in the "classical" stabilization procedure with increasing $N$. In this case the character of the stabilization is more obvious than in the case of increasing dimension of the basis $N$. Indeed, the resonance scattering function $\psi(E_R, r)$ is localized in the interaction region and its amplitude in this region is substantially in excess of the amplitude in its asymptotic part. Thus, as soon as the magnitude of $R_0$ approaches the first node $R_1$ of the wave function $\psi(E_R, r)$ lying in the asymptotic region, the problem in the box becomes coincident with the problem of scattering in the interval $(0, R_0)$ because the scattering wave function in this interval, $\psi(E_R, r)$, satisfies the same boundary conditions (4.3) as do the basis functions. Therefore, in this case the solution of the problem in the box reproduces, up to normalization, the basic (inner) part of the resonance scattering function $\psi(E_R, r)$. The location of the first positive eigenvalue at $R_0$ close to $R_1$ is stabilised near the energy $E_R$. As $R_0$ increases further, the first eigenvalue and the first eigenfunction become destabilized, but at $R_0 \sim R_2$, ($R_2 > R_1$ is the location of the next asymptotic node of $\psi(E_R, r)$), the second eigenvalue and the second eigenfunction become stabilized, etc. In this case $\partial \varepsilon_p / \partial R_0$ is analogous with $\Delta \varepsilon_p^{(N)} / \Delta N$ in the "classical" stabilization procedure and is also proportional to the resonance width $\Gamma$.

### 7.5. Complex Scaling Method

At present this method, which is also called the method of complex coordinate rotation or the dilatation method, is one of the most frequently used methods for calculating the resonances in atomic and molecular physics. Its accuracy is high and the resonance state width and energy are determined simultaneously. This method was recently reviewed in detail by Ho [36].

The method is based on the observation that in the case of complex coordinate transformation

$$r \rightarrow r \, e^{i\vartheta} \tag{5.1}$$

where $\vartheta$ is some real number, the wave function of the resonance state turns into a quadratically integrable function. Indeed, at high values of $r$ the wave function corresponding to the $S$-matrix pole behaves as

$$\psi \sim e^{ikr}. \tag{5.2}$$

In the case of bound states, $\operatorname{Im} k > 0$ $(\operatorname{Re} k = 0)$ and this function becomes quadratically integrable. In the case of resonance states, $\operatorname{Im} k < 0$ $(\operatorname{Re} k \neq 0)$ and the function diverges exponentially. Let us denote $k = |k| \cdot e^{-i\alpha}$. Thus, the argument $\alpha$ for resonances is within the interval $(0, \pi)$. Substituting (5.1) in (5.2), we obtain

$$\psi(r) \to \psi_\vartheta(r) \sim e^{ir|k|\cos(\vartheta-\alpha)} \cdot e^{-|k|r\sin(\vartheta-\alpha)}. \tag{5.3}$$

It is clear that, if the angle is selected so that

$$\sin(\vartheta - \alpha) > 0 \qquad \qquad .$$

the wave function $\psi_\vartheta(r)$ decreases at infinity and thus it behaves as a bound-state wave function, i.e. it is quadratically integrable. However, it is immediately evident that this function is much more complicated than the bound-state function. The first term on the right-hand side of (5.3) makes it an oscillating function of coordinates in the asymptotic region too. Nevertheless, the boundary condition (2.3) imposed on the resonance wave function is transformed using (5.1) into the condition for bound states and, therefore, the resonances and the bound states may be treated on the same footing.

The idea of complex coordinate transformation (5.1) is by no means new and was used earlier under other circumstances in scattering theory [37–39].

Due to the transformation (5.1) the Hamiltonian $H$ of the system of particles will change into $\mathcal{H}(\vartheta)$ which is evidently non-Hermitian, so the question arises as to what its spectrum will be. This problem has been studied in the case of the so-called dilatation analytic potentials in several works [40–42]. We shall not discuss this problem in detail from the mathematical point of view (for details see [40–44]), but mention only that the Coulombic interaction and the sum of the Yukawa potentials are among the dilatation analytic operators. We shall show how the operator spectrum of the system $H^-(p + e^- + e^-)$ changes as a result of the transformation (5.1). This system has only one bound state $H^-$ with energy $-0.754$ eV. The continuous spectrum consists of cuts lying on the real axis, starting from the bound-state energies $B_i$ of the hydrogen atom. The Hamiltonian of this system (in the case of an infinitely "heavy" proton) is

$$\mathcal{H} = \mathcal{T} + \mathcal{V}$$

where

$$\mathcal{T} = -\frac{\hbar^2}{2m}\Delta_1 - \frac{\hbar^2}{2m}\Delta_2, \tag{5.4}$$

$$\mathcal{V} = -\frac{e^2}{r_1} - \frac{e^2}{r_2} + \frac{e^2}{|r_1 - r_2|}. \tag{5.5}$$

The transformation (5.1) changes this Hamiltonian into

$$\mathscr{H}(\vartheta) = e^{-2i\vartheta}\mathscr{T} + e^{-i\vartheta}\mathscr{V} \tag{5.6}$$

i.e. the kinetic and potential energies are only multiplied by certain complex numbers. This is an excellent property because the matrix elements of the kinetic (5.4) and potential (5.5) energies may well be calculated (at real $r$) using the

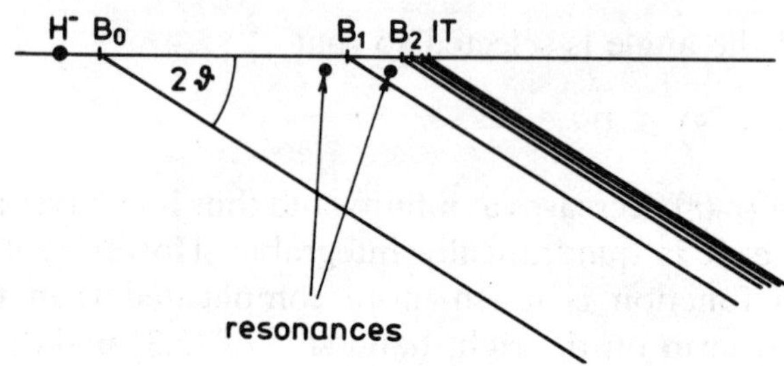


Fig. 7.3
Spectrum of the rotated Hamiltonian $\mathscr{H}(\vartheta)$ for $H^-$.

available very effective programs for calculating bound states which need not be modified for the given case. The spectrum of the operator $\mathscr{H}(\vartheta)$ is shown in Fig. 7.3 and may be characterized as follows:

(1) In the transformation (5.1) the location of the three-particle bound state $H^-$ does not change.

(2) The continuous spectrum splits into several branches which come out of the individual bound states of the hydrogen atom. The corresponding cuts are rotated to the lower half-plane of complex energy and make an angle of $2\vartheta$ with the real axis.

(3) When $\vartheta$ increases and reaches the value at which one of the branches of the continuous spectrum goes through the resonance energy, this energy becomes an eigenvalue of the operator $\mathscr{H}(\vartheta)$.

(4) As soon as a resonance has appeared, i.e. as soon as a branch of the continuous spectrum has gone through it, its energy becomes independent of the angle $\vartheta$ and remains as such until another branch of the continuous spectrum goes through it. Once this happens, this eigenvalue disappears.

### 7.5.1. Analytic Model

Let us illustrate the complex-scaling method by an example of a simple analytically solvable model [45]. The potential of this model consists of two parts, namely, the attractive part $-\gamma/r^2$ and the repulsive Coulombic barrier (the atomic units [46] are used in this section):

$$V(r) = \frac{1}{r} - \frac{\gamma}{r^2}. \tag{5.7}$$

This model can describe a system composed of an electron approaching a negative ion which has nearly degenerate resonances whose angular momentum differ by unity from each other. When the electron approaches the ion, the degenerate ion resonances may be expected to mix and to produce an attractive $r^{-2}$ potential. The simplest ion of the type is $H^-$. The Hamiltonian $\mathcal{H}$ of this system

$$\mathcal{H} = -\frac{1}{2}\frac{1}{r^2}\frac{\mathrm{d}}{\mathrm{d}r}\left(r^2\frac{\mathrm{d}}{\mathrm{d}r}\right) + \frac{1}{r} - \frac{\gamma}{r^2}$$

is changed by the complex scaling transformation (5.1) into

$$\mathcal{H}(\vartheta) = -\frac{e^{-2i\vartheta}}{2r^2}\frac{\mathrm{d}}{\mathrm{d}r}\left(r^2\frac{\mathrm{d}}{\mathrm{d}r}\right) + \frac{e^{-i\vartheta}}{r} - \gamma\frac{e^{-2i\vartheta}}{r^2}.$$

The Schrödinger equation

$$\mathcal{H}(\vartheta)\psi_\vartheta = E(\vartheta)\psi_\vartheta \tag{5.8}$$

changes after the standard substitution

$$\psi_\vartheta(r) = \frac{R_\vartheta(r)}{r}$$

into the equation

$$\frac{\mathrm{d}^2 R_\vartheta}{\mathrm{d}r^2} + \left(\frac{b}{r} - \frac{c}{r^2} - a\right)R_\vartheta = 0 \tag{5.9}$$

where

$$a = -2e^{2i\vartheta}E(\vartheta)\,,$$

$$b = -2e^{i\vartheta}\,,$$

$$c = -2\gamma\,.$$

On making another substitution:

$$R_\vartheta(r) = r^\alpha\, e^{\mu r}P(r)$$

where $\alpha$ and $\mu$ satisfy the equations

$$\alpha(\alpha - 1) = c\,,$$

$$\mu^2 = a$$

we may use (5.9) to obtain the equation for the confluent hypergeometric function

$$\frac{d^2 P}{dr^2} + \frac{dP}{dr}\left(2\mu + \frac{2\alpha}{r}\right) + P\,\frac{2\alpha\mu + b}{r} = 0$$

which has a solution (regular at the origin)

$$P(r) = {}_1F_1\left(\alpha + \frac{b}{2\mu}, 2\alpha, -2\mu r\right).$$

Thus, the wave function $\psi_\vartheta$ is of the form

$$\psi_\vartheta(r) = N r^{\alpha-1} e^{\mu r}\, {}_1F_1\left(\alpha + \frac{b}{2\mu}, 2\alpha, -2\mu r\right) \tag{5.10}$$

where $N$ is the normalization factor. The hypergeometric function on the right-hand side of (5.10) reduces to a polynomial when

$$\alpha + \frac{b}{2\mu} = -n$$

where $n = 0, 1, 2, \dots$ . The energy corresponding to these states is

$$E_n = -\frac{2}{\left(1 + \sqrt{1 - 8\gamma} + 2n\right)^2}$$

and does not depend on the angle $\vartheta$. However, the coefficient $\mu$ determining the asymptotic behaviour of the wave function $\psi_\vartheta$ does depend on $\vartheta$

$$\mu = \frac{2e^{i\vartheta}}{1 + 2n + \sqrt{1 - 8\gamma}}.$$

If the attractive part of the interaction is sufficiently weak, i.e.

$$8\gamma < 1$$

the energies $E_n$ are real and the wave function $\psi_\vartheta$ contains the exponential term

$$\exp\left(\frac{2r(\cos\vartheta + i\sin\vartheta)}{1 + 2n + \sqrt{1 - 8\gamma}}\right).$$

These states correspond to virtual states and their wave function will be normalizable if we select $\vartheta$ such that $\cos\vartheta < 0$ .

However, a far more interesting situation arises at

$$8\gamma > 1$$

because in this case the attractive interaction is so strong that the resonances with energy

$$E_n = 2\,\frac{8\gamma - 4n^2 - 4n - 2 - i(4n + 2)\sqrt{8\gamma - 1}}{(4n^2 + 4n + 8\gamma)^2} \qquad (5.11)$$

appear. This energy is again independent of $\vartheta$. At $n = 0$ and $\gamma = 1/2$ ( which corresponds to a resonance with the greatest real part), we get

$$E_0 = \frac{1 - i\sqrt{3}}{4}$$

and

$$R_\vartheta(r) = r^{\frac{1+i\sqrt{3}}{2}}\,\exp\left[-2r\sin\left(\vartheta - \frac{\vartheta_R}{2}\right)\right]\exp\left[2ir\cos\left(\vartheta - \frac{\vartheta_R}{2}\right)\right] \qquad (5.12)$$

where $\vartheta_R$ is the angle between the real axis and a straight line connecting the resonance energy with the origin. The wave function (5.12) depends strongly on the angle $\vartheta$. At $\vartheta < \frac{1}{2}\vartheta_R$ it increases exponentially and at $\pi + \frac{1}{2}\vartheta_R > \vartheta > \frac{1}{2}\vartheta_R$ it is quadratically integrable. As compared with the wave function of the bound state which has no node at $n = 0$, the resonance wave function (5.12) is much more complicated and has an infinite number of nodes. A fraction of the oscillations are due to the term $r^{i\sqrt{3}/2}$ which originates from the potential $\gamma r^{-2}$ and another fraction arises from the last term on the right-hand side of (5.12), which is inherent to the method of complex scaling. From this it follows that problems will appear in each method which approximates the wave function $\psi_\vartheta(r)$ by a finite sum of quadratically integrable functions.

This model may easily be generalized to the case of $l \neq 0$ [34]. If we denote

$$l' = -\frac{1}{2}\left(1 + i\sqrt{8\gamma - (2l + 1)^2}\right) \qquad (5.13)$$

we obtain the $S$-matrix as

$$S_l(k) = \frac{\Gamma\left(\frac{1}{2}[1 + \sqrt{(2l + 1)^2 - 8\gamma}] + ik^{-1}\right)}{\Gamma\left(\frac{1}{2}[1 + \sqrt{(2l + 1)^2 - 8\gamma}] - ik^{-1}\right)}. \qquad (5.14)$$

This $S$-matrix has poles at the points

$$k_n = -\frac{i}{n + \frac{1}{2}[1 + \sqrt{(2l + 1)^2 - 8\gamma}]}, \qquad n = 0, 1, 2, \dots$$

when

$$8\gamma < (2l + 1)^2$$

and at the points

$$k_n = \frac{2\left[\pm\sqrt{8\gamma - (2l + 1)^2} - i(2n + 1)\right]}{(2n + 1)^2 - 8\gamma - (2l + 1)^2}$$

in the case of strong attraction

$$8\gamma > (2l + 1)^2 .$$

The wave function at these points is

$$R_{nl\vartheta}(r) =$$

$$= C_{l'}\, e^{i|k|r}\, e^{i(\vartheta-\beta)}\, [|k|r\, e^{i(\vartheta-\beta)}]^{l'}\; {}_1F_1(-n;\, 2l' + 2;\, -2i|k|r\, e^{i(\vartheta-\beta)})$$

where

$$\beta = -\tan^{-1}\frac{2n + 1}{8\gamma - (2l + 1)^2}$$

and $l'$ is determined by (5.13)

### 7.5.2. *Computational Aspects of Complex Scaling*

The model analysed in the previous section, i.e. the one-particle problem in the potential field, can be solved analytically. The realistic systems with three or more particles involved in the scattering process cannot be solved analytically, so they have to be solved numerically. This gives rise to a number of new problems. First of all, the variational principle has to be modified. In applying the complex scaling method, one proceeds most frequently from the Ritz variational principle

$$\delta[W] = \delta\frac{\int \psi^{*+}_\vartheta\, \mathcal{H}(\vartheta)\psi_\vartheta\, d\tau}{\int \psi^{*+}_\vartheta\, \psi_\vartheta d\tau} = 0 \tag{5.15}$$

but must take into consideration that at $\vartheta \neq 0$ the operator is not Hermitian. As a result, the equation (5.15) provides only for a stationary principle without upper or lower bounding properties. The stationarity condition for the functional (5.15) leads to the equations

$$\mathcal{H}(\vartheta)\psi_\vartheta = W\psi_\vartheta ,$$

$$\mathcal{H}^+(\vartheta)\psi^+_\vartheta = W^*\psi^+_\vartheta .$$

The function $\psi^{*+}_\vartheta$ is defined to be the complex conjugate of the time-reversed solution and is obtainable in spherical coordinates by taking the complex conjugate of the angular part and leaving the radial functions unchanged [24].

As shown in Section 7.5, the resonance position does not depend on $\vartheta$ once the resonance has been exposed. This, of course, is strictly true provided the resonance energy is calculated exactly. In the approximate calculations which are necessary for all real systems, the resonance position proves to depend (very strongly in many cases) on the angle so it is impossible to determine the exact resonance energy. Instead we obtain a certain domain within which the resonance can be found and its magnitude is the measure of usability of the trial function $\psi$. The $^1S$ resonance in the $H^-$ system is a touchstone of each of the methods for calculating resonances in atomic physics. Theoretically, this three-body system is one of the simplest nontrivial problems for which not a single exact analytic solution has been found. One of the first applications of the complex scaling method was also aimed at calculating this resonance. Doolen [47] has chosen the trial function $\psi$ to be of the Hyleraas form

$$\psi = e^{-a(r_1+r_2)/2} \sum_{l+m+n\leq N} c_{lmn} r_{12}^m \left( r_1^l r_2^n + r_1^n r_2^l \right) =$$

$$= \sum_{l+m+n\leq N} c_{lmn} u_{lmn} \tag{5.16}$$

so that the problem reduced to finding the eigenvalues from the modified secular equation:

$$\det \left( \mathcal{H}_{ij} - W \mathcal{N}_{ij} \right) = 0$$

where the Hamiltonian matrix elements are

$$\mathcal{H}_{ij} = \int u_{l'm'n'} \mathcal{H}(\vartheta) u_{lmn} \, d\tau$$

and the overlap matrix is

$$\mathcal{N}_{ij} = \int u_{l'm'n'} u_{lmn} \, d\tau \,.$$

Fig. 7.4 shows the results of his calculations made in the case of $N = 8$ (i.e. 95 Hyleraas functions) at three values of the parameter $a$.

Examination of these eigenvalue trajectories for different nonlinear parameters indicates that they converge in what is expected to be the vicinity of the exact eigenvalue position. If the basis set is broad enough to simulate the resonance wave function, the resonance eigenvalue trajectories will meet each other or change their directions in the vicinity of the exact resonance position. Also for a wide range of rotation angles $\vartheta$, the eigenvalue remains near the resonance location.

If the number $N$ of the expansion terms in (5.16) increases, the eigenvalue trajectory spirals round the resonance position. This situation is shown in Fig. 7.5 [48].

From the two dependences we see that, for calculation of resonances by the complex scaling method to be fruitful, two assumptions must be satisfied, namely, we must have a good trial function $\psi$ and a reliable estimate for

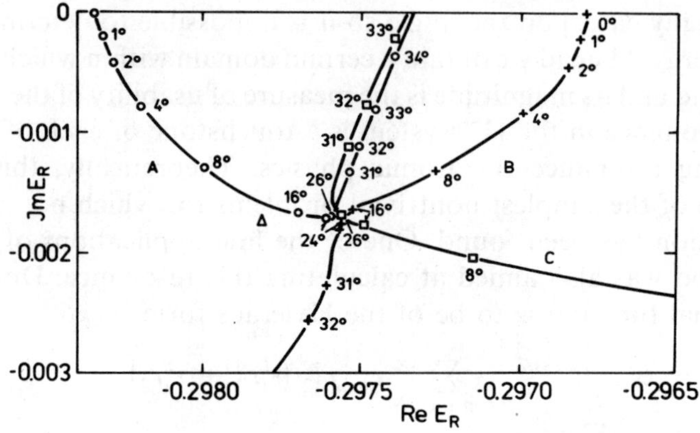

Fig. 7.4

Energy of the $H^-$ $^1S$ resonance calculated by the complex scaling method as a function of the rotation angle $\vartheta$ at three values of the variational parameter $a$.

selecting the angle $\vartheta$. In practice, the first condition is satisfied most frequently, so the trial wave function $\psi$ is determined first by the stabilization method which, at the same time, gives an approximate estimate of the resonance energy; and then this trial function is used to minimize the functional $W$ (5.15). When examining the case of the analytic model we saw that the introduction of the angle $\vartheta$ leads to oscillations in the function $\psi$, thereby resulting in a slow convergence of the variational estimate. It appears [49] that, to obtain good

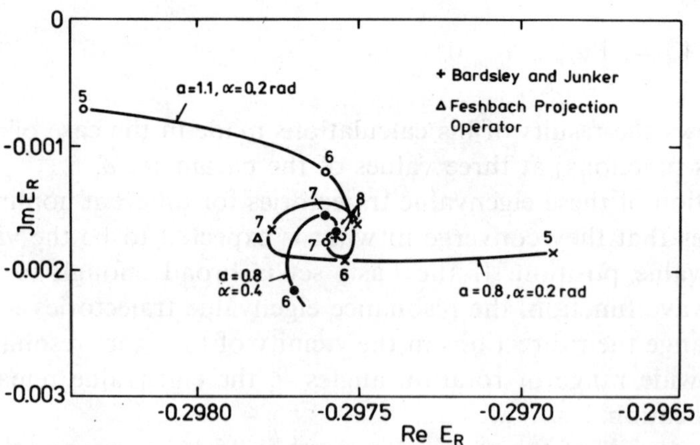


Fig. 7.5

Energy of the $H^-$ $^1S$ resonance as a function of the number $N$ of terms in (5.16).

results at arbitrary $\vartheta$, it is necessary to include approximately $10^N$ terms ($N$ is the number of electrons). As $\vartheta$ increases, oscillations are introduced into all of the bound state orbitals and even the representation of a hydrogenic 1s orbital requires many terms.

To satisfy the second condition, Winkler and Yaris [50] have proposed to generalize the virial theorem. This theorem says that in the case of the bound--state wave functions $\psi$ for the system of particles with Coulombic interaction the following relation holds:

$$2 \langle \psi | T | \psi \rangle + \langle \psi | V | \psi \rangle = 0 .\tag{5.17}$$

We shall show that this relation is also valid in the case of the transformation (5.1). The resonance state energy is determined by the equation

$$W(\vartheta) = \langle \psi_\vartheta | \mathcal{H}(\vartheta) | \psi_\vartheta \rangle$$

if the function is normalized so that $\langle \psi_\vartheta | \psi_\vartheta \rangle = 1$ .

In the case of the exact wave function $\psi_\vartheta$, the resonance energy $W$ does not depend on $\vartheta$. Thus

$$W(\vartheta) = W(\vartheta + \mathrm{d}\vartheta) .\tag{5.18}$$

After the transformation (5.1), the Hamiltonian $\mathcal{H}$ is of the form

$$\mathcal{H}(\vartheta) = \mathcal{T}(\vartheta) + \mathcal{V}(\vartheta) = \mathrm{e}^{-2\mathrm{i}\vartheta} T + \mathrm{e}^{-\mathrm{i}\vartheta} V .$$

For minor variations of the angle $\vartheta$, we may write

$$\mathcal{H}(\vartheta + \mathrm{d}\vartheta) - \mathcal{H}(\vartheta) = -\mathrm{i}\,\mathrm{d}\vartheta(2\mathcal{T}(\vartheta) + \mathcal{V}(\vartheta)) + \text{higher-order terms.}$$

Thus, when the variation $\vartheta \to \vartheta + \mathrm{d}\vartheta$ occurs, the resonance energy varies as

$$W(\vartheta + \mathrm{d}\vartheta) - W(\vartheta) =$$

$$= -\mathrm{i}\,\mathrm{d}\vartheta(2 \langle \psi_\vartheta | \mathcal{T}(\vartheta) | \psi_\vartheta \rangle + \langle \psi_\vartheta | \mathcal{V}(\vartheta) | \psi_\vartheta \rangle) .$$

The condition that the resonance energy should be independent of the angle $\vartheta$, i.e. when eq. (5.18) is valid, gives rise to a generalized (complex) virial theorem

$$2 \langle \psi_\vartheta | \mathcal{T}(\vartheta) | \psi_\vartheta \rangle + \langle \psi_\vartheta | \mathcal{V}(\vartheta) | \psi_\vartheta \rangle = 0 .\tag{5.19}$$

Winkler and Yaris [50] suggest to choose a value of $\vartheta$ for which the relation (5.19) is satisfied for a given trial function.

In practice, however, it is more suitable to choose the angle $\vartheta$ in such a way that $W(\vartheta)$ would depend on $\vartheta$ as little as possible, i.e. for example, in the region where the eigenvalue trajectories meet each other (see Fig. 7.4).

After the dilatation transformation (5.1) the Hamiltonian $\mathcal{H}(\vartheta)$ is no longer Hermitian and, therefore, the variational principle (5.15) does not provide for the bounding properties. However, the complex scaling method can be formulated on the basis of Hermitian operators. This makes it possible to obtain the bounds for the resonance position and width. The complex Schrödinger equation (5.8) obtained by rotating the coordinates to the complex plane

$$[(H_r + iH_i) - E_r - iE_i] (\psi_r + i\psi_i) = 0$$

(where $H_r$, $E_r$, and $\psi_r$ are the real parts, and $H_i$, $E_i$ and $\psi_i$ the imaginary parts of $\mathscr{H}(\vartheta)$, $E$, and $\psi$, respectively) can be split into the set of two equations [51]

$$\left[ \begin{pmatrix} -H_i, & H_r \\ H_r, & H_i \end{pmatrix} - E_i \begin{pmatrix} 1, & 0 \\ 0, & -1 \end{pmatrix} + E_r \begin{pmatrix} 0, & -1 \\ -1, & 0 \end{pmatrix} \right] \begin{pmatrix} \psi_i \\ \psi_r \end{pmatrix} = 0$$

which can be formally written in the matrix form as

$$\hat{\mathscr{H}}(\vartheta, E_r, E_i)\psi = 0$$

where

$$\psi \equiv \begin{pmatrix} \psi_i \\ \psi_r \end{pmatrix}.$$

The operator $\hat{\mathscr{H}}(\vartheta, E_r, E_i)$ is Hermitian and contains the resonance position and width as parameters. The eigenvalues $\lambda$ of this Hamiltonian $\hat{\mathscr{H}}\psi = \lambda\psi$ can be both positive and negative. The zero eigenvalue corresponds to the exact values of the resonance parameters $E_r$ and $E_i$.

Thus, we must go over to the operator $\hat{\mathscr{H}}^2$ which has non-negative eigenvalues:

$$\hat{\mathscr{H}}^2\varphi = \lambda^2\varphi .$$

By applying variational methods to this equation, the upper and lower bounds of the resonance position and width can be obtained. The exact position and width of the resonance is a point on the surface of a ring, i.e.

$$E_r + \lambda_0 \geq E_r \text{ (exact)} \geq E_r + \lambda$$

or

$$E_r - \lambda \geq E_r \text{ (exact)} \geq E_r - \lambda_0$$

and the width

$$2E_i + 2\lambda_0 \geq \Gamma \text{ (exact)} \geq 2E_i + 2\lambda$$

or

$$2E_i - 2\lambda \geq \Gamma \text{ (exact)} \geq 2E_i - 2\lambda_0$$

where $\lambda$ and $\lambda_0$ are certain expectation values [51].

### 7.5.3. *The Stark Effect*

The complex scaling method discussed in the previous part of this Chapter cannot be used directly to calculate the Stark effect because the corresponding potential is not an analytic dilatation potential. The potential $\boldsymbol{F}.\boldsymbol{r}$ of the electrostatic field alters the whole spectrum substantially in the case of even a very weak field. The potential of an atomic electron in an electric field is schematically shown in

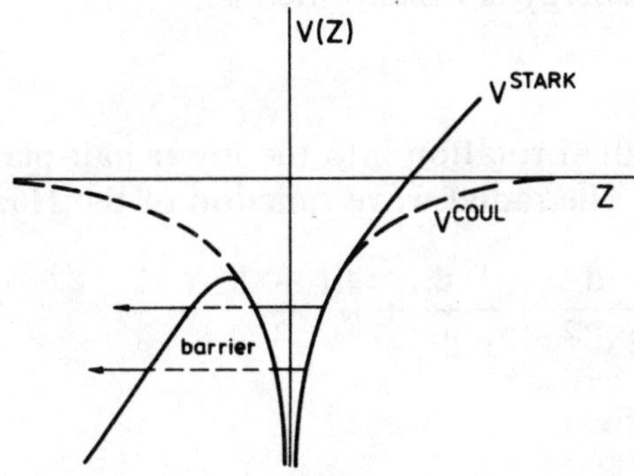


Fig. 7.6

Schematic plot of a potential consisting of the Coulombic and electric field interactions.

Fig. 7.6. It is clear that bound states cannot exist in this case. After the electric field is included, all bound states change into resonances. Moreover, the continuous spectrum also changes. It will occupy the whole axis $(-\infty, \infty)$, so no room will be left for the spectrum to begin. We have seen that in the complex scaling transformation (5.1) the individual branches of the continuous spectrum rotate about their origins to the lower halfplane at angle $2\vartheta$. What will happen, however, to the continuous spectrum, if no origins exist? In spite of all these obscurities, the method of complex scaling was used successfully to calculate the Stark effect [52–56], as a result of which the theory was elaborated step-by-step [57–59]. Let us discuss the transformation of the continuous spectrum in the coordinate transformations of the (5.1) type in more detail. We shall follow [60]. The effect of various coordinate transformations on the operator spectrum is determined by the principle of preservation of the asymptotic form of the continuum eigenfunctions. Generally, the principle may be stated as follows: given a coordinate transformation, the continuous spectrum of an operator shifts in such a way as to preserve the asymptotic form of eigenfunction. Let us examine the validity of this theorem using several examples.

(1) *Free particle.* The Hamiltonian of a free particle

$$H = -\frac{1}{2}\frac{\mathrm{d}^2}{\mathrm{d}r^2}$$

has the eigenfunctions $e^{ikr}$, $e^{-ikr}$ and the eigenvalues $k^2/2$. The expression (5.1) transforms these functions as follows:

$$e^{\pm ikr} \rightarrow e^{\pm ikr \, e^{i\vartheta}} .$$

For the asymptotic behaviour to be unchanged, k must be transformed as

$$k \rightarrow k \, e^{-i\vartheta} . \tag{5.20}$$

This means that the energy is transformed as

$$k^2 \rightarrow k^2 \, e^{-2i\vartheta} .$$

which is none other than rotation into the lower half-plane through angle $2\vartheta$.

(2) *Coulomb field.* The radial wave function of the Hamiltonian $H$

$$H = -\frac{1}{2}\frac{d^2}{dr^2} - \frac{1}{r}\frac{d}{dr} + \frac{l(l+1)}{r^2} - \frac{1}{r}$$

is of the asymptotic form

$$A\frac{1}{r}\sin kr\left(1 + \frac{\ln 2kr}{k^2 r}\right) + B\frac{1}{r}\cos kr\left(1 + \frac{\ln 2kr}{k^2 r}\right) .$$

The transformation (5.20) ensures the invariance of the spectrum again, because $\ln(2kr)/r \rightarrow 0$ at $r \rightarrow \infty$ .

(3) *Stark Hamiltonian.* In this case the one-dimensional Hamiltonian is of the form

$$H = -\frac{1}{2}\frac{d^2}{dx^2} - Fx , \qquad x \in (-\infty, \infty) \tag{5.21}$$

and its eigenfunctions are the Airy functions $A_i(\xi)$ and $B_i(\xi)$ ,where

$$\xi = \frac{2}{3}(E + Fx)^{3/2} .$$

At high values of $x$, these functions behave as $\sin \xi$ or $\cos \xi$, respectively [61]. This means that the wave function behaves as

$$\sin\left[\frac{2}{3}(E + Fx)^{3/2}\right] . \tag{5.22}$$

It is clear that none of the $x$-independent transformations of $E$ can compensate for the transformation

$$x \rightarrow x \, e^{i\vartheta} .$$

As Herbst [62] has shown, in this case the continuous spectrum is empty for $\vartheta \neq 0$.

(4) *Stark Hamiltonian. Coordinate translation.* As the last example we present the Stark Hamiltonian (5.21) again but instead of the rotation, we shall carry out the translation

$$x \rightarrow x + iq .$$

In this case it is easy to preserve the asymptotic behaviour (5.22) using the transformation

$$E \rightarrow E - iFq$$

which means that the continuous spectrum is shifted by the value $Fq$ to the lower half-plane. Using this method, the resonances were calculated with the model Hamiltonian [60]

$$H = -\frac{1}{2}\frac{d^2}{dx^2} - Fx - A \exp\left(-\sigma x^2\right) .$$

The resultant spectrum is shown in Fig. 7.7. Both transformations, i.e. the complex rotation and the translation, yield the same results, but the translation

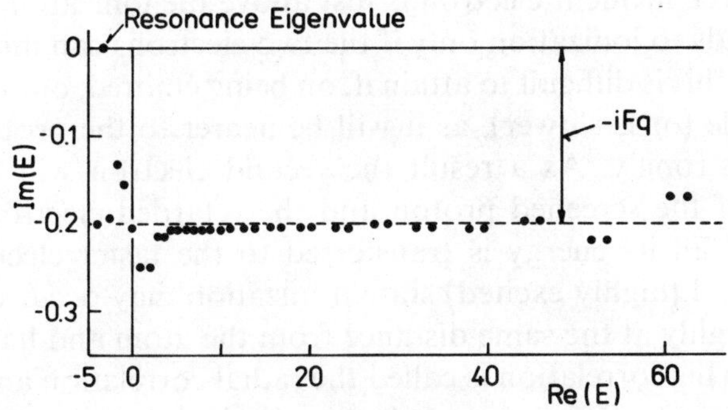


Fig. 7.7

Spectrum distortion of the Stark model Hamiltonian by the complex coordinate translation.

method seems to converge more slowly. From this example the interesting conclusion follows that transformations other than the complex rotation may also be used to calculate resonance energies and widths.


## 7.6. Post-Collision Interaction Model

In this section we shall study again the autoionizing states which we discussed briefly in Section 7.1. The autoionizing states are of great interest from both theoretical and practical points of view. For example, they play an important role in solar corona where the temperatures exceed one million K, thereby making their excitation possible. From a theoretical point of view, they are of interest in a study of the correlation effects which arise in the collective interactions of atomic electrons with each other and of atomic electrons with an incident particle. We shall treat here only the correlation effects between an incident particle and atomic electrons. If the velocity of incident particles is too high, atomic electrons cannot respond and the influence of these effects is inconsiderable. If the incident particles are slow, however, the atom has enough time to react to the presence of an incident particle and the correlation effects will play an important role. They can markedly influence the values and the energy behaviour of various collision cross sections. A similar situation arises also when the projectile leaves the atom. The situation gets considerably more complicated, however, when ionization occurs. In this case the projectile, which is assumed to be charged, can interact with the emitted electron even at a great distance from the atom. Phenomena of this type are typically exemplified by the ionization of a hydrogen atom by electron impact:

$$e^- + H \rightarrow p + e^- + e^- . \tag{6.1}$$

If the energy of the incident electron is just above the ionization threshold, the reaction (6.1) leads to ionization only if the two electrons can move far away in space. However, this is difficult to attain if, on being emitted, one of the electrons is retarded a little (or is slower), as it will be nearer to the proton and will be attracted more strongly. As a result the second electron will be affected by a weaker field of the screened proton and the retarded electron will even be decelerated until all its energy is transferred to the faster electron so that it finishes in a bound (highly excited) state. Ionization may occur only if the two electrons are roughly at the same distance from the atom and have roughly the same velocities. This correlation is called the radial correlation and proves to be very unstable. Because of their equal charges, both electrons tend to be emitted in opposite directions. This is the angular correlation which proves to be stable. Based on these correlations, Wannier [3] predicted the behaviour of the ionization cross section near threshold (1.2). In the context of the present monograph, however, it is far more interesting to study the correlation effects in those cases where the ionization is from a certain electronic state whose energy and, to a certain extent, wave function are known. These are just the autoionizing states. The energy of the autoionizing states is calculated to within a high accuracy by, for example, the complex scaling method and is measured to a high accuracy [8, 63] in photoabsorption experiments of the type of

$$\hbar\omega + He \rightarrow He^{**} \rightarrow He^+ + e^-_{fast} .$$

Experiments in which the incident (i.e. exciting) particles are fast ions or electrons yield the same results. For example,

$$\text{He}^+ + \text{He} \rightarrow \text{He}^+ + \text{He}^{**} \rightarrow \text{He}^+ + \text{He}^+ + \text{e}^-_{\text{fast}}.$$

$$(6.2)$$

The situation changes substantially however when the energy of the incident ions decreases. Barker and Berry [64] measured the spectrum of the emitted electrons in the reaction (6.2) in the 30–40 eV range where several sharp peaks should occur (see Table 7.1). Instead, they observed a single very broad and asymmetric peak whose position and width depend on the incident ion energy according to the law

$$E = 35.78 - \frac{37.9}{E_0} \quad (\text{eV}),$$

$$\Gamma = 0.41 - \frac{94.4}{E_0} \quad (\text{eV})$$

where $E_0$ is the incident ion energy. The results of the experiment were used by Barker and Berry to construct a model which was called the post-collision interaction (PCI) model. If the energy of an incident ion is low, the ion will be located at sufficiently small distances from the excited atom during a period within the autoionizing state lifetime which is about $10^{-14}$ s, so the emitted electron will be strongly affected by the ion. It will be attracted more strongly and energy (and angular momentum) transfer will occur between the electron and the ion. In terms of this model Baker and Berry inferred the probability distribution for the energy transfer $\Delta E$

$$P(\Delta E)\, \mathrm{d}(\Delta E) = \frac{b}{(\Delta E)^2} \exp\left(-\frac{b}{\Delta E}\right) \mathrm{d}(\Delta E)$$

where

$$b = \frac{e^2 \Gamma}{4\pi\varepsilon_0 \hbar \sqrt{2 M_{\text{He}} E_0}}$$

and $\Gamma$ is the autoionization level width. The resulting line shape of the electron peak is asymmetric with a long tail towards higher energies.

In more precise experiments where the incident particles were electrons [7, 65] individual peaks could be discriminated. In this case their energy and width are also dependent on the incident electron energy and the positions of the emitted-electron peaks are shifted to higher energies when the incident electron energy approaches the excitation threshold. This situation is clearly seen in Fig. 7.8

where the left-hand peak corresponds to the electrons emitted from the $(2s^2)^1S$ state whose lifetime $\tau = 5 \times 10^{-15}$s, and the right-hand peak to the electrons emitted from the $(2s, 2p)^3P$ state whose lifetime $\tau = 4 \times 10^{-14}$s. The lifetime

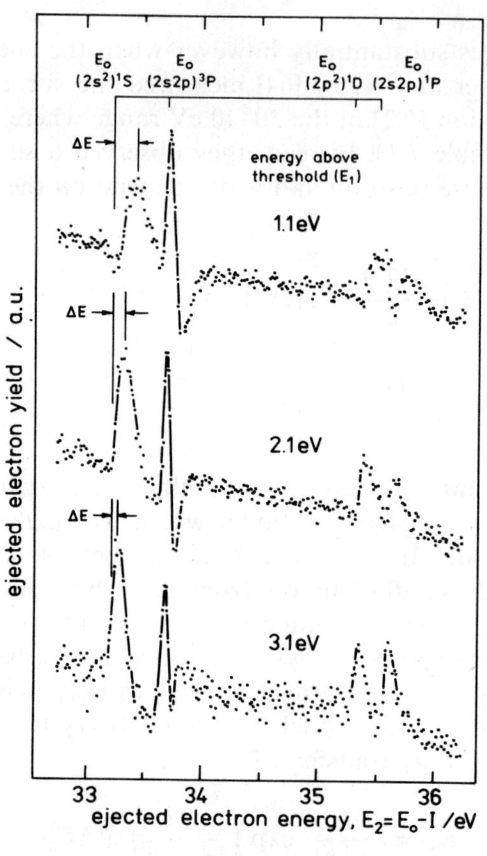



Fig. 7.8

The ejected electron spectrum for electron excitation of autoionizing states of He. The shift of the left peak manifests the presence of the post collison interaction.

of the first state is about an order shorter therefore, the PCI manifests itself in the first state more strongly than in the second.

The post-collision interaction not only manifests itself in the spectrum of fast electrons but also markedly affects the slow electrons. If the energy of the incident electron is just above the excitation threshold of the autoionizing state, it will exhibit a certain energy excess after excitation. Because of the PCI, it will transfer a fraction of the excess energy to the fast electron. This means that we shall fail to observe the slow electrons until the energy of the incident electrons exceeds the autoionizing state energy by a certain energy $\varepsilon$, i.e. the PCI manifests itself as a displaced emission threshold of the slow electron. This can be seen in

Fig. 7.9 showing the spectrum of slow electrons $(E < 10\ \mathrm{meV})$ as a function of incident ion energy [65]. The feature at 57.22 eV is due to the He$^-$ resonance state $(2s^2 2p)^2 P$, but there is no apparent yield of the threshold electrons at the

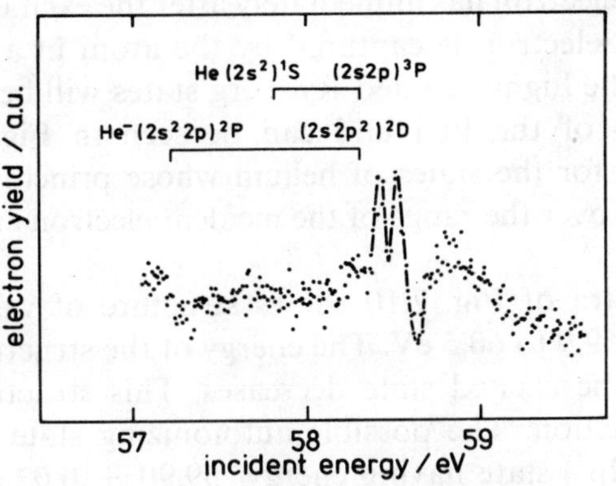


Fig. 7.9

Spectrum of the slow electrons $(E < 10\ \mathrm{meV})$ ejected after the excitation of an autoinizing state of He. The electrons should appear at an energy 57.82 eV, but the post collision interaction shifts the threshold to higher energies.

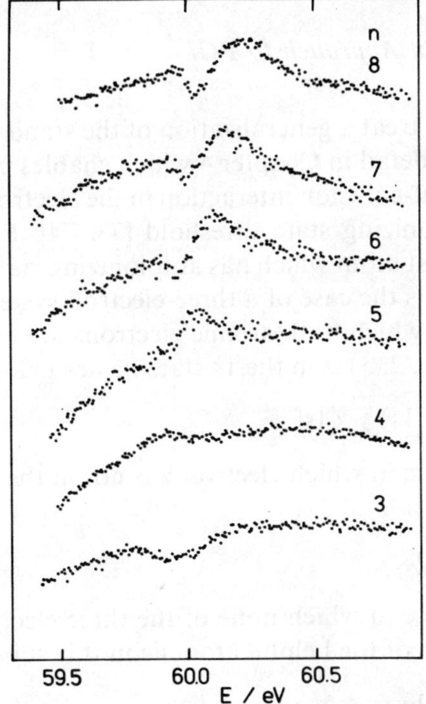


Fig. 7.10

Excitation function of He$^*$. The structure indicates the excitation of high Rydberg states caused by the post collision interaction.

57.82 eV energy of the $(2s^2)^1S$ state of He. Prominent features are two sharp peaks at 58.42 eV and 58.52 eV.

If the energy $\Delta E$ transferred to the fast electron exceeds the energy excess $E_1$ which the scattered electron has immediately after the excitation of the autoionizing state, the slow electron is captured by the atom in a bound state. Thus, owing to the PCI, the highly excited Rydberg states will be formed. This is the third manifestation of the PCI and can be seen in Fig. 7.10 showing the excitation functions for the states of helium whose principal quantum number $N$ varies from 3 to 8 over the range of the incident electron energies from 59.3 eV to 60.80 eV [66].

Each of the spectra of Fig. 7.10 shows structure of various shapes in the energy region from 59.9 to 60.5 eV. The energy of the structures increases as the binding energy of the excited state decreases. This structure is attributed to post-collision interaction. The possible autoionizing state responsible for the structure is the $^1D(2p^2)$ state having energy $59.90 \pm 0.02$ eV.

To describe these phenomena, several models have been constructed which include, apart from the classical model [7, 64, 67], the semiclassical [68–70] and the quantal shake-down [66] models. Angular momentum exchange between scattered and emitted electrons is discussed in [71–73].

### 7.6.1. Optical-Potential Approach to PCI

In this section we shall treat a generalization of the standard Feshbach resonance theory discussed in detail in Chapter 4 which enables us to include the effects associated with the post-collision interaction in the electron-impact excitation of atoms near the autoionizing state threshold [73, 74]. For simplicity we limit ourselves to the simplest atom which has autoionizing states, namely, the helium atom [73, 74]. Thus, it is the case of a three-electron system whose autoionizing states are the states in which both atomic electrons are excited. Let the state in which electron $\alpha(\alpha = 1, 2, 3)$ is in the 1s state be designated $|1s\rangle_\alpha$. The operator

$$Q_\alpha = 1 - |1s\rangle_{\alpha\ \alpha}\langle 1s|$$

projects on the subspace in which electron $\alpha$ is not in the 1s state and, thus, the operator $Q$

$$Q = Q_1Q_2Q_3 \tag{6.3}$$

projects on the subspace in which none of the three electrons is in the 1s state. The autoionizing states of the helium atom lie in this subspace. The operator $P$

$$P = 1 - Q$$

projects on the subspace of states in which at least one electron is in the $1s$ state and to which the initial and final states of the excitation process of the autoionizing states belong. Once the projectors are constructed, we can find the optical potential

$$V_{\text{opt}} = PHQ \frac{1}{E + i\varepsilon - QHQ} QHP \qquad (6.4)$$

and the effective Hamiltonian

$$H_{\text{eff}} = PHP + V_{\text{opt}} . \qquad (6.5)$$

The optical potential $V_{\text{opt}}$ contains the effect of autoionizing states, whereas the operator $PHP$ includes the excitation and the direct ionization without excitation of the autoionizing states. Let us denote by $|ki; \alpha\rangle$ a state in which electron $\alpha$ is free and electrons $\beta$ and $\gamma$ are in the $i$-th bound state. For instance, $|k0; 1\rangle$ is the state with the atom in its ground state $\varphi_0$ and the first electron moving freely, i.e.

$$\langle r_1 r_2 r_3 \mid k0; 1 \rangle = \exp\left(ikr_1\right)\varphi_0(r_2, r_3) .$$

This state is an eigenfunction of the operator $H_1$ defined as

$$H_1 = H - V_1$$

where $V_1$ is the interaction of the first electron with the atom. Analogously, we define

$$H = H_0 + V = H_\alpha + V_\alpha , \qquad \alpha = 1, 2, 3 . \qquad (6.6)$$

The final states of the excitation process of the autoionizing state are the eigenfunctions of the $PHP$ operator. Assume that electron 1 collides with a helium atom in the ground state. The solution $|k0; 1^{P+}1\rangle$ describing the outgoing state and satisfying the equation

$$|k0; 1^{P+}\rangle = P |k0; 1\rangle + \frac{1}{E + i\varepsilon - PH_1P} PV_1P |k0; 1^{P+}\rangle \qquad (6.7)$$

is the eigenfunction of the $PHP$ operator; the function $P |k0; 1^+\rangle$ given by the equation

$$P |k0; 1^+\rangle = |k0; 1^{P+}\rangle + \frac{1}{E + i\varepsilon - PHP} V_{\text{opt}}P |k0; 1^+\rangle \qquad (6.8)$$

is the eigenfunction of the effective Hamiltonian (6.5). The amplitude of the scattering into the singly excited state $|i\rangle$ in the arrangement $\alpha$ is determined by the asymptotic behaviour of the scattered wave of $P |k0; 1^+\rangle$

$$\langle r\, i; \alpha| \, P |k0; 1^+\rangle \underset{r\to\infty}{\longrightarrow} \frac{e^{ik_i r}}{r} f(k_i\, i; \alpha \leftarrow k0; 1) \qquad (6.9)$$

where the wavenumber of outgoing $k_i$ wave is defined by energy conservation

$$E = \frac{\hbar^2 k_i^2}{2m} + \varepsilon_i \tag{6.10}$$

in which $\varepsilon_i$ is the energy of the state $|i\rangle$.

From the relations (6.7), (6.8), (6.9) and after some operations we obtain [71] the following expression for the scattering amplitude:

$$f(k_i \text{ i; } \alpha \leftarrow k0; 1) = -\frac{m}{2\pi\hbar^2}(T_p + T_{\text{opt}})$$

where

$$T_p = \langle k_i \text{ i; } \alpha|\, PV_\alpha P\, |k0; 1^{P+}\rangle \tag{6.11}$$

describes the excitation of the singly excited state $|i\rangle$ with the scattering mechanism confined to the $P$-space and

$$\bar{T}_{\text{opt}} = \langle k_i \text{ i; } \alpha^{P-}|\, PHQ\, \frac{1}{E + i\varepsilon - QHQ}\, QHP\, |k0; 1^+\rangle \tag{6.12}$$

contains the effect of the autoionizing states on the excitation process. In this equation

$$\langle k_i \text{ i; } \alpha^{P-}| = \langle k_i \text{ i; } \alpha|\left(1 + PV_\alpha P\,\frac{1}{E + i\varepsilon - PHP}\right).$$

The relations (6.8)–(6.10) are exact and the term $T_{\text{opt}}$ contains the contribution of all the autoionizing states. In what follows we shall assume that only one autoionizing state is important for the excitation of the singly excited states and that the contribution of all other autoionizing states is neglected. Let us denote by $|k_a a; \beta\rangle$ a state in which electron $\beta$ is free with the wave vector $k_a$ and the atom is in the autoionizing state $|a\rangle$. The state $|k_a a; \beta\rangle$ is the eigenfunction of the operator $QH_\beta Q$ and its total energy is

$$E_a = \frac{\hbar^2 k_a^2}{2m} + \varepsilon_a$$

where $\varepsilon_a$ is the autoionizing state energy.

In conformity with our intuitive understanding of the first step of the PCI mechanism which has the autoionizing state $|a\rangle$ as its final state, we define the state $|k_a a; \beta^{Q-}\rangle$ to be an incoming-wave eigenstate of $QHQ$ which has the state $|k_a a; \beta\rangle$ as an outgouing wave:

$$|k_a a; \beta^{Q-}\rangle = |k_a a; \beta\rangle + \frac{1}{E - i\varepsilon - QHQ}\, QV_\beta Q\, |k_a a; \beta\rangle. \tag{6.13}$$

The effect of a single autoionizin g state $|a\rangle$ on the scattering amplitude may be studied by introducing the approximate closure

$$\left(\frac{1}{2\pi}\right)^3 \sum_\beta \int \mathrm{d}k_a \, |k_a a; \beta^{Q-}\rangle\langle ka; \beta^{Q-}| \tag{6.14}$$

in the optical potential (6.4). This expression would be equal to the projection operator $Q$ if it were summed over all eigenstates $|a\rangle$ of $QHQ$. Substituting (6.14) in (6.12), we obtain

$$T_{\mathrm{opt}} = \left(\frac{1}{2\pi}\right)^3 \int \mathrm{d}k_a \, \langle k_i \, i; \alpha^{P-}| \, PHQ \, |k_a a; \beta^{Q-}\rangle \, \times$$

$$\times \, \frac{1}{E + i\varepsilon - E_a} \, \langle k_a a; \beta^{Q-}| \, QHP \, |k0; 1^+\rangle \, . \tag{6.15}$$

Using the relations (6.8) and (6.14), we can write the integral equation for the last term on the right-hand side of (6.15)

$$\langle k_a a; \beta^{Q-}| \, QHP \, |k0; 1^+\rangle \, =$$

$$= \langle k_a a; \beta^{Q-}| \, QHP \, |k0; 1^{P+}\rangle \, +$$

$$+ \left(\frac{1}{2\pi}\right)^3 \sum_\gamma \int \mathrm{d}k'_a \, \langle k_a a; \beta^{Q-}| \, QHP \, \frac{1}{E + i\varepsilon - PHP} \, PHQ \, |k'_a a; \gamma^{Q-}\rangle \, \times$$

$$\times \, \frac{1}{E + i\varepsilon - E'_a} \, \langle k'_a a; \gamma^{Q-}| \, QHP \, |k0; 1^+\rangle$$

which determines this term through a simpler term

$$\langle k_a a; \beta^{Q-}| \, QHP \, |k0; 1^+\rangle \, .$$

In what follows we assume that the decay of the autoionizing state $|a\rangle$ is not affected strongly by the receding electron, so it may be written approximately:

$$\langle k_a a; \beta^{Q-}| \, QHP \, \frac{1}{E + i\varepsilon - PHP} \, PHQ \, |k'_a a; \gamma^{Q-}\rangle \, =$$

$$= \left(\varDelta - \frac{i\varGamma}{2}\right) (2\pi)^3 \delta(k_a - k'_a) \delta_{\beta\gamma}$$

where the real quantities $\Delta$ and $\Gamma$ are assumed to be the properties of the autoionizing state $|a\rangle$ which are independent of $\boldsymbol{k}_a$. Then, from (6.13) we obtain

$$\langle \boldsymbol{k}_a a; \beta^{\,Q-}| \, QHP \, |k0; 1^+\rangle =$$

$$= \frac{E + i\varepsilon - E_a}{E - E_a - \Delta + \dfrac{i\Gamma}{2}} \langle \boldsymbol{k}_a a; \beta^{\,Q-}| \, QHP \, |k0; 1^{P+}\rangle \,. \tag{6.16}$$

This equation shows that the state $|a\rangle$ has an effective threshold $\varepsilon_a + \Delta$ and a width $\Gamma$. The expressions (6.15) and (6.16) lead to the approximate expression

$$T_{\mathrm{opt}} \approx \left(\frac{1}{2\pi}\right)^3 \sum_\beta \int \mathrm{d}\boldsymbol{k}_a \, \langle \boldsymbol{k}_i i; \alpha^{P-}| \, PHQ \, |\boldsymbol{k}_a a; \beta^{\,Q-}\rangle \times$$

$$\times \frac{1}{E - \varepsilon_a - \Delta + \dfrac{i\Gamma}{2}} \langle \boldsymbol{k}_a a; \beta^{\,Q-}| \, QHP \, |k0; 1^{P+}\rangle \,.$$

From right to left, this equation describes the successive steps of the PCI. First, the atom is excited from the ground state to the autoionizing state $|a\rangle$, whereupon the scattered electron moves with wave vector $\boldsymbol{k}_a$. The possible decay channels of the autoionizing state give rise to the finite width $\Gamma$ which is expressed as an imaginary term in the propagator. Finally, the scattering state $|\boldsymbol{k}_a a; \beta^{\,Q-}\rangle$, in which the outgoing electron $\beta$ leaves the atom in the state $|a\rangle$, decays to the scattering state in which the ejected electron leaves the atom in the singly-excited state $|i\rangle$.

The optical potential description of the post-collision interaction described above was used in [75]. After some drastic but intuitively plausible approximations a model was obtained which leads to practical calculations. Results have been obtained on: i) angular momentum exchange during PCI, ii) angular distribution of ejected electrons and iii) lineshapes of PCI structures in the excitation of Rydberg states. In this model the scattered electron is captured into a singly excited state. A model for the case that the scattered electron remains in a continuum state was formulated in [76].

### References

1. BROWN G. E., JACKSON A. D., The Nucleon-Nucleon Interaction, North Holland, Amsterdam 1976.
2. O'MALLEY T. F., SPRUCH L., ROSENBERG L., J. Math. Phys. *2* (1961) 491.
3. WANNIER G. H., Phys. Rev. *90* (1953) 817.
4. TEMKIN A., Phys. Rev. Lett. *49* (1982) 365.
5. BRAUM A. J., J. Phys. *B14* (1981) 4 377.

6. Ho Y. K., Bhatia A. K., Temkin A., Phys. Rev. *A15* (1977) 1 423.
7. Hicks P. J. at al., Vacuum *24* (1974) 573.
8. Quéméner J. J., Paquet C., Marmet P., Phys. Rev. *A4* (1971) 494.
9. Marchand P. D., Can. J. Phys. *51* (1973) 814.
10. Burke P. G., Smith K., Rev. Mod. Phys. *34* (1962) 458.
11. Burke P. G., Adv. Phys. *14* (1965) 521.
12. Smith K., Rep. Prog. Phys. *29* (1966) 373.
13. Burke P. G., Adv. At. Mol. Phys. *4* (1968) 173.
14. Bardsley J. N., Mandl F., Rep. Prog. Phys. *31* (1968) 472.
15. Andrick D., Adv. At. Mol. Phys. *9* (1973) 207.
16. Schultz G. J., Rev. Mod. Phys. *45* (1973) 378, *45* (1973) 423.
17. Golden D. E., Adv. At. Mol. Phys. *14* (1978) 1.
18. Burke P. G., Schey H. M., Phys. Rev. *126* (1962) 147.
19. Schultz G. J., Phys. Rev. Lett. *13* (1964) 583.
20. Gailitis M. K., Damburg R.: Proc. Phys. Soc. London *82* (1963) 192.
21. Nicolaides C. A., Beck D. R., Int. J. Quant. Chem. *14* (1978), 457.
22. Siegert A. F. J., Phys. Rev. *56* (1939) 750.
23. Bardsley J. N., Int. J. Quant. Chem. *14* (1978) 343.
24. Herzberg A., Mandl F., Proc. Roy Soc. (London) *A274* (1963) 253.
25. Moiseyev N., Certain P. R., Weinhold F., Mol. Phys. *36* (1978) 1 613.
26. Rittby M., Elander N., Brandas E., Phys. Rev. *A24* (1981) 1636, Mol. Phys. *45* (1982) 553.
27. Giraud B. G., et. al., Preprint DLINUCI *P134T* (1981).
28. Gazdy B., Phys. Lett. *64A* (1977) 193, J. Phys. *A9* (1976) L39.
29. Mayer H. D., Walter O., J. Phys. *B15* (1982) 3 647.
30. Hazi A. U., Taylor H. S., Phys. Rev. *A1* (1970) 1 109.
31. Hazi A. V., in Electron-Atom and Electron-Molecule Collisions, ed. J. Hinze, Plenum Press, New York and London 1983, p. 103 and references cited therein.
32. Ho Y. K., Phys. Rev. *A19* (1979) 2 347.
33. McCurdy C. W., Layderable J. G., Mourey R. C., J. Chem. Phys. *75* (1981) 1 835.
34. Junker B. R., Adv. At. Mol. Phys. *18* (1982) 207.
35. Maier C. H., Cederbaum L. S., Domcke W., J. Phys. *B13* (1980) L119.
36. Ho Y. K., Phys. Reports *99* (1983)1.
37. Lovelace C., In: Strong Interactions and High Energy Physics, ed. R. G. Moorhouse, Oliver and Boyd, London 1964.
38. Nuttal J., Cohen H. L., Phys. Rev. *188* (1969) 1 542.
39. Nuttal J., Int. J. Quant. Chem. *14* (1978) 519.
40. Aguilar J., Combes J. M., Comm. Math. Phys. *22* (1971) 269.
41. Balslev E., Combes J. M., Comm. Math. Phys. *22* (1971) 280.
42. Simon B., Ann. Math. *97* (1973) 247.
43. Simon B., Int. J. Quantum Chem. *14* (1978) 529.
44. Reed M. C., Simon B., Methods of Modern Mathematical Physics Vol. IV: Analysis of Operators, Academic Press, New York, 1978.
45. Doolen G., Int. J. Quantum Chem. *14* (1978) 523.
46. Bethe H. A., Salpeter E. E., Quantum Mechanics of One-and Two-Electron Atoms, Springer, Berlin, 1957.
47. Doolen G. D., J. Phys. *B8* (1975) 525.
48. Doolen G. D., Nuttal J., Stagat R. W., Phys. Rev. *A10* (1974) 1 612.
49. Bardsley J. N., Int. J. Quantum Chem. *14* (1978) 343.
50. Winkler P., Yaris R., J. Phys. *B11* (1978) 1 475.
51. Moiseyev N., Springer Lecture Notes in Physics *211* (1984) 235.
52. Reinhardt W. P., Int. J. Quant. Chem. *10* (1976) 359.

53. CERJAN C., REINHARDT W. P., Avron J. E., J. Phys. *B11* (1978) L201.
54. CHU S. I., Chem. Phys. Lett. *58* (1978) 462.
55. CHU S. I., Chem. Phys. Lett. *64* (1979) 178.
56. CHU S. I., REINHARDT W. P., Phys. Rev. Lett. *77* (1977) 1 195.
57. HERBST I. W., Comm. Math. Phys. *75* (1980) 197.
58. HERBST I. W., SIMON B., Comm. Math. Phys. *80* (1981) 181.
59. HERBST I. W., In: Rigorous Atomic and Molecular Physics, ed. G. Vels, Plenum Press 1981, p. 131.
60. CERJAN C. at al., Int. J. Quant. Chem. *14* (1978) 393.
61. ABRAMOWITZ M., STEGAN I. A., Handbook of Mathematical Functions, Dover, New York, 1978.
62. HERBST I., Comm. Math. Phys. *64* (1979) 279.
63. MADDEN R. P., CODLING K., Astrophys. J. *141* (1965) 364.
64. BARKER R. B., BERRY H. W., Phys. Rev. *151* (1966) 14.
65. READ F. H., Radiation Research *64* (1975) 23.
66. KING G. C., READ F. H., BARDFORD R. C., J. Phys. *B8* (1975) 2 210.
67. HEIDEMAN H. G. M., NIENHUIS G., Van ITTERSUM T., J. Phys. *B7* (1974) L493.
68. MORGENSTERN R., NIEHAUS A., THIELMAN U., Phys. Rev. Lett. *37* (1976) 199.
69. MORGENSTERN R., NIEHAUS A., THIELMAN U., J. Phys. *B9* (1976) L363.
70. MORGESTERN R., NIEHAUS A., THIELMAN U., J. Phys. *B10* (1977) 1 039.
71. van de WATER W. at al., J. Phys. *B11* (1978) L465.
72. van de WATER W., HEIDEMAN H. G. M., J. Phys. *B14* (1980) 1 065.
73. van de WATER W., Thesis Rijks University, Utrecht 1981.
74. NIENHUIS G., HEIDEMAN H. G. M., J. Phys. *B12* (1976) 2 053.
75. van de WATER W., HEIDEMAN H. G. M., NIENHUIS G., J. Phys. *B14* (1981) 2 935
76. van der BURGT P. M. J., van ECK J., HEIDEMAN H. G. M., NIENHUIS G., Proceedings ICPEAC 1985.

# Conclusion, Open Problems

In this book we have examined a relatively small fraction of the material from the sea of the literature devoted to resonance processes in physics. The aim of the book was to describe some general concepts and methods used in calculating the resonance and other long-lived states in one- and many-particle quantum systems.

When preparing the book, we laid special emphasis on those problems relevant to determining the resonance-state parameters in few-body system studying the pole trajectories as the model parameters change and deriving the parameters of metastable and near-threshold states from experimental data. However, because of the obvious limitations on the volume of the book a great number of interesting problems related to this field were omitted. Below, we shall outline some of these problems and give the minimum references bearing in mind the interests of those readers who would like to familiarize themselves with actual unsolved problems in the examined field. We shall also mention other problems the solution of which is advanced to a certain extent, but for which some basic elements of practical importance are still missing.

(1) In the theory of many-particle resonances in systems of strongly interacting particles, the following important elements are developed insufficiently:

– a general connection between the parameters of two-particle interactions and the parameters of three- (or four-)-particle near-threshold and resonance states. To the best of our knowledge there exists only a single estimate [1–3] for a very simple three-body model system (a one-dimensional problem of three bodies with a $\delta$-interaction) and some general theory has been constructed [4]. However, no reliable estimates for realistic three-particle problems have been obtained; the decay dynamics of few-body quasistationary states has practically not been studied with due allowance for the strong resonance interaction of particles in the final state. In particular, the problem of separation of the direct (i.e. instantaneous) three-particle decay from the decay through *several* resonance rescatterings in the final state is not very clear. To resolve these problems is important with a view to correct derivation of the parameters of two-particle resonances and generally, of unstable fragments from the three-particle decay data;

– studies of threshold singularities in the coupling constant in the problem of three and more particles are almost absent. In particular, the evolution of the

342

three-particle Efimov levels *above* the three-particle threshold has not been studied.

(2) The status of the near-threshold long-lived states introduced by Baz' has not been established yet. Undoubtedly, this idea is fruitful because such states are present in many systems. However, no clear conditions for their appearance in various systems have been formulated;

(3) A very interesting and evidently little studied class of problems is closely associated with the specific mechanism of the metastable-state decay and with nuclear reactions of medium and heavy ions fusion and includes the problems concerning the calculations of the potential barrier penetrability allowing for internal degrees of freedom of colliding (or decaying) particles [5–9], especially in the presence of long-range (that is Coulombic and dipole) forces and also the investigation of the penetration of multidimensional barriers [8, 9]. These problems play an important role in studying the new types of nuclear radiactive decays discovered in recent years, in particular two-proton radioactivity [10], carbon radioactivity (i.e. the emission of $^{14}C$ nuclei from the Ra nuclei) [11, 12], etc. Many related problems are closely associated with the subject matter of this book. Unfortunately, it is mainly only strongly idealized problems that have been investigated so far [1, 3].

(4) Numerous problems relevant to the decays of highly excited quasi-stationary states are undoubtedly worth studying further, especially in cases of strong locking of these states in the collective degree of freedom of a system. They include problems relevant to the decay of giant resonances in nuclei, where the observed widths have been only qualitatively explained in many cases, and also interesting problems concerning the decay of deep-hole states in nuclei and other Fermi systems, the damping of quasi-particle excitations in nuclei, etc.

Finally, some interesting problems arise from experiments with long-lived states from direct measurements of very short lifetimes $(10^{-20}$–$10^{-21}$ s) of nuclear states to a study of the dynamic of the many-particle resonance state decays. Such experiments are common in high-energy physics, but they are far from being used extensively in other branches.

"De rebus omnibus et quibusdam aliis" (about all and still about many other things).


*References*

1. BRAYSHAW D. D., PEIERLS R. F., Phys. Rev. *177* (1969) 2539.
2. BADALYAN A. M., SIMONOV Yu. A., Yad. Fiz. *21* (1975) 458. [Sov. J. Nucl. Phys. *21* (1975) 239].
3. SIMONOV Yu. A., Nucl. Phys. *A266* (1976) 163.
4. SIMONOV Yu. A., The Resonances in the Three-Particle System.
   In: Proceeds. 1977 Europ. Symp. on Few-Particles Problems in Nucl. Phys., Potsdam 1977, p. 81;
   SIMONOV Yu. A., GRACH I. L., SHAMATIKOV M. Zh., Nucl. Phys. *A334* (1980) 80.
5. TANG H., NEGELE J. W., Nucl. Phys. *A406* (1983) 205.

6. DASSO C. H., LANDOWNE S., WINTER A., Nucl. Phys. *A407* (1983) 221.
7. LINDAY R., ROWLEY N., J. Phys. G.: Nucl. Phys. *10* (1984) 805.
8. CARLSON B. V., McKOY K. W., NEMETS M. C., Nucl. Phys. *A331* (1979) 117.
9. RING P., RASMUSSEN J. O., MASSANN H., EChAYa *7* (1976) 916 (in Russian) [Sov. J. Part. Nucl. *7* (1979) 366].
10. GOLDANSKY V. I., Nucl. Phys. *19* (1960) 482; ZhETF *39* (1960) 497 [Sov. Phys. JETP *12* (1961) 348];
    GABLE M. D., HOUKANEN H., PARRY R. F., ZHOU S. H., CERNY J., Phys. Rev. Lett. *50* (1983) 404; Phys. Lett. *B12* (1983) 25.
11. ALEKSANDROV D. V., BELYATSKY A. F., GLUKOV Yu. A., NIKOLSKY Yu. E., NOVATSKY B. G., DGLABLIN A. A., STEPANOV D. V., Pisma v ZhETF *40* (1984) 152. [JETP Lett. *40* (1984) 909].
12. ALEKSANDROV D. V., GANZA E. A., GLUKHOV Yu. A., NOVATSKY B. G., OGLOBLIN A. A., STEPANOV D. V., Yad. Fiz. *39* (1984) 513. [Sov. J. Nucl. Phys. *39* (1984) 323].

# Appendix A

# Rigged Hilbert Spaces and the Properties of Self-Adjoint Operators in them[1])

A Hilbert space in quantum mechanics, as well as in analysis, arises from completing the space $\Phi$ of 'sufficiently good' (for example, smooth and decreasing in infinity) functions with respect to a norm defined by a scalar product. The RHS theory assumes that $\Phi$ is the nuclear[2]) space in which the convergence is defined by a countable system of norms (the $\tau_\Phi$-convergence). If the scalar product $(\varphi, \psi)$ is introduced in $\Phi$ continuous relative to the $\tau_\Phi$-convergence, then $\Phi$ will not be complete relative to the new convergence in the norm $\|\varphi\| = = \sqrt{(\varphi, \varphi)}$ (the $\tau_{\mathcal{H}}$-convergence). However, $\Phi$ can be completed with respect to this new $\tau_{\mathcal{H}}$-convergence to the Hilbert space $\mathcal{H}$. The space $\mathcal{H}^*$ of all antilinear functionals on $\mathcal{H}$ is isomorphic to $\mathcal{H}$ itself. The functionals from $\mathcal{H}^*$ are also continuous linear functionals on the space $\Phi \subset \mathcal{H}$. The space $\Phi^*$ of all linear functionals on $\Phi$ turns out to be broader and to include the Hilbert space $\mathcal{H}$. The set of the three spaces embedded densely into each other

$$\Phi \subset \mathcal{H} \subset \Phi^* \tag{A.1}$$

is called the rigged Hilbert space (or the Gel'fand triad).

Henceforth in the Appendix we shall use the following notation: small letters $g, h, \ldots$ denote the vectors from the Hilbert space $\mathcal{H}$; the Greek letters $\varphi, \psi, \ldots$ denote the vectors from $\Phi$; the capital letters $F, R, \ldots$ denote the vectors from $\Phi^*$.

The notation for the scalar product in the sense of RHS (A1) should be understood as follows:
$\langle h \mid g \rangle$ is the conventional scalar product in $\mathcal{H}$; $h, g \in \mathcal{H}$; $\langle \varphi \mid F \rangle = \overline{\langle F \mid \varphi \rangle}$ is the value of the functional $F \in \Phi^*$ on the vector $\varphi \in \Phi$ (the horizontal bar indicates the complex conjugation). Generally, the quantity $\langle F \mid G \rangle$ (where $F$, $G \in \Phi^*$, but neither $F$ no $G$ belong to $\Phi$) is not defined. In particular, the norm of the Gamow state in the RHS formalism is not defined.

Let a linear operator $A$ be given in $\mathcal{H}$ with the definition domain $D(A) \supset \Phi$ such that the operator $A^*$, adjoint of $A$, is defined in the domain $D(A^*) \supset \Phi$.

---

[1]) Since Appendix A pertains mainly to Chapter 4, the references here are those from that chapter.

[2]) For the definition see [22].

Here, $A$ and $A^*$ do not map outside the space $\Phi$ (i.e. $\Phi$ is invariant with respect to $A$ and $A^*$):

$$A\Phi \subset \Phi, \qquad A^*\Phi \subset \Phi.$$

It should be reminded that the operator $A^*$ adjoint of $A$ in the Hilbert space is called such an operator that

$$\langle A^*h \mid g \rangle = \langle h \mid Ag \rangle \tag{A2}$$

at arbitrary $g \in D(A)$ and $h \in D(A^*)$. If $A = A^*$ provided $D(A) = D(A^*)$, the operator is called self-adjoint.

Throughout the space $\Phi^*$ the extension $\hat{A}$ of the operator $A$ is defined by the relation

$$\langle \hat{A}F \mid \varphi \rangle = \langle F \mid A^*\varphi \rangle$$

or

$$\langle \varphi \mid \hat{A}F \rangle = \langle A^*\varphi \mid F \rangle \tag{A3}$$

which must be satisfied for all $\varphi \in D(A^*)$ and $F \in \Phi^*$. The extension $A^+$ of the adjoint operator $A^*$ may be defined analogously:

$$\langle A^+F \mid \varphi \rangle = \langle F \mid A\varphi \rangle, \quad \varphi \in D(A), \; F \in \Phi^*. \tag{A4}$$

For these extensions it is possible to formulate the generalized problems for the eigenvalues:

$$\hat{A}R(\lambda) = \lambda R(\lambda), \tag{A5a}$$

$$A^+L(\mu) = \bar{\mu}L(\mu) \tag{A5b}$$

where the complex number $\lambda$ is called the right eigenvalue of the operator $A$ corresponding to the right generalized eigenvector (GEV) $R(\lambda) \in \Phi^*$; $\mu$ and $L(\mu)$ are the left eigenvalue and the left GEV of the operator $A$, respectively. Using the Dirac notation, we can write equations (A5) in a more common manner, whence the meaning of the terms 'right' and 'left' becomes clear:

$$A \mid R(\lambda)\rangle = \lambda \mid R(\lambda)\rangle, \tag{A6a}$$

$$\langle L(\mu)\mid A = \mu \langle L(\mu)\mid. \tag{A6b}$$

However, it should be remembered that the meaning of these equations must conform to the definitions (A3) and (A4):

$$\langle A^*\varphi \mid R(\lambda)\rangle = \lambda \langle \varphi \mid R(\lambda)\rangle, \; \varphi \in D(A^*); \tag{A7a}$$

$$\langle L(\mu) \mid A\varphi \rangle = \mu \langle L(\mu) \mid \varphi \rangle, \; \varphi \in D(A). \tag{A7b}$$

If $R(\lambda)$ is the right GEV of the operator $A$ belonging to the eigenvalue $\lambda$, then it proves to be the left GEV of the operator $A^*$ corresponding to the eigenvalue $\bar{\lambda}$. Indeed, we have

$$\langle A^*\varphi \mid R(\lambda)\rangle = \lambda \langle \varphi \mid R(\lambda)\rangle$$

or

$$\overline{\langle R(\lambda) \mid A^*\varphi \rangle} = \lambda \overline{\langle R(\lambda) \mid \varphi \rangle}$$

i.e.

$$\langle R(\lambda) \mid A^*\varphi \rangle = \bar{\lambda} \langle R(\lambda) \mid \varphi \rangle .$$

For the self-adjoint operator $A = A^*$, the extensions $\hat{A}$ and $A^+$ coincide with each other, so any GEV is simultaneously the right vector with the eigenvalue $\lambda$ and the left vector with the eigenvalue $\bar{\lambda}$. For the real eigenvalues, in particular, the right and left GEV of the self-adjoint operator are the same. Nevertheless, we retain the difference in the notation ($R$ and $L$) bearing in mind the continuation of GEV to the complex plane. Furthermore, the right GEV will usually appear in the position of a ket, and the left GEV in the position of a bra.

Generally, the self-adjoint operators in the Hilbert space, e.g. such as the Hamiltonians of physical systems, have not only a discrete, but also a continuous, spectrum; only the eigenvectors of the discrete spectrum belong to $\mathcal{H}$, however. If we use the RHS, we may also treat the (generalized) eigenvectors of the continuous spectrum belonging to $\Phi^*$. Moreover, a self-adjoint operator in the RHS has a complete GEV system. The following (Gel'fand-Maurin) nuclear spectral theorem holds [18, 23]: for any self-adjoint operator $A$ in the complex separable Hilbert space $\mathcal{H}$ with the definition domain $D(A)$ there exists a nuclear space $\Phi \in D(A)$ densely and continuously imbedded in $\mathcal{H}$ which is invariant with respect to $A$ such that the system of generalized eigenvectors of the operator $A$ corresponding to the real eigenvalues from the spectrum $\sigma(A)$ is complete, i.e. for all $\varphi$, $\psi \in \Phi$ we get

$$\langle \varphi \mid \psi \rangle = \int_{\sigma(A)} d\mu(\lambda) \sum_i \langle \varphi \mid R^i(\lambda)\rangle \langle L^i(\lambda) \mid \psi \rangle . \tag{A8}$$

In fact, this is the conventional spectral expansion of unity expressed in terms of RHS. The sum in (A8) allows for the multiplicity of the spectrum (for example, in the case of a one-particle Hamiltonian, $i$ is a set of quantum numbers of the angular momentum $l$ proper and of its projection $m$) and the Stieltjes integral taken over the spectrum of the operator $A$. The physical Hamiltonians have usually a discrete and absolute continuous spectrum

$\Lambda$. Now, having introduced the Lebesgue measure $d\mu_{ac}(\lambda) = h(\lambda)d\lambda$ , we can write (for simplicity we omit the multiplicity indices

$$\langle \varphi \mid \psi \rangle = \sum_j \langle \varphi \mid f_j \rangle \langle f_j \mid \psi \rangle +$$

$$+ \int_\Lambda \langle \varphi \mid R(\lambda) \rangle \langle L(\lambda) \mid \psi \rangle \, d\lambda \tag{A9}$$

where $f_j \in \mathcal{H}$ are the proper eigenvectors of the discrete spectrum; $R(\lambda) = L(\lambda)$ are the GEV of the continuous spectrum. Going over from (A8) to (A9), we change their normalization in such a way that the weight function $h(\lambda)$ disappears. It should be remembered again that $R(\lambda)$ in (A9) coincides with $L(\lambda)$ because $\lambda \in \Lambda$ is real.

Apart from the GEV entering the expansion (A9) and corresponding to the real eigenvalue $\lambda \in \sigma(A)$, there exist other GEV with eigenvalues located both inside and outside the spectrum. In fact, the GEV of the continuous spectrum entering (A9) form an analytic GEV family, i.e. a vector-valued function with the value in $\Phi^*$, analytic in the region $\Omega$ (which includes $\Lambda$) such that at every $\lambda \in \Omega$ $R(\lambda)$ is a GEV and $\lambda$ is the corresponding eigenvalue. The analyticity of $R(\lambda)$ means that the function $\langle \varphi \mid R(\lambda) \rangle$ is analytic at any $\varphi \in \Phi$ .

Let us examine an isolated singular point $\lambda_0$ of the function $R(\lambda)$ . From the conventional Laurent expansion of the function $\langle \varphi \mid R(\lambda) \rangle$ the expansion for the GEV arises:

$$R(\lambda) = \sum_{n=-\infty}^{\infty} C_n(\lambda - \lambda_0)^n \tag{A10}$$

whose coefficients are

$$C_n = \frac{1}{2\pi i} \int_C \frac{R(\lambda)}{(\lambda - \lambda_0)^{n+1}} \, d\lambda \tag{A11}$$

where $C$ is a simple closed and positively orientated curve inside the analyticity region $\Omega$ of the function $R(\lambda)$. The integral (A11) defines the vector $C_n \in \Phi^*$ . The properties of the Laurent series coefficients $C_n$ in the expansion of GEV $R(\lambda)$ of the operator $A$ are similar to the properties of the same Laurent series coefficients in the expansion of the resolvent near its singular points, thereby resulting in

$$\hat{A}C_n = \lambda_0 C_n + C_{n-1} . \tag{A12}$$

This follows immediately from (A11) and from the weak continuity of $\hat{A}$ in $\Phi^*$. Thus, if $C_{n-1} = 0$ , then $C_n$ is GEV of the operator $A$ corresponding to $\lambda_0$. In particular, if $\lambda_0$ is a first-order pole, the first non-zero coefficient is $C_{-1}$ and

$C_{-2} = 0$. Therefore, $C_{-1}$ is a GEV corresponding to the eigenvalue $\lambda_0$. Thus, the first-order-pole residue of the GEV $R(\lambda)$ belonging to continuous spectrum and continued to the complex plane $\lambda$ is also the GEV of the operator $A$. This residue in the resonance pole is a Gamow state.

Consider the simple example of the Schrödinger equation for spinless particles with central local potential $V(r)$ which behaves 'properly' at the origin and at infinity: $V(r) \underset{r \to 0}{\sim} \mathcal{O}(r^{-3/2+\varepsilon})$, $\varepsilon > 0$; $V(r) \underset{r \to \infty}{\sim} \sim \mathcal{O}(r^{-3-\delta})$, $\delta > 0$. Any function $\psi(r) \in L^2(\mathbb{R}^3)$ can be expanded in spherical functions:

$$\psi(r) = \frac{1}{r} \sum_{lm} \psi_{lm}(r) Y_{lm}(\hat{r}) \tag{A13}$$

and the scalar product is defined as

$$\langle \psi \mid \varphi \rangle = \sum_{lm} \int_0^\infty dr \, \bar{\psi}_{lm}(r) \varphi_{lm}(r) = \sum_{lm} \langle \psi_{lm} \mid \varphi_{lm} \rangle \,.$$

The functions $\psi_{lm}(r)$ belong to the Hilbert space $L_2(R)$ which is denoted as $\mathcal{H}_{lm}$. A formal expansion of the Hamiltonians $H$ and $H_0$ corresponds to the expansion (A13), namely, the self-adjoint operators $H_l^0$ and $H_l$ act in each $\mathcal{H}_{lm}$

$$H_l^0 = -\frac{1}{2\mu} \frac{d^2}{dr^2} + \frac{l(l+1)}{2\mu r^2} \,,$$

$$H_l = H_l^0 + V(r) \,.$$

Following [20], we shall now construct an example of the nuclear subspace $\Phi$ convenient for 'rigging' the Hilbert space of the examined problem. First, we shall costruct the nuclear space in each partial Hilbert space $\mathcal{H}_{lm}$.

Examine the space of the complex-valued infinitely-differentiable finite functions $\varphi(x)$ on the straight line with the carrier $[a, b]$, i.e. vanishing outside $[a, b]$. The convergence on this space is usually prescribed using a countable set of norms; for example,

$$\|\varphi\|_\sigma = \max_{\alpha \le \sigma} \sup_x |\varphi^\alpha(x)|$$

i.e. the convergence in this space (denoted as $C_0^\infty([a, b])$) means a uniform on $[a, b]$ convergence of $\varphi$ together with a convergence of all its derivatives. Such a space is called countably normed and it is a metrizable, but not normalizable, complete nuclear space.

Let us examine further an increasing sequence of such spaces, for example $\mathcal{D}_n = = C_0^\infty([1/n, n])$ and treat their union. As a result, we obtain a set of all infinitely differentiable functions along the semistraight line $(0, \infty)$ with a finite carrier. This set will be denoted $C_0^\infty(0, \infty)$. Given such a union, we may introduce the topology of the inductive limit; as a result the linear manifold $C_0^\infty(0, \infty)$ turns into a locally-convex nonmetrizable topological space $\mathscr{D}(0, \infty)$ which is known in the theory of generalized functions to be the space of the trial Schwartz functions (along a ray) [27]. The convergence of the sequence $\{\varphi_k\}$ in the space $\mathscr{D}(0, \infty)$ means that there exists such a number $n$ that all $\varphi_k$ vanish outside $[1/n, n]$ and the sequence $\{\varphi_k\}$ converges in the space $C_0^\infty([1/n, n])$, i.e. converges uniformely on $[1/n, n]$ together with all the derivatives. In its capacity of being the inductive limit of nuclear spaces, the space $\mathscr{D}(0, \infty)$ is nuclear (but is not countably-normalized). At each value of $l, m$ such space $\mathscr{D}(0, \infty)$ will be denoted $\mathscr{D}_{lm}$. The operators $H_l^0$ and $H_l$ are essentially self-adjoint in $\mathscr{D}_{lm}$ at $l \neq 0$. At $l = 0$ the self-adjointness requires the additional conditions

$$\varphi_{00}(0) = 0 , \quad \varphi_{00}(r) = 0 \quad \text{at} \quad r > R .$$

Let us construct now the complete nuclear space $\Phi$ imbedded densely in $\mathscr{H} = L^2(\mathbb{R}^3)$. For this purpose we shall again use the same technique, i.e. we take the direct sum of spaces $\mathscr{D}_{lm}$ at $l \leq L$:

$$\mathscr{D}_L = \sum_{l=0}^{L} \sum_{m=-l}^{l} \oplus \, \mathscr{D}_{lm}$$

and then construct the union of the increasing sequence of spaces $\mathscr{D}_L (\mathscr{D}_{L+1} \supset \mathscr{D}_L)$ and introduce the inductive-limit topology in the union. The so constructed space $\Phi$ is a closed subspace of the space of three-dimensional Schwartz functions $\mathscr{D}(\mathbb{R}^3)[27]$. It comprises only the vectors with a *finite* number of nonvanishing components $\varphi_{lm}$ each of which is an infinitely differentiable function with a finite carrier.

Let us determine now the GEV of the operators $H^0$ and $H$. First, we shall examine $H^0$. The normalized Ricatti-Bessel functions

$$\hat{j}(x) = x^{1/2} J_{l+1/2}(x)$$

which are regular solutions for the free radial Schrödinger equation $H_l^0 \hat{j}_l(kr) = \dfrac{k^2}{2\mu} \hat{j}_l(kr)$ ,define the continuous functionals on $\mathscr{D}_{lm}$. Therefore, the functionals

$$\langle klm|$$

$$\langle klm \,|\, \chi \rangle = \int_0^\infty \mathrm{d}r \hat{j}_l(kr) \chi_{lm}(r) , \quad \chi \in \Phi$$

are continuous on $\Phi$ and are the GEV of the operator $H^0$ with eigenvalue $k^2/2\mu$. Moreover, they are analytic throughout the $k$-plane. The completeness condition is

$$\langle \chi \mid \varphi \rangle = \int_0^\infty dk \sum_{lm} \langle \chi \mid klm \rangle \langle klm \mid \varphi \rangle .$$

Consider now the total Hamiltonian. The radial Schrödinger equation is known [3] to have the regular solution $\varphi_{l,\,k}(r)$

$$H_l \, \varphi_{l,\,k}(r) = \frac{k^2}{2\mu} \, \varphi_{l,\,k}(r)$$

which behaves as $(kr)^{l+1}/(2l+1)!!$ at $r \to 0$. Under the adopted assumptions concerning $V(r)$, this regular solution is an entire function throughout the $k$-plane at any value of $r$ [3]. The regular solution $\varphi_{l,k}(r)$ is related to the normalized physical solution corresponding to an incoming wave as

$$\psi_{l,\,k}^{(+)}(r) = \sqrt{\frac{2}{\pi}} \, \frac{\varphi_{l,k}(r)}{f_l(k)} , \tag{A14}$$

where $f_l(k) = \bar{f}_l(-\bar{k})$ is the Jost function which is analytic in the upper halfplane and continuous along the real axis (may be, except for the point $k = 0$).

The functionals $|R_{lm}^{(+)}(k)\rangle$ defined on $\Phi$ by the relation

$$\langle \varphi \mid R_{lm}^{(+)}(k) \rangle = \int_0^\infty dr \, \bar{\varphi}_{lm}(r) \psi_{l,\,k}^{(+)}(r) \tag{A15}$$

are the GEV of the operator $H$ and form an analytic family in the upper $k$-halfplane. The functionals $\langle L_{lm}^{(+)}(k)|$:

$$\langle L_{lm}^{(+)}(k) \mid \chi \rangle = \int_0^\infty dr \, \bar{\psi}_{l,\,k}^{(+)}(r) \chi_{lm}(r) \tag{A16}$$

are analytic in the lower $k$-halfplane. Thus, the formula for continuing the resolvent is

$$G^{\mathrm{II}}(z) = G(z) - 2\pi \, i \sum_{lm} |R_{lm}^{(+)\,\mathrm{II}}(k)\rangle \langle L_{lm}^{(+)}(k)| ,$$

$$k = \sqrt{2mz} , \quad \mathrm{Im} \, k < 0 \tag{A17}$$

where $R_{lm}^{(+)\text{II}}(k)$ is the continuation of the functionals $R_{lm}^{(+)}(k)$ to the lower $k$-halfplane. This continuation is possible to the same region to which the continuation of the Jost function $f_l(k)$ for a given potential is also possible. The singularities of the continued GEV $R_{lm}^{(+)}(k)$ coincide with the zeros of the Jost function and, hence, with the $S$-matrix poles

$$S_l(k) = \frac{\bar{f}_l(k}{f_l(k)} \, .$$

The functionals $\langle L_{lm}^{(+)}(k)|$ do not have singularities in the lower $k$-halfplane (except for the poles on the imaginary axis which correspond to bound states). Instead of the functions $\psi_{l,k}^{(+)}$ we may use the functions $\psi_{l,k}^{(-)}$ corresponding to the out-states. In this case the right and left GEV switch roles because $\psi_l^{(-)} = \bar{\psi}_l^{(+)}$ .

# Index